W9-DFH-605

MATHEMATICS IN DAILY LIFE

MAKING DECISIONS AND SOLVING PROBLEMS

ON OUR COVER

Koenigsberg is perhaps the only city that has ever become a mathematical problem or inspired a branch of mathematics. The Koenigsbergers' fascination with trying to find a continuous route which would cross each of its seven bridges once and only once gave the world a famous puzzle. Two centuries ago, their problem attracted the interest of Leonhard Euler, a Swiss mathematician; and finding a solution stimulated his development of the theory of graphs. (Readers are invited to tackle the puzzle in Chapter 14.)

The engraving on our cover (dating from about 1590) shows six of the renowned bridges of Koenigsberg; the seventh is hidden behind the woman in the foreground.

For those who are interested in old texts, the German inscription says: "The ducal capital of Koenigsberg in Prussia." The Latin says: "Koenigsberg, Prussia (or Borussia); finest maritime city; ducal residence."

MATHEMATICS IN DAILY LIFE
MAKING DECISIONS AND SOLVING PROBLEMS

JOANNE SIMPSON GROWNEY
BLOOMSBURG UNIVERSITY

McGRAW-HILL BOOK COMPANY
NEW YORK ST. LOUIS SAN FRANCISCO AUCKLAND BOGOTÁ HAMBURG JOHANNESBURG
LONDON MADRID MEXICO MONTREAL NEW DELHI PANAMA PARIS SÃO PAULO
SINGAPORE SYDNEY TOKYO TORONTO

Cover credit
View of the city of Koenigsberg. Colored copperplate engraving by Braun and Hogenberg, c. 1590. Kunstbibliothek, Bildarchiv Preussicher Kulturbesitz, Berlin.

Part-opening photograph credits
Part One: From *Heavy Love*, a Joe Rock production with Fatty Arbuckle. Springer/Bettmann Film Archive.
Part Two: Stan Laurel and Oliver Hardy. Culver Pictures, Inc.
Part Three: Duncan Wilmore and Juanin Clay in *WarGames*. © 1983 United Artists Corporation. All rights reserved.
Part Four: Ray Bolger and Judy Garland in *The Wizard of Oz*. Springer/Bettmann Film Archive; © 1939, Loew's Inc.
Part Five: W. C. Fields. Bettmann Archive, Inc.
Part Six: Ivan Triesault in *Barabbas*. Culver Pictures, Inc.; © 1962 Columbia Pictures Corporation. All rights reserved.

MATHEMATICS IN DAILY LIFE: MAKING DECISIONS AND SOLVING PROBLEMS

1 2 3 4 5 6 7 8 9 0 VNHVNH 8 9 8 7 6 5

ISBN 0-07-025015-4

This book was set in Helvetica by Better Graphics.
The editors were Peter R. Devine and Susan Gamer;
the designer was Scott Chelius;
the production supervisor was Diane Renda.
The photo editor and cover researcher was Randy Matusow.
The drawings were done by Danmark & Michaels, Inc.
Von Hoffmann Press, Inc., was printer and binder.

Library of Congress Cataloging in Publication Data

Growney, JoAnne Simpson.
Mathematics in daily life.

Includes bibliographical references and index.

1. Problem solving. 2. Decision-making. I. Title.
QA63.G76 1986 510 85-9999
ISBN 0-07-025015-4

ABOUT THE AUTHOR

JoAnne Growney was born on a farm in Indiana, Pennsylvania. She grew up there, milking cows, making hay, and solving problems. She went to Keith Laboratory School and Indiana Joint High School, then to Westminster College in New Wilmington, Pennsylvania, for her bachelor of science degree. Her graduate degrees came from Temple University and the University of Oklahoma. Her debt to her teachers is great, especially to Laura Church, who offered early and enthusiastic proof that women can excel in mathematics; to Elinor Blair, who showed that all of life can be a classroom for learning; to T. K. Pan, who insisted that mathematical intuition was at least as important as rigor; and to Miriam C. Ayer, who expected good mathematics to be expressed in good English. The author of a number of articles presenting applications of elementary mathematics, Dr. Growney recently has focused her scholarly interests on decision making. Presently a professor of mathematics and computer science at Bloomsburg University, she resides with four teenagers on an otherwise quiet street. Decent weather finds her jogging; when it rains, she tries her hand at the piano. And sometimes she just sits around.

TO MY MOTHER,
ADDA B. STEPHENS,
WITH GRATITUDE AND LOVE

CONTENTS

IN BRIEF

PREFACE *xvii*

PART ONE

BUILDING A FRAMEWORK FOR PROBLEM SOLVING

CHAPTER 1

KEEP THE GOAL IN MIND *3*

CHAPTER 2

LEARN FROM THE WORLD AROUND YOU *16*

CHAPTER 3

LEARN FROM CONSTANTS *34*

CHAPTER 4

OBSERVE WIDELY; SPECULATE FREELY; TEST CAREFULLY *47*

PART TWO

APPLYING MEASUREMENT AND ARITHMETIC TO PROBLEM SOLVING

CHAPTER 5

USING NUMERICAL MEASURES TO PROMOTE PRECISE THINKING *75*

CHAPTER 6

WHEN IS ADDITION APPROPRIATE? *92*

CHAPTER 7

USING MEASUREMENT IN DECISION MAKING: MATRICES AND WEIGHTED AVERAGES *99*

CHAPTER 8

MAXIMIZING SATISFACTION; MINIMIZING COST *115*

PART THREE

ORGANIZING COMPLEX INFORMATION

CHAPTER 9

ORGANIZING INFORMATION IN MUTUALLY EXCLUSIVE GROUPS *135*

CHAPTER 10

CALCULATING NUMBERS OF POSSIBILITIES: A MULTIPLICATION RULE *146*

CHAPTER 11

VISUALIZING THE STRUCTURE OF INFORMATION WITH A TREE DIAGRAM *162*

CHAPTER 12

SUMMARIZING INFORMATION USING A FEW NUMBERS *183*

PART FOUR

VISUALIZING RELATIONSHIPS AND ROUTES

CHAPTER 13

PICTURING RELATIONSHIPS *201*

CHAPTER 14

FINDING ROUTES *227*

PART FIVE

PROBABILITY ESTIMATION AND INDIVIDUAL DECISION MAKING

CHAPTER 15

MEASURING CERTAINTY 245

CHAPTER 16

EXPECTED VALUE 275

CHAPTER 17

DECISION RULES: DECIDING HOW TO DECIDE 294

CHAPTER 18

PROBABILITIES AND ESTIMATION 309

CHAPTER 19

REDUCING THE EFFECTS OF INTERRUPTIONS 318

PART SIX

GROUP DECISION PROCESSES

CHAPTER 20

INDIVIDUAL THRESHOLDS AND GROUP BEHAVIOR 337

CHAPTER 21

METHODS OF RECOGNIZING CONSENSUS 361

CHAPTER 22

THE PRISONER'S DILEMMA: A MODEL OF THE CONFLICT BETWEEN INDIVIDUAL AND GROUP BENEFITS

AFTERWORD 413

REFERENCES 415

APPENDIX

ANSWERS AND SUGGESTIONS FOR SELECTED EXERCISES 419

INDEX 447

CONTENTS

PREFACE *xvii*

PART ONE

BUILDING A FRAMEWORK FOR PROBLEM SOLVING

CHAPTER 1

KEEP THE GOAL IN MIND *3*

A Game of Nim 3

Nim: Analysis and Generalizations 4

Exercises 1 to 4 6

More Nim Games 8

Exercises 5 to 13 8

Goal-Directed Problem Solving 12

Exercises 14 to 19 13

CHAPTER 2

LEARN FROM THE WORLD AROUND YOU *16*

Observe Yourself 16

Exercises 1 to 3 17

Observe Others 19

Exercises 4 to 9 20

Learn from Mistakes 23

Exercises 10 to 15 23

Take Control 25

Exercises 16 to 19 26

The Role of Values 27

Exercises 20 to 30 30

CHAPTER 3

LEARN FROM CONSTANTS *34*

Look for Background Constants 34

Exercises 1 to 6 35

Percents: Confusing Constants 37

Exercises 7 to 21 39

CHAPTER 4

OBSERVE WIDELY; SPECULATE FREELY; TEST CAREFULLY *47*

Believe It or Not? *47*

Testing by Calculation: Speculation about Population Growth *48*

Exercises 1 to 13 *50*

Studying Random Events by Simulation *55*

Exercises 14 to 26 *66*

PART TWO

APPLYING MEASUREMENT AND ARITHMETIC TO PROBLEM SOLVING

CHAPTER 5

USING NUMERICAL MEASURES TO PROMOTE PRECISE THINKING *75*

The Value of Measurement Scales *75*

Types of Measurement Scales *76*

Exercises 1 to 14 *80*

Utility: Measuring Strengths of Preferences *86*

Exercises 15 to 22 *88*

CHAPTER 6

WHEN IS ADDITION APPROPRIATE? *92*

What Mathematical Process Suits a Given Problem? *92*

When Should We Add? *93*

Exercises 1 to 11 *94*

CHAPTER 7

USING MEASUREMENT IN DECISION MAKING: MATRICES AND WEIGHTED AVERAGES *99*

Forming Matrices *99*

Assigning Weights and Computing Weighted Averages *101*

Exercises 1 to 14 *107*

CHAPTER 8

MAXIMIZING SATISFACTION; MINIMIZING COST *115*

Introduction *115*

A Problematic Situation *116*

A Utility Matrix *116*

Exercises 1 to 4 *119*

Getting the Most from a Fixed Budget *121*

More on Marginal Utility *123*

Shortcuts *125*

Summary *126*

Exercises 5 to 18 *127*

PART THREE

ORGANIZING COMPLEX INFORMATION

CHAPTER 9

ORGANIZING INFORMATION IN MUTUALLY EXCLUSIVE CATEGORIES *135*

Addition and Mutually Exclusive Groups *135*
Organizing Information with Venn Diagrams *136*
Exercises 1 to 14 140

CHAPTER 10

CALCULATING NUMBERS OF POSSIBILITIES: A MULTIPLICATION RULE *146*

How Many License Plates? *146*
What Is the Multiplication Rule? *148*
A Solution to the License Bureau's Problem *148*
Use of the Multiplication Rule in Solving Problems *149*
Exercises 1 to 21 151
Binary Coding of Numbers *157*
Exercises 22 to 31 158

CHAPTER 11

VISUALIZING THE STRUCTURE OF INFORMATION WITH A TREE DIAGRAM *162*

An Introduction to Tree Diagrams *162*
Exercises 1 to 10 163
Numbered Tree Diagrams *171*
Summary *178*
Exercises 11 to 24 178

CHAPTER 12

SUMMARIZING INFORMATION USING A FEW NUMBERS *183*

Locating the "Center" of a Collection of Data Items *183*
Exercises 1 to 19 185
Measuring Variation from the "Center" *191*
Exercises 20 to 30 194

PART FOUR

VISUALIZING RELATIONSHIPS AND ROUTES

CHAPTER 13

PICTURING RELATIONSHIPS *201*

Friends, Enemies, and Communication *201*
Exercises 1 to 15 206

Finding an Efficient Order for Activities 211
Exercises 16 to 19 215
Developing a Project Schedule 217
Exercises 20 to 30 221

CHAPTER 14

FINDING ROUTES 227
The Rabbit Under the Bush 227
Graphs and Routes 227
Exercises 1 to 14 229
An Application of Euler Circuit Techniques 235
Exercises 15 to 25 236

PART FIVE

PROBABILITY ESTIMATION AND INDIVIDUAL DECISION MAKING

CHAPTER 15

MEASURING CERTAINTY 245
The Meaning of Probability 245
Exercises 1 to 8 246
A Mathematical Framework for Probability 249
Exercises 9 to 22 253
Determining Probability Assignments 258
Exercises 23 to 49 262

CHAPTER 16

EXPECTED VALUE 275
Gambling Games 275
Expected Value 276
Exercises 1 to 24 283

CHAPTER 17

DECISION RULES: DECIDING HOW TO DECIDE 294
Introduction 294
Deciding What to Study 294
Which Rule Should I Use? 299
Exercises 1 to 19 299

CHAPTER 18

PROBABILITIES AND ESTIMATION 309
An Almanac Survey 309
Evaluating the Survey Results 311
Confidence Intervals in Statistics 313
Exercises 1 to 6 315

CHAPTER 19

REDUCING THE EFFECTS OF INTERRUPTIONS 318

Introduction 318
Simulation of Interruptions 319
Exercise 1 322
More Interruption Problems—And a Formula 322
Exercises 2 to 4 323
Organizing Tasks to Reduce the Effect of Interruptions 326
Exercises 5 to 17 329

PART SIX

GROUP DECISION PROCESSES

CHAPTER 20

INDIVIDUAL THRESHOLDS AND GROUP BEHAVIOR 337

Chain Reactions 337
Exercises 1 to 3 339
Threshold Values and Possible Behaviors 339
Exercises 4 to 9 343
Role of Environment in Threshold Analysis 347
References 350
In Conclusion 351
Exercises 10 to 20 352

CHAPTER 21

METHODS OF RECOGNIZING CONSENSUS 361

Introduction 361
Exercises 1 and 2 363
Sequential Pairwise Voting—Avoid It If You Can 365
Exercises 3 to 6 366
Different Voting Procedures Lead to Different Decisions—Beware 368
Five Voting Decision Methods 371
Exercises 7 to 12 376
Criteria for a Voting Decision Method 379
Exercises 13 to 30 380

CHAPTER 22

THE PRISONER'S DILEMMA: A MODEL OF THE CONFLICT BETWEEN INDIVIDUAL AND GROUP BENEFITS 389

Introduction 389
What Is "Rational" Decision Making? 390
Prisoner's Dilemma 390
Exercise 1 392
Prisoner's Dilemma as a Model for Other Decision Problems 394
Exercises 2 to 12 395
Circumventing the Dilemma 400

Suggestions for Further Reading *404*
Exercises 13 to 23 *405*
Multiperson Prisoner's Dilemma Situations *408*
Exercises 24 to 29 *410*

AFTERWORD *413*

REFERENCES *415*

APPENDIX
ANSWERS AND SUGGESTIONS FOR SELECTED EXERCISES *419*

INDEX *447*

PREFACE

USING MATHEMATICS FOR DAY-TO-DAY DECISION MAKING AND PROBLEM SOLVING

Mathematics is useful. We all know that this mathematics is useful in accounting, economics, engineering, and physics. But it is also useful at a personal level—in managing time, organizing activities, evaluating priorities, and deciding among alternatives. There are many powerful ways to use the ability to calculate and reason that we have been developing since childhood. *Mathematics in Daily Life* shows how you can apply familiar mathematical ideas to the analysis and solution of day-to-day problems.

FOR WHOM IS THIS TEXT WRITTEN?

Mathematics in Daily Life is for students who have these characteristics:

- Skill with arithmetic operations (addition, subtraction, multiplication, and division) for integers, fractions, and decimals. (Familiarity with algebra is helpful but not necessary.)
- Curiosity about how mathematics is useful; willingness to experiment with mathematics.
- Patience to develop the habit of careful analysis of problems and critical evaluation of proposed solutions.
- An inclination toward independent thinking, insistence on full understanding of ideas, and acceptance of only those with merit.

Mathematics in Daily Life is designed primarily for use in college or university "general education" mathematics courses that develop students' abilities in quantitative and logical reasoning. Its emphasis is not on mathematical topics but on the *use* of elementary mathematics (mainly arithmetic) in decision making and problem solving.

Elementary and secondary school teachers provide another audience for this text. Its exercises serve as a model of what these teachers might create for their own teaching. Furthermore, many of its topics and strategies are adaptable for precollege use.

Mathematics is
not
a spectator sport.

A MESSAGE TO THE STUDENT

The most important attribute you can bring to your study of *Mathematics in Daily Life* is curiosity—a willingness to ask questions and to experiment. Read the text with a pencil in hand so that you can not only underline but make marginal notes—questions, comments, points of disagreement, and examples stimulated by your reading. Also, keep a notebook with complete solutions to all assigned exercises. Expressing a solution in detail, in your own words, is a sure way to confirm your understanding. When an exercise is difficult or confusing, attempt a solution anyhow. Even the experiment of developing an incorrect solution has value, for it provides an opportunity to learn to detect errors and a basis from which to improve. If you obtain a correct solution with the help of another problem solver, your next step should be to write out the solution in your own words, to verify it, and to implant it in your own understanding.

To write down a
thought focuses our
full attention on it
automatically.
Few of us can write
one thought and
think another
at the same time.

Discuss your reactions to reading assignments and your own attempted solutions with fellow students. Sharing hunches about solutions can serve as a stimulus for developing them further. Two heads are better than one when both are thinking carefully about the faults and merits of a proposed solution.

Use a pocket calculator, as needed, for the exercises. Especially in Chapters 3 and 4, some of the exercises require considerable calculation. Even though you may not mind calculating by hand, you will find that a calculator makes it easier to experiment with trial solutions. In addition, a calculator can free you to think more about "Does this answer make sense?" and less about "Did I calculate correctly?"

Truth will sooner
come out of error
than from
confusion.

Francis Bacon
(1561–1626)

The most important ingredients in a course using this text are your own solutions to the exercises. Willingness to participate actively is the key to success in learning to use mathematics to solve real problems.

Most of the text exercises fall into these general categories:

- Questions that require a brief application of a specific concept or method.
- Questions that require discussion and evaluation of the merits of a given idea or procedure.
- Questions that require multistep application and evaluation of several ideas or procedures.

I hear and I forget.
I see and I remember.
I do and I understand.

Chinese proverb

For all the exercises, an important step is verifying the solution. Ask questions such as, "Is this answer a good one?" "Is it the best possible one?" "How can I be sure?" Because real-world problems do not come with an "answer key," verification of one's own solutions is both necessary and useful. The Appendix, which offers answers for a few exercises and solution suggestions for a few others, is a portion of this book that I hope you won't get too familiar with; the exercises will be more fun if you insist on thinking independently and finding your own solutions. On the other hand, it is a good idea to treat the Appendix as you might treat a fellow student—consulting it occasionally but not considering it the final authority. I have tried to offer helpful advice but also to emphasize that finding correct solutions is primarily *your* task. Taking my word for things won't do for people who want to learn to think for themselves.

There are many special exercises called "invent" or "journal." Each "invent" exercise asks you to make up an example—describing a realistic problem—that

> When you really take problem solving seriously, problems and solutions will be at the back of your mind all the time. While you are walking to class, jogging, or cleaning your room, ideas will pop into your head. Jot them down on a piece of paper for later use. Once continuous problem solving becomes a habit, it will also be fun. Make it a habit by spending a few minutes daily thinking about special problems (such as those suggested in the "journal" exercises) and fixing them in your mind; then, as you move through routine tasks, your mind can be coming up with solutions.

applies mathematical ideas presented in the text. The "journal" exercises suggest ways that you might apply text ideas to personal problem situations. (Because of their personal nature, you may want to use a separate section of your course notebook for these exercises.)

Creation of an example is one of the best ways to demonstrate that you understand a concept; for this reason, "test yourself" exercises at the end of a chapter ask you to supply examples of the terms and ideas introduced in that chapter.

A warning Learning to use mathematics for effective decision making and problem solving may not be easy. Many of the ideas we will consider together are not difficult to understand, but it is hard to make their use a habit. Developing the overall habit of applying mathematical reasoning to diverse situations is a project that will need your persistent attention if it is to succeed. It requires a commitment—both to careful thinking and to trying new ways of doing things.

A guarantee The effort required to improve decision making and problem solving by learning to think systematically—identifying goals, organizing information, measuring attributes, and evaluating alternatives—will be amply rewarded. People who develop the ability to apply mathematical reasoning improve their ability to take charge of their lives and achieve their goals.

A MESSAGE TO THE TEACHER

The most valuable class activity for a course using *Mathematics in Daily Life* is discussion (including evaluation) of solutions to the exercises. Because the text explains concepts in patient detail, the student can read about most topics without an introduction by the instructor. On the other hand, for some classes, many of the exercises should be introduced (and some completed) in class. A number of the exercises ask for discussion of the merits of an idea or an attempted solution; students inexperienced in evaluation of solutions to mathematical problems may need to see this take place in the classroom before they realize the possible scope of such evaluation and the value of their own contributions to it.

For exercises which are good sources of discussion or those in which you anticipate that students may get bogged down on one of many steps, in-class solution may be preferable to assignment as homework. In such cases, you might serve as consultant—asking questions, stimulating and praising students' efforts, and encouraging evaluation of solutions. When class activity is devoted to problem solving, homework assignments can consist of preparatory reading, follow-up review and analysis, and writing up of solutions.

Division of a class into small groups to discuss solutions to "invent" and "journal" exercises motivates students to take these exercises seriously. Requiring students to submit an end-of-semester report summarizing their responses to "journal" exercises (possibly entitled "How I use mathematics in personal problem solving") encourages their participation in this activity. Some students resist completing the text exercises that request discussion instead of short answers; students may also resist the "invent" and "journal" exercises but, if pushed to solve them, will later consider them valuable. Your students need your continual support in their effort to increase the worth of their mathematical knowledge by learning to use it.

Probably you will enjoy using this text only if the primary role that you play is that of "inquiry stimulator" (rather than "knowledge disseminator"). The text is addressed to an audience of students whose primary interests are nonmathematical; on the other hand, most of them know more mathematics than they are able to use effectively. Teaching them is an experience that requires a high level of energy. The continuing effort to convince them of what many others take for granted—that some knowledge of mathematics and confident use of its processes can make them better prepared for life—can be an exhausting process. But when a student ends the term with a statement like, "This course helped me discover what I wanted to do and to see better ways of doing it," then the effort becomes unquestionably worthwhile.

In your effort to stimulate inquiry, your successful use of the text probably will find you switching with dexterity among the following roles:

- Questioner and gentle skeptic. Probe with questions such as, "What's the main idea here?" "Does it make sense?" "How can it be useful to us?" "Is this a good solution?" "What criteria shall we use to decide?" "How can we improve it?"
- Curious problem solver. Do the exercises yourself; experience the difficulties; resolve the ambiguities; prepare to discuss and evaluate solutions and not just give answers.
- Inventor. Invent problems that apply text ideas to your own day-to-day situations. These examples, more than any other contribution you can make, testify to the usefulness of mathematics.
- Organizer. Schedule class activities, assignments, quizzes, and tests. Lead class discussions; draw out and coordinate students' responses.
- Consultant. Serve as an advisor for both group and individual problem solving during class. Encourage students to gain information (from you and from other students) by asking questions. Prod, question, and praise as the students develop and evaluate solutions.

On quizzes and tests, in addition to problems, it is effective to use discussion questions, including questions that offer solutions to be evaluated. Evaluation of solutions is not only an important part of each exercise but also an important activity on its own. Many students who do not aspire to use mathematics independently as a problem-solving tool still want to be informed consumers of quantitative information. Interest in evaluating solutions deserves encouragement, and class discussions and test questions can stimulate and reward it.

Available on request from the publisher is an *Instructor's Manual* with chapter overviews, supplementary examples and ideas for class discussion, references, sample test questions, and solutions (or suggestions) for all text exercises.

SELECTION OF TOPICS FOR A COURSE

Many chapters of *Mathematics in Daily Life* are independent of the others; thus varied lists of topics can be chosen, depending on the interests of instructors and students. Chapters 1 and 2 (and probably also Chapters 3 and 4) should begin the course: they "set the stage" by introducing attitudes and strategies that are applied throughout the text. In addition, the content of Chapter 5 is drawn on in a number of later chapters. Certain class groups may be disposed to linger in or move quickly through certain chapters. The text allows either approach. Most chapters permit detailed study through challenging exercises that further develop text ideas; on the other hand, prerequisites for later chapters usually will be satisfied by a cursory coverage.

The diagram below shows dependencies among chapters. Any chapter may be covered after those that lie on the path or paths above it. Most of the dependencies

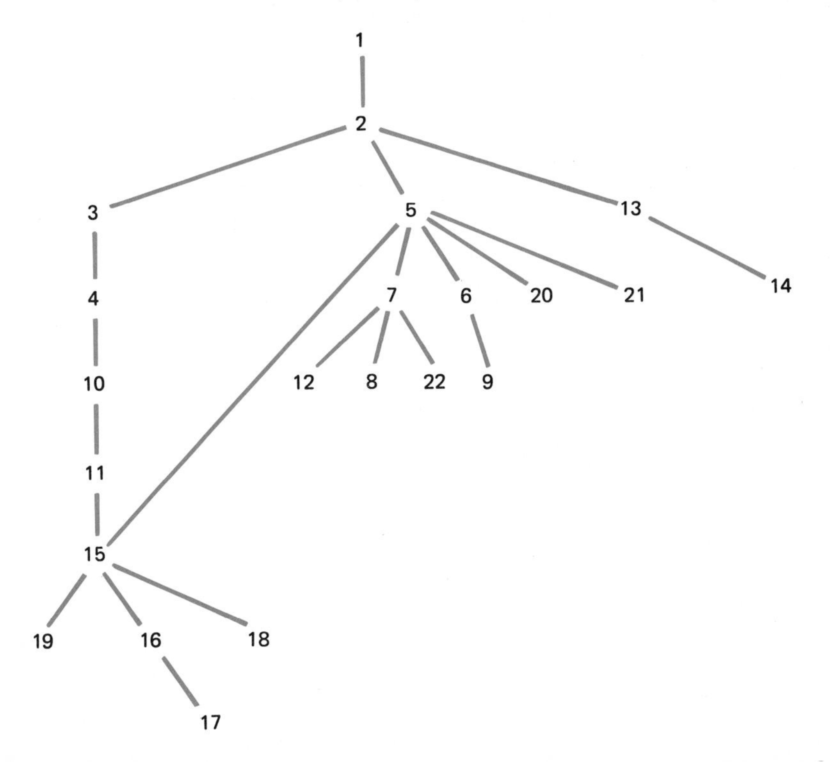

Diagram of chapter dependencies

are rather weak, however, and the need for prerequisite chapters can be avoided with a bit of careful supplementation by the instructor.

Comments from students who have used *Mathematics in Daily Life* have characterized certain of the chapters as noted below. All these characterizations are based on responses repeated in a number of classes; but they are, of course, subjective and are dependent on the length of time taken to cover the chapter, the mood of the class group, the enthusiasm of the instructor, and a multitude of other factors.

Chapters that are least difficult: 1, 5, 6, 10, 11

Chapters that are students' favorites: 9, 11, 19, 20

Chapters that students think will be most useful to them after the course has ended: 8, 11, 13, 15, 19, 20

Not only do the chapters have different levels of difficulty but, within a chapter, the exercises vary considerably in difficulty. This encourages students to tackle an exercise without preconceived notions about how hard it is.

If a class tackles a challenging chapter and finds it rough going, then it may choose to move through that chapter slowly, solving many of the exercises as a group rather than individually. Chapters 4 and 19 are, for example, ones that some students have found difficult; but these same chapters have been well liked by many when covered on a schedule that includes solving problems together in class, sharing the labor of computation, checking each other's solutions, and working together toward understanding.

HOW TO CONTACT THE AUTHOR

I will be delighted to hear from any student or instructor using *Mathematics in Daily Life*. If you have a question, a suggestion, or a point of disagreement to discuss, write to me at this address:

JoAnne S. Growney
Department of Mathematics and Computer Science
Bloomsburg University
Bloomsburg, PA 17815

I will respond. Writing *Mathematics in Daily Life* has been a valuable learning experience for me; your correspondence can help to keep me learning from it.

ACKNOWLEDGMENTS

This book has come into being as a result of the inspiration and hard work of many people. Family, friends, and colleagues patiently helped me even before I could explain my goals clearly; thus my vague ideas took shape and my expression of them improved.

Those who deserve special mention include:

Richard V. Andree and Moshe F. Rubinstein, both outstanding problem solvers who make good use of mathematics.

Joseph Malkevitch and Walter Meyer, whose text *Graphs, Models and Finite Mathematics*—which I enjoyed teaching for a number of years—provided a springboard for my own writing.

Eileen M. Peluso, who was the first reader of the initial draft and who provided detailed and valuable comments.

Stephen D. Beck, Deborah Frantz, Wallace J. Growney, Carol N. Harrison, John E. Kerlin, Anneli Lax, Thomas L. Ohl, James C. Pomfret, and Peter Sternberg—who taught students from preliminary versions of the text and provided feedback helpful for revision.

Ray H. Simpson, who offered valuable advice—avuncular and other—throughout the writing and publishing process.

Marjorie Clay, Richard Brook, and Seymour Schwimmer, whose interested, critical reading helped keep me from saying more than I know.

Scores of students—most of them from Bloomsburg University, but also some from Moravian College and Susquehanna University—who reacted thoughtfully to preliminary versions and helped me see the strengths and weaknesses.

My family, who—once I had begun this project—never doubted that I would complete it, though they wondered why it took so long.

I would also like to express my thanks for the many useful comments and suggestions provided by colleagues who reviewed this text during the course of its development, especially to Thomas Bartlow, Villanova University; Duane Deal, Ball State University; Joel Haack, Oklahoma State University; Robert Hafer, Brevard Community College; Robert Hoburg, Western Connecticut State University; Peter Lindstrom, North Lake College; Maurice Monahan, South Dakota State University; John Novosel, Richard J. Daley College; James Stasheff, University of North Carolina; Daniel Wachter, College of the Desert; and Frederick Ward, Boise State University.

JoAnne Simpson Growney

PART ONE

BUILDING A FRAMEWORK FOR PROBLEM SOLVING

PART CONTENTS

CHAPTER 1

KEEP THE GOAL IN MIND

A Game of Nim
Nim: Analysis and Generalizations
More Nim Games
Goal-Directed Problem Solving

CHAPTER 2

LEARN FROM THE WORLD AROUND YOU

Observe Yourself
Observe Others
Learn from Mistakes
Take Control
The Role of Values

CHAPTER 3

LEARN FROM CONSTANTS

Look for Background Constants
Percents: Confusing Constants

CHAPTER 4

OBSERVE WIDELY; SPECULATE FREELY; TEST CAREFULLY

Believe It or Not?
Testing by Calculation: Speculation about Population Growth
Studying Random Events by Simulation

CHAPTER 1

KEEP THE GOAL IN MIND

> If one advances confidently in the direction of his dreams, and endeavors to live the life which he has imagined, he will meet with a success unexpected in common hours.
>
> —Henry David Thoreau

A GAME OF NIM

We begin with a game. In part, solving mathematics problems is like playing Checkers or Monopoly or any other game: becoming good at it requires learning certain rules and then engaging in a great deal of practice. Usually one plays badly at first, but persistence pays—and as one becomes skillful, problem solving also becomes fun.

In the game of Nim, which is described below, it is possible to devise a strategy that will *guarantee* winning. To discover a winning strategy, a player must look ahead to the goal (winning) and think carefully about what moves to make to get there.

Here is the game:

> Nim calls for 2 players and 12 objects (say, matchsticks, pennies, paper clips, or marks on a piece of paper). The objects are placed on a table between the players. Each player, in turn, removes either 1 or 2 of the objects. To win, a player must leave the opponent with the final object.

For example, Judy and Rudy might play the game this way:

Start: | | | | | | | | | | | |
Judy plays first and takes 2: +—+ | | | | | | | | | |
Rudy plays next and takes 2: +—+ +—+ | | | | | | | |
Judy plays again and takes 1: +—+ +—+ + | | | | | | |
Rudy plays again and takes 2: +—+ +—+ + +—+ | | | | |
Judy plays again and takes 2: +—+ +—+ + +—+ +—+ | | |
Rudy plays again and takes 1: +—+ +—+ + +—+ +—+ + | |
Judy plays again and takes 1 and the last mark is left for Rudy.
He loses; she wins.

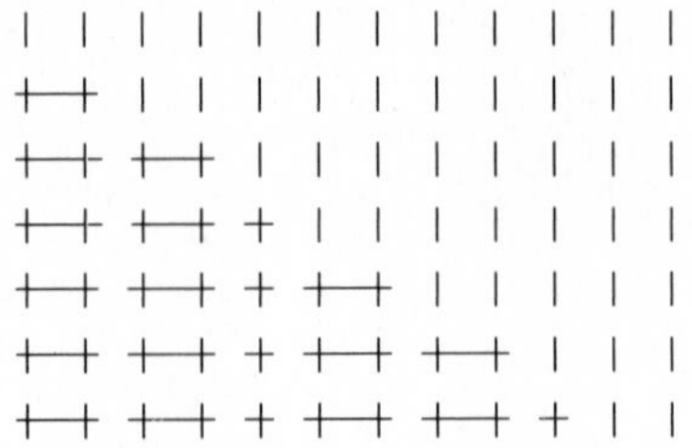

Now, you try it. With an opponent, play a few rounds of Nim. As you play, look ahead and try to predict who will win. How far ahead can you predict? Attempt to formulate a strategy that will guarantee you a win each time you play. *Summarize* your strategy in a brief written paragraph. *Test* your strategy against opponents. Is it foolproof?

NIM: ANALYSIS AND GENERALIZATIONS

The Nim game you have played and analyzed will from now on be referred to as (12;1, 2) Nim, so that we can later distinguish it from other Nim games in which the total number of objects and the number that may be taken at each turn will differ.

Below is an analysis of (12;1, 2) Nim. Although this approach may differ from yours, you will find it worth considering. The analysis begins with the goal—leaving one object for the opponent's last turn—and works backward to see how that goal can be achieved. Organizing the information in a table—like Table 1-1—is helpful.

Summarizing Table 1-1, we have:

If 1 object is left at your turn, you *lose.*

If 2 or 3 objects are left at your turn, you can make a choice that will *win.*

If 4 objects are left at your turn, no matter what you do, your opponent can make a countermove that will make you *lose*.

Thus, playing to win requires avoiding having 1 or 4 objects left when it is your turn. To see how to accomplish this, go on to Table 1-2.

Summarizing Table 1-2, we have:

If 5 or 6 objects are left at your turn, you can make a choice that will leave your opponent in a bad position—with 4 objects left. From there you can *win.*

If 7 objects are left at your turn, no matter what you do, your opponent can make a countermove that will leave you with 4 objects. Unless your opponent is not thinking, you will *lose.*

Thus, playing to win requires avoiding having 7 objects left when it is your turn. To see how to accomplish this, go on to Table 1-3.

TABLE 1-1
Analysis of moves in Nim

NUMBER OF OBJECTS LEFT AT YOUR TURN	YOU TAKE THIS MANY	EVALUATION OF RESULT
1	1	Bad. You lose. Avoid this situation.
2	1	Good. You win; opponent is left with last object.
	2	Bad. You lose—but you didn't have to.
3	1	Bad. You have given opponent the chance to win.
	2	Good. You win; opponent is left with last object.
4	1	Bad. Opponent is left with 3 objects and a chance to win.
	2	Bad. Opponent is left with 2 objects and a chance to win.

TABLE 1-2
Analysis of moves in Nim (continued)

NUMBER OF OBJECTS LEFT AT YOUR TURN	YOU TAKE THIS MANY	EVALUATION OF RESULT
5	1	Good. Opponent is left with 4 objects—a bad number. You can win.
	2	Bad. Opponent is left with 3 objects and a chance to win.
6	1	Bad. Opponent can take 1 object and leave 4 for you.
	2	Good. Opponent is left with 4 objects—a bad number. You can win.
7	1	Bad. Opponent can take 2 objects and leave 4 for you.
	2	Bad. Opponent can take 1 object and leave 4 for you.

TABLE 1-3
Analysis of moves in Nim (continued)

NUMBER OF OBJECTS LEFT AT YOUR TURN	YOU TAKE THIS MANY	EVALUATION OF RESULT (YOU SUPPLY THE REASON)	
8	1	Good.	Why?
	2	Bad.	Why?
9	1	Bad.	Why?
	2	Good.	Why?
10	1	Bad.	Why?
	2	Bad.	Why?

TABLE 1-4
Analysis of moves in Nim (continued)

NUMBER OF OBJECTS LEFT AT YOUR TURN	YOU TAKE THIS MANY	EVALUATION OF RESULT (YOU SUPPLY THE REASON)	
11	1	Good.	Why?
	2	Bad.	Why?
12	1	Bad.	Why?
	2	Good.	Why?

Summarizing Table 1-3, we have:

> If 8 or 9 objects are left at your turn, you can make a choice that will leave your opponent in a bad position—with 7 objects left. From there you can *win.*
>
> If 10 objects are left at your turn, no matter what you do, your opponent can make a countermove that will leave you with 7 objects. Unless your opponent is not thinking, you will *lose.*

Thus, playing to win requires avoiding having 10 objects left when it is your turn. To complete this analysis, consider Table 1-4.

The step-by-step analysis of Nim now can be synthesized into a strategy for play. Looking back at the summaries and Tables 1-4, 1-3, 1-2, and 1-1, we see that winning strategies require avoiding having 10, 4, 7 or 1 objects left at your turn. To accomplish this, seek to play first and remove 2 objects. This leaves your opponent with 10 objects. No matter what the opponent's move is (taking 1 object or 2), you can make a countermove (taking 2 objects or 1) to reduce the total to 7. From 7, the opponent's move (taking 1 object or 2) and your countermove (taking 2 objects or 1) can reduce the pile to 4. Then after another pair of moves you can leave the opponent with 1.

A more concise version of this strategy is:

> To win at (12; 1, 2) Nim, insist on playing first. On your first turn, remove 2 objects. On subsequent moves, counter your opponent's preceding move x with a countermove y so that $x + y = 3$. (S)

Exercise 1

a Using what you have learned from playing and analyzing (12; 1, 2) Nim, devise a strategy for winning at (15; 1, 2) Nim.* A complete strategy will

*In (15; 1, 2) Nim there are 15 objects at the start; each player may take 1 or 2 objects at a turn. To designate a Nim game, first list the starting number of objects followed by a semicolon; after the semicolon list the numbers of objects that a player may take at a turn.

specify whether or not you want to play first and how many objects you will take at each step.

b Describe a strategy for winning at (14; 1, 2) Nim.

c Describe a strategy for winning at (13; 1, 2) Nim.

Hint Sometimes it's a useful strategy to be "generous" and insist that your opponent play first.

Exercise 2

a Describe a strategy for winning at (28; 1, 2) Nim.

b Describe a strategy for winning at (29; 1, 2) Nim.

c Describe a strategy for winning at (30; 1, 2) Nim.

d Describe a strategy for winning at (31; 1, 2) Nim.

Exercise 3

The number of objects left at your turn is called *winning* if a careful choice of moves can lead you to win from that play; it is called *losing* if, no matter what you do, your opponent can win. Make a list of the *losing* numbers for (50; 1, 2) Nim. (Your list should begin 1, 4, 7, 10, ...)

Exercise 4

a If you are trying to win a game of (n; 1, 2) Nim, for what values of n would you insist that your opponent play first? If some very large value of n is given, what calculation can you perform to decide this?

b If you are trying to win a game of (n; 1, 2) Nim, for what values of n would you insist on playing first and removing 1 object on your first turn?

c If you are trying to win a game of (n; 1, 2) Nim, for what values of n would you insist on playing first and removing 2 objects on your first turn?

d In *a, b,* and *c,* what rule would you follow for moves after the first move?

e Apply the results of *a, b, c,* and *d* to describe how you would play to win for (999; 1, 2) Nim, (1000; 1, 2) Nim, and (1001; 1, 2) Nim.

Note The Appendix contains complete answers for Exercises 1*a, b,* and *c* and 2*a* and *d.* It contains partial solutions or hints for Exercises 3 and 4.

MORE NIM GAMES

The discussion of how to win at (12; 1, 2) Nim involved several important steps that are useful in many problem-solving situations—both mathematical and nonmathematical. These steps are:

- Become familiar with the details of the problem situation. (This was accomplished by playing Nim several times and observing what happened.)
- Focus on the goal of the problem and think backward from the goal, devising steps to achieve it. [The goal of (12; 1, 2) Nim was to leave the opponent with the last object. Step-by-step we worked from that goal backward to the beginning of the game and were thereby able to see a winning strategy. In the process, we classified possible steps as *good* or *bad,* depending on whether or not they helped achieve the goal.]
- Apply the results of one analysis to other similar situations, to exend the power and usefulness of the analysis. (The exercises constituted such an application.)

The exercises that follow continue the application of Nim analysis by asking you to devise winning strategies for still other Nim games.

Exercise 5

Another version of Nim allows each player, in turn, to remove 1, 2, or 3 objects from the starting pile. Again the winner is the player who can leave a single object for the opponent's last turn. Suppose that you and an opponent are going to play this game, starting with 10 objects. Keeping the goal (leaving the last object for your opponent) in mind, analyze the game of (10; 1, 2, 3) Nim by completing Table 1-5.

Exercise 6

Using the results of Exercise 5, write a complete strategy for winning (10; 1, 2, 3) Nim. Your strategy should specify whether or not you should play first and, if you do play first, how many objects to take on the first move. It should then specify what countermove to make after each of your opponent's moves. (Strategy S, given on page 6, illustrates a concise format.)

Test your strategy. Does it result in a *winning* number of objects each time your turn comes and a *losing* number of objects for each of your opponent's turns?

TABLE 1-5
Analysis of moves in (10; 1, 2, 3) Nim

NUMBER OF OBJECTS LEFT AT YOUR TURN	YOU TAKE THIS MANY	EVALUATION OF RESULT (GOOD OR BAD? REASON?)
1	1	
2	1	
	2	
3	1	
	2	
	3	
4	1	
	2	
	3	
5	1	
	2	
	3	
6	1	
	2	
	3	
7	1	
	2	
	3	
8	1	
	2	
	3	
9	1	
	2	
	3	
10	1	
	2	
	3	

Exercise 7

Extend the results of Exercises 5 and 6 to other, similar situations:

a Devise a winning strategy for (11; 1, 2, 3) Nim.
b Devise a winning strategy for (12; 1, 2, 3) Nim.
c Devise a winning strategy for (13; 1, 2, 3) Nim.
d Devise a winning strategy for (14; 1, 2, 3) Nim.
e Devise a winning strategy for (15; 1, 2, 3) Nim.

Exercise 8

a List the *losing* numbers—the numbers of objects that you would like to avoid at your turn—for (51; 1, 2, 3) Nim.

b What calculation can you perform on a number to test whether it is a *losing* number in (1000; 1, 2, 3) Nim?

c In the Nim games presented so far, all numbers of objects are either *winning* or *losing*. Following are two possible strategies for winning Nim games. Are they equivalent? Which is easier to follow? Why?

Play so that there is always a *winning* number of objects left when it's your turn. (S1)

Play so that there is always a *losing* number of objects left when it's your opponent's turn. (S2)

Exercise 9

Suppose that you want to win a game of (n; 1, 2, 3) Nim.

a For what values of n would you insist that your opponent play first? If some very large value of n is given, what calculation can you perform to decide this?

b For what values of n would you insist on playing first and removing 1 object on your first turn?

c For what values of n would you insist on playing first and removing 2 objects on your first turn?

d For what values of n would you insist on going first and removing 3 objects on your first turn?

e In *a, b, c,* and *d,* what rule would you follow for moves after the first one?

f Apply the results of *a–e* to describe how you would play to win for

(750; 1, 2, 3) Nim
(800; 1, 2, 3) Nim

(943; 1, 2, 3) Nim
(991; 1, 2, 3) Nim

Exercise 10

A popular pocket calculator game called Wipe-Out pits person against machine. The game begins with the calculator providing a starting number in its display. The person may decide who takes the first turn, and then the two alternate in subtracting from the displayed value any whole number from 1 to 9. If 1 is left at a player's turn, that player loses. Analyze Wipe-Out by drawing on your experience with Nim games. Devise and state a strategy that will enable you to win no matter what starting number n is provided by the calculator.

Exercise 11

The best-laid plans sometimes encounter the unexpected.

a Suppose you meet a friend, Sam, who wants to play a game that is like (12; 1, 2) Nim in every respect except one: in Sam's game the person who takes the final object wins. Devise a strategy for beating Sam at his own game.

b Your little brother, Sid, whom you have beaten badly at (12; 1, 2) Nim and (12; 1, 2, 3) Nim, proposes a new game—(12; 1, 3) Nim. Devise a strategy to beat Sid at his game.

Exercise 12

The original Nim game is believed to have originated in Asia several thousand years ago. This version of the game is still popular. It starts with 12 objects (originally these were stones) placed in heaps of 3, 4, and 5. In turn, each player may choose any one heap and remove any number of objects from it—at least one object and at most the entire heap. Again the goal is to force the opponent to take the last object.

a Play this Nim game and try to win.

b Apply the method of working backward from the goal to analyze this Nim game. (Analysis of this game is more complex than others we have considered because the number of possible moves is larger.)

c Devise a strategy that will enable you to win each time you play.

Exercise 13: Invent

Think of possible variations of Nim other than those already discussed. Analyze these games and formulate strategies for play. Test your strategies in actual play against an alert opponent.

Note The Appendix contains complete answers for Exercises 8*a* and *b* and 9*c*. It contains partial solutions or hints for Exercises 6, 7*c*, 10, 11, 12, and 13.

GOAL-DIRECTED PROBLEM SOLVING

"Keep the goal in mind" is an important problem-solving strategy that extends to many situations other than winning Nim games.

In solving mathematical "word problems," for example, the first step in making sense of a confusing array of information is to look for a goal: What do I want to find out? The value of identifying the goal is illustrated by a well-known English rhyme that poses a problem:

As I was going to Saint Ives
I met a man with seven wives;
Every wife had seven sacks;
Every sack had seven cats;
Every cat had seven kits.
Kits, cats, sacks and wives—
How many were going to Saint Ives?*

A practical example of goal-directed problem solving is saving money. Consider the problem faced by Kevin, a college senior who has just inherited $5000. In about 10 years Kevin sees himself settling down and purchasing a home; at that time he would like to have $10,000 available for a down payment. Present financial opportunities would permit Kevin to invest the money at 9 percent interest, a rate guaranteed over the 10-year period. Kevin wonders if he could invest his inheritance and have $10,000 in 10 years; he also wonders if he would need to invest the whole amount.

To see what could happen if he invested the money, Kevin starts calculating. He multiplies $5000 by 0.09 to determine the first year's interest. His calculations for the 10-year period are shown in Table 1-6. The values in Table 1-6 reveal that Kevin will not need to invest the entire $5000 to achieve his goal. Exercises 14 and 15, below, continue the problem of how much he should invest.

---

*Only one person, the speaker, was going *to* Saint Ives.

TABLE 1-6
Investment of $5000 at 9 percent interest

YEAR	INTEREST EARNED DURING YEAR	VALUE OF $5000 INVESTMENT AT YEAR'S END
1	$450.00	$ 5450.00
2	490.50	5940.50
3	534.64*	6475.14
4	582.76	7057.90
5	635.21	7693.11
6	692.38	8385.49
7	754.69	9140.18
8	822.62	9962.80
9	896.65	10859.45
10	977.35	11836.80

*When rounding values to the nearest cent, if the digits to be discarded are a 5 followed by zeroes, round to the nearest *even* digit. Thus, for example, $3.4550 becomes $3.46 and $2.34500 becomes $2.34. This rounding procedure is common in scientific calculations. Financial institutions use a number of different rounding procedures. Some round all half cents to the next larger cent. Others discard all fractions of a cent; for example, they would round $3.458 to $3.45.

Defining a goal and thinking backward from it to find strategies for achieving it is a process that applies in many situations. For example:

- Georgia has just received a B on her first statistics test. She is disappointed, for she wants to earn an A in the course. Focusing on that goal, she wonders, "What do I need to do right now to achieve it?"
- Mark needs to lose 15 pounds in 2 months to be ready for the start of wrestling season. What steps does he need to take right now to achieve that goal?
- Wendy has a term paper due in 5 weeks. Her typist will need to have the handwritten draft at least a week ahead of the deadline. Synthesizing her notes into a good paper will require at least a week. Thinking backwards from her deadline, Wendy sees that she has only 3 weeks in which to do her research.

Exercise 14

Complete Table 1-7 to determine whether Kevin can achieve his goal (accumulating $10,000 in 10 years) by investing only $4500 at 9 percent interest.

TABLE 1-7
Investment of $4500 at 9 percent interest

YEAR	INTEREST GAINED DURING YEAR	VALUE OF $4500 INVESTMENT AT YEAR'S END
1	$405.00	$4905.00
2	441.45	
3		
4		
5		
6		
7		
8		
9		
10		

TABLE 1-8
Discovering how much to invest to yield $10,000 in 10 years (at 9 percent interest)

YEAR	VALUE OF INVESTMENT AT YEAR'S END	VALUE OF INVESTMENT AT BEGINNING OF YEAR
10	$10,000.00	$9174.31
9	9,174.31	
8		
7		
6		
5		
4		
3		
2		
1		

Exercise 15

After completing Table 1-7, Kevin knows that he can afford to spend at least $500 of his inheritance money on a vacation to celebrate his graduation from college, and still have enough to invest toward his goal. This time, work backward from the goal ($10,000 in ten years) to see exactly how much Kevin needs to invest to achieve it; complete Table 1-8. (As you work backward, you can obtain each beginning-of-year value by dividing the desired year-end value by 1.09. Why does this work?)

Exercise 16

Some people like to begin each day by reading a horoscope or an inspirational passage. Is this a useful tool for selecting a goal for the day? What are the limitations of such a practice?

Exercise 17: Journal

Single out a goal that you would like to achieve in a week or less. Think backward from that goal and work out strategies that can lead to it step by step.

Exercise 18: Journal

Single out an important goal that you would like to achieve within the next 6 months. Think backward from the goal; what effect does it have on your activities today and in the near future?

If you don't aim
at a target,
you won't hit it.

Warning posted in
archery range

Exercise 19: Test Yourself

Develop a list of new terms and key ideas presented in this chapter. Supply an example of each item on your list. (Providing your own example of an idea is a major step toward your own independent use of that idea.)

Note The Appendix contains partial solutions or hints for Exercise 14 through Exercise 18.

CHAPTER 2

LEARN FROM THE WORLD AROUND YOU

It was six men of Indostan
To learning much inclined,
Who went to see the Elephant
(Though all of them were blind),
That each by observation
Might satisfy his mind.

—John Godfrey Saxe

OBSERVE YOURSELF

Television, books, and newspapers—even college courses—are important sources of new information and ideas for solving problems; but *you* are your own most important resource. Your own accumulation of knowledge and habits provides the basis for everything new you try.

It's appropriate, therefore, to devote some attention to analyzing yourself as a problem solver. First, ask yourself the following question: "When I have a problem to solve, what strategies work best for me?" Then consider some additional questions:

Do I need to be alone?
Do I need to be in a quiet place?
Are some settings (at my desk? out of doors?) better than others?
Does it help if I discuss the problem with someone else?
Do I accomplish more in frequent short intervals or in a few long intervals?
At what time of day am I most effective?
How do fatigue, worry, haste, or hunger affect my problem-solving ability?

How do I combat frustration and discouragement?

Do I waste valuable time and energy by being disorganized? (Or do I devote too much time to organizing?)

Do I focus on goals, to supply direction for problem solving?

Finally, ask yourself: "Are the answers to these questions different for different types of problems?" Answering all these questions will reveal your own characteristics as a problem solver; and once you are aware of them, you will be in a better position to capitalize on your strengths and deal with your weaknesses.

Exercise 1

To observe how you work with problems, it can be helpful to select a difficult problem to solve and observe yourself in the process. Select one of the following problem-solving activities—*a, b, c,* or *d*—and observe yourself while you solve it. As you work, ask yourself the questions given in the text.

Warning Each of these problems can become tedious, but don't give up. Many real-life problems also require long and patient examination. Observing how you react to difficulty and frustration is an important part of this assignment. Remember that few important problems can be solved in a few minutes.

a Put together a jigsaw puzzle of at least 500 pieces.

b Create a "magic square" by filling in the numbers 1, 2, 3, . . . , 16 in a four-by-four square array so that each horizontal, vertical, and diagonal line of numbers has the same sum.

c A *cryptarithm* is a puzzle in which letters of the alphabet have been substituted for the digits of a simple arithmetic problem. Each letter represents the same digit throughout the puzzle, and no digit is represented by more than one letter. A "leading" (leftmost) digit may not be zero. For example, the cryptarithm

$$\begin{array}{r} \text{AT} \\ +\ \text{A} \\ \hline \text{TEE} \end{array}$$

has the solution

$$\begin{array}{r} 91 \\ +\ 9 \\ \hline 100 \end{array}$$

Here is another example. The cryptarithm

$$\begin{array}{r} \text{AT} \\ \times\ \text{A} \\ \hline \text{PAD} \end{array}$$

has the solution

$$\begin{array}{r} 85 \\ \times\ 8 \\ \hline 680 \end{array}$$

Solve two of the following cryptarithmetic puzzles:

$$\begin{array}{r} \text{DONALD} \\ +\ \text{GERALD} \\ \hline \text{ROBERT} \end{array} \qquad \begin{array}{r} \text{CROSS} \\ +\ \text{ROADS} \\ \hline \text{DANGER} \end{array} \qquad \begin{array}{r} \text{SEND} \\ +\ \text{MORE} \\ \hline \text{MONEY} \end{array} \qquad \begin{array}{r} \text{DIETS} \\ -\ \text{FAD} \\ \hline \text{FOOD} \end{array}$$

Hint In the first of these, D = 5.

d Find the double agent:

Next week's duty roster at headquarters is causing difficulty. Two of the five secret agents will be needed each day. Each is to have 3 days' duty except for the double agent, who will have only 2. No pair of agents are to work together more than once. For reasons too secret to disclose, agent 1 is to work Tuesday, with one of Saturday's pair. Agent 4 is to work Thursday, with one of Friday's pair. Agent 5 is to work Monday, with one of Wednesday's pair. No one is to work both Thursday and Saturday; no one is to work both Wednesday and Sunday. Agents 1 and 3 may not work together, nor may agents 2 and 5. Agent 3 must be assigned to Wednesday.

Who is the double agent, and what days shall he or she be on duty?

Exercise 2: Journal

Analyze yourself as a problem solver by answering the questions given as guidelines in the text. Write several paragraphs that summarize your answer to the question, "When I have a problem to solve, what strategies work best for me?" In Exercise 1 you were asked to observe yourself working with a puzzle problem; now expand your view and examine your personal problem solving as well. For example, making an important decision—such as choosing a college, a major subject, a career, or a mate—is a situation in which your most effective problem-solving strategies can be useful.

Be careful not to confuse the way you presently solve problems with the way you would like to solve them. After you have honestly identified your present characteristics as a problem solver, list one or more *goals* for your problem solving. Include in this list problem-solving strategies that you would like to acquire.

Exercise 3: Journal—Imitate Your Successes

Think of something you are good at. Is it tennis or sketching or playing the flute? Is it writing or typing or algebra? Is it working with children or telling jokes at a party or organizing a budget?

Now stop and analyze this. *Why* are you good at it? Do you practice a lot? Are you well-organized? Do you have some special talent? Do you find it especially satisfying? Do you earn approval from others, or income?

If you can identify reasons why you are good at something, you may be able to use that information to help you become good at something else. For example:

Perhaps being well-organized has made you a good history student. Why not try the same approach for French?

Hours of disciplined practice may have made you a good tennis player. Can you apply the same strategy to physics?

Perhaps you are good at sketching cartoons because you have a special way of singling out the most important feature of a situation. Can you do that in problem solving?

Perhaps you jog or bicycle or do some other regular exercise, even though you find it tedious and unpleasant, because it contributes to your overall well-being. Can you motivate yourself to acquire other good habits because of their eventual benefits?

What habits and skills would you like to acquire? List these in your journal. Then answer this question: "What insights have I gained—from examining the things I do well—that can help me achieve these new goals?

Note The Appendix contains brief suggestions for Exercises 1, 2, and 3.

OBSERVE OTHERS

The people around us can provide a rich variety of ideas for problem solving. But to learn the most from others we must focus on the *processes* they use rather than on the answers they find. To illustrate this point, consider the following problem:

Tennis tournament A tennis pro is organizing a single-elimination tournament. Initial matches are determined by drawing entrants' names, in pairs, from a hat. After the initial matches the losers retire from the tournament, the winners' names are put back into the hat, and again pairs are drawn and the second round takes place. If at any time there is an odd number of names in the hat, the name left over after the drawing of pairs remains in the hat until the next round (that player gets a "bye" in the round). The pro wants to know in advance how to determine how many matches will have to be played for any number n of people who enter the tournament. That is, what will be the total number of matches in the tournament if there are n entrants?

Give a man
a fish
and he eats
for a day;
teach him
to fish
and he eats
for a
lifetime.

Chinese proverb

Suppose that this problem has been assigned for a mathematics course. At your first reading of the problem, you are stumped, and you ask your friend Dave for help. If Dave simply gives you the correct answer—which is $n-1$—you have not gained much; you need to know how he got the answer. A description of the process that Dave used to get the answer is of more value than the answer itself. Now we will listen in as Dave describes the analysis he used.

Here is Dave's account:

"First I picked a specific value, $n = 4$," Dave says, "to see how the tournament worked. In that case, I figured out that 3 matches would need to be played.

"Next I tried several other values of n. For $n = 5$, the number of matches turned out to be 4; this was harder than the first case because I had to figure out how to take care of 'byes.' I then tried $n = 10$ and $n = 15$. By this time I was pretty sure that the number of matches was 1 less than the number of entrants and I was just checking that hunch. Sure enough, for $n = 10$ the number of matches needed was 9 and for $n = 15$ the number of matches was 14.

"I then stopped to think about my hunch," Dave continues. "Would the number of matches always be $n - 1$? I thought so—but I couldn't supply a good reason.

"I asked Marge, who has played in a lot of tennis tournaments. She assured me that my answer was right. She said, 'In a single-elimination tournament, every match has one loser. The number of matches will be the same as the number of losers. Thus the tournament will have one fewer matches than the number of entrants.' "

Dave's process of solving a general problem by working out several special cases is one that requires patience. Working out the details of the special cases also calls for considerable care and organization. And note that Dave took time to verify his answer; this too requires patience. Exercise 4 below provides several chances for you to test your patience and try Dave's process.

FIGURE 2-1

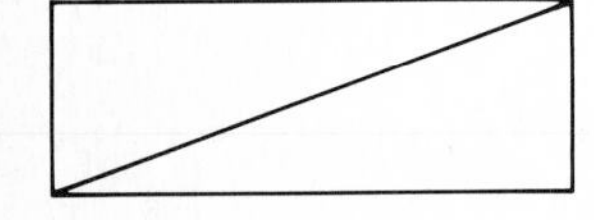

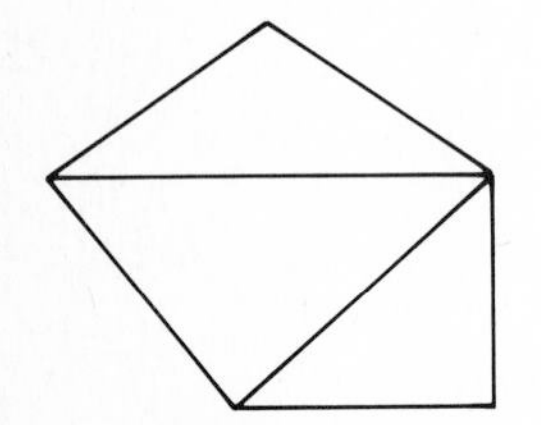

Exercise 4

Imitate the process that Dave used for the tennis-tournament problem and solve each of the following:

a In geometry, the study of triangles is emphasized because polygons with more than three sides can be divided into triangles by inserting of diagonals.* For example, by adding diagonals to a rectangle and a pentagon, we obtain the triangulations shown in Figure 2-1. For a polygon with n sides, into how many triangles may it be divided by insertion of noncrossing diagonals?

b Katy observes, in doing her mathematics assignments, that the first and easiest problem generally requires 1 minute, the second problem requires 3 minutes, the third requires 5 minutes, and so on. If Katy has n problems to do, give a formula that will predict in advance how long it will take her to solve all of them.

*A *diagonal* of a polygon is a line segment joining two of its nonadjacent vertices.

c A soccer league of n teams requires that each team play each other team twice (once on each home field) during the season. What is the total number of league games that will be played during a given season?

d In certain city school districts, parent volunteers serve as crossing guards at each intersection. If the district is a rectangular region composed of square blocks, and if the region is m blocks by n blocks in size, how many intersections require crossing guards? (For example, consider the 2-block by 3-block region shown in Figure 2-2; the 8 intersections marked with asterisks require crossing guards.)

FIGURE 2-2

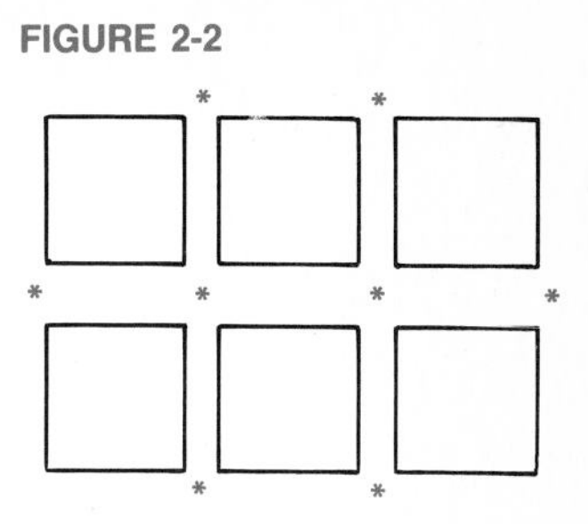

Exercise 5

Marilyn was having trouble with a puzzle problem. She asked her roommate Lisa how to solve it. Sidestepping Marilyn's request, Lisa said, "When I first read the problem, I didn't know how to solve it either. I began by making up a table to organize the information, but I still couldn't see how to do it—so I started guessing. I checked each guess and kept track of my trials until I found a correct answer." Use Lisa's process to solve the problem, given below, that she and Marilyn were discussing.

What does Bigfoot look like? Probably the first real evidence of the existence of Bigfoot is a photograph taken by the renowned woodsman Bunyan Paul and his exploration team. Before the photo was developed, Bunyan was asked to describe the beast.

"It is over 12 feet tall, with long white fur and a long tail," he replied.

"He's a liar," objected his guide, Babe. "Bigfoot has no fur at all, is under 5 feet tall, and has a short tail."

The descriptions given by Bunyan Paul's three brothers, all members of the exploration party, also disagreed.

Brother Jonah Paul said, "Bigfoot is 8 feet tall and has long white fur and a short tail."

Brother Horatio Paul said that Bigfoot is over 12 feet tall and has brown fur and no tail at all.

Brother Carter Paul estimated the beast's height at under 5 feet and said that it had brown fur and a long tail.

The five explorers' descriptions of Bigfoot probably disagreed because the beast was sighted during a severe snowstorm. When the photograph was developed, it revealed that each of the explorers was correct about one aspect of Bigfoot's appearance. What does Bigfoot look like?

Exercise 6

Terry has D averages in physics, psychology, and French, but he intends to take positive action. He estimates that

He can improve his physics grade by one letter for each additional 3 hours per week that he can spend studying physics.

He can improve his psychology grade by one letter for each additional 2 hours per week that he can spend studying psychology.

He can increase his French grade by one letter for each additional hour per week that he can spend studying French.

If Terry has up to 12 extra hours per week to spend on physics, psychology, and French, how should he invest his time to maximize his average grade?

Hint Try to adapt Lisa's strategy for Exercise 5 to this problem: organize the possibilities using a table; then evaluate each possibility and see which is best. Calculate the average grade using A = 4, B = 3, C = 2, and D = 1 and dividing the point total for all three course grades by 3.

Exercise 7: Invent

Make up a problem similar to those in Exercise 4 and solve it.

Exercise 8: Journal

Consider one of the problems in Exercises 4, 5, or 6 or find a new problem—either mathematical or personal—whose solution interests you. Observe how someone else solves the problem, and enter in your journal the insights you gain from your observation. Look for ways that you can apply those insights to other situations.

Exercise 9: Journal

In the following story, Arthur has trouble getting things done. If you have the same problem, you may find Arthur's observation of Alfred useful. Adapt Alfred's process to your own situation and record the results in your journal. (Note that Alfred is observing the motto of Chapter 1: Keep the goal in mind.)

Arthur has trouble getting all the things he wants to do done in a given day. He observes that Alfred, although basically a lazy fellow, always gets the important things done. Finally Arthur brings himself to ask Alfred what his secret is. Alfred is embarassed because he really has no great secret to share, but he confides his process: each evening he makes a list, of the things he'd like to get done the next day, in order of priority. The next morning he reviews the list, fixes it in his mind, and accomplishes as much of it as he can.

Note The Appendix contains partial solutions or hints for Exercises 4, 5, 6, and 7.

LEARN FROM MISTAKES

> Experience is the worst teacher;
> it gives the test before presenting the lesson.
>
> Vernon Law

When I took a computer programming course, on the first day the instructor said, "The best way to learn to program the computer is to make a lot of mistakes." I can recall laughing at that statement; I thought it was ridiculous and remembered it primarily because it seemed so absurd. I still do not entirely agree with it; but I have begun to recognize that making mistakes is one of the opportunities I can count on. Despite all efforts to avoid them, mistakes will occur. Because of this, I have tried to develop ways of learning from them. The following exercises illustrate some ways to learn from mistakes.

Exercise 10

Sometimes a problem calls for a process that we don't know how to do. For example, suppose we have a large lot and 200 feet of fence, and we wish to use the fencing to enclose the largest possible rectangular area for our dog. What should the dimensions be?

To answer this question, we can experiment. We sketch a possible region, as shown in Figure 2-3*a*.

Can we do better? A second rectangular region is shown in Figure 2-3*b*. But obviously the second guess is worse than the first—that is, it is a "mistake."

a Continue experimenting. "Guess" dimensions of rectangles until your guesses (possibly "mistakes") give you enough information to decide what is the largest area you can enclose with 200 feet of fence.

b What are the dimensions of the largest rectangular area you can enclose with 100 feet of fence? With 500 feet of fence?

c What is the smallest amount of fence needed to enclose a rectangular region whose area is 1000 square feet?

FIGURE 2-3
(a) A = 20 feet × 80 feet = 1600 square feet; (b) A = 10 feet × 90 feet = 900 square feet

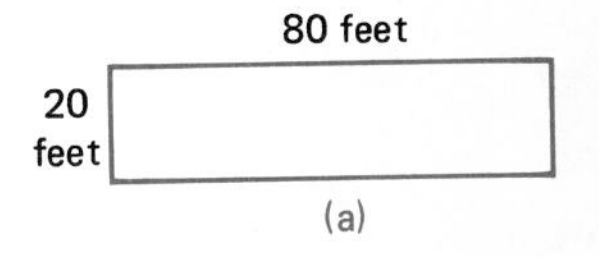

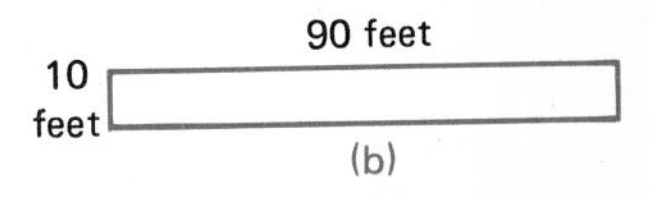

Exercise 11

Use trial and error (guessing and learning from mistakes) to solve the following.

a Find two counting numbers whose sum is 10 and whose product is as large as possible. [For example: $3 + 7 = 10$ and $3 \times 7 = 21$; $2 + 8 = 10$ and $2 \times 8 = 16$. Thus the pair (3, 7) is a better guess than the pair (2, 8) because its product is larger. What are still better guesses?]

b Find three counting numbers whose sum is 10 and whose product is as large as possible.

c Find four counting numbers whose sum is 10 and whose product is as large as possible.

d Find five counting numbers whose sum is 10 and whose product is as large as possible.

e What is the largest possible product that can be obtained from a list of counting numbers whose sum is 10?

f What is the largest product that can be obtained from multiplying the members of a list of counting numbers whose sum is 20?

g What is the largest product that can be obtained by multiplying the numbers from a list of counting numbers whose sum is 100?

Exercise 12

Repeat Exercise 11, with "largest" replaced by "smallest."

Exercise 13

Choosing numbers from anywhere on the number line, find two numbers whose product is 36 and whose sum is as small as possible.

Exercise 14

Sometimes students face confusing and difficult assignments which they feel unable to complete: "I just don't know enough to answer these questions." One of my students provided an interesting rationale for attempting such assignments despite the probability of making mistakes:

"I always try to solve confusing and difficult problems because those are the ones that provide practice for dealing with confusing and difficult situations in real life. Also, since I know I'm bound to make some mistakes, I would rather experiment on an assignment than on a test or a real situation. When I try and succeed, I feel proud. But even when I get something wrong, I learn. My instructors and my fellow students are quick to spot my errors. Somehow it's easier for others to offer helpful advice about my mistakes than to respond when all I can say is, 'I don't understand.' "

Do you agree with this reasoning? Do assignments such as homework offer an opportunity to make low-cost mistakes from which one can learn and to avoid later high-cost mistakes? Is it true that other people offer more helpful advice in response to mistakes than to pleas of ignorance? What are your own reasons for doing or not doing certain assignments?

Exercise 15: Journal

Make two lists: (1) a list of situations in which you can derive greater benefit from taking action that might result in mistakes than from doing nothing; (2) a list of situations in which you can derive greater benefit from doing nothing than from taking action that might result in mistakes. First consider situations related to courses you are presently taking; then consider more general situations of personal interest to you. What insights do you gain from your lists about your own attitude toward making mistakes? How can you utilize mistakes to your advantage?

Note The Appendix contains complete answers for Exercises 11*c*, *d*, and *f*. It contains partial solutions or hints for Exercises 10*b* and *c*, 11*g*, and 14.

TAKE CONTROL

When you face a problem, do you treat it as a demand on you or do you take charge of it, insisting on asking questions to gain more information and restating the problem in a form that makes sense to you? Success in problem solving requires a certain attitude: the problem solver must take control of the problem rather than being controlled by it.

Consider the question, "When did you stop cheating on your income tax?" Acceptance of the question puts the honest respondent in a difficult position—for the answer "Never!" suggests continual cheating. A proper response for an honest person is, "Your question contains a trick, and it should be restated before it is answered. I have never cheated on my income tax."

Most real problems do not involve tricks. Nevertheless, we may need to impose some control on them rather than timidly responding to them as to a command. The given information may need reorganization to be readily understood; and the sense of the problem may need to be questioned. Solving a problem may depend on our willingness to take control and restate it in a way that makes better sense to us. When we have done this, we are ready to apply solution procedures.

> "When I use a word," Humpty Dumpty said, "it means just what I choose it to mean—neither more nor less."
> "The question is," said Alice, "whether you can make words mean so many different things."
> "The question is," said Humpty Dumpty, "which is to be the master—that's all."
>
> Lewis Carroll (Charles Lutwidge Dodson, 1832–1898)

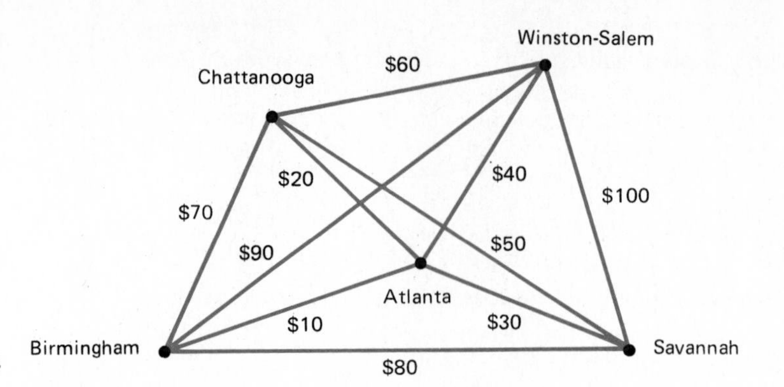

FIGURE 2-4

Exercise 16

For each of the following problems,* "take control" and restate the problem so that it is more sensible and leads to a better solution than is requested by the original version.

a A problem for travelers Figure 2-4 shows (obviously fictitious) costs of travel between pairs of cities. What round trip—starting and ending at Winston-Salem and going to each other city once—will result in the least total cost?

b Redecorating the basement Suppose you have a spacious area in your basement that you want to convert into a recreation room. You decide to panel the walls, install an acoustical ceiling, carpet the floor, and assemble a pool table, and you plan to have all this done professionally. The estimated times for the activities are:

Activity	Time required (hours)
Assemble pool table	4
Carpet floor	6
Panel walls	6
Install ceiling	8

The floor should be carpeted before the pool table is assembled, but everything else can be done at the same time. What is the minimum time required for the entire project? If the carpet layers can arrive at 8 A.M. next Monday, when should you schedule the other workers to get the whole job done on Monday? Will the work be completed by suppertime?

*These exercises are adapted from problems in Gary Chartrand, *Graphs as Mathematical Models*, Prindle, Weber, and Schmidt, Boston, 1977, pp. 98 and 76.

Exercise 17: Journal

Taking a test is a "problem" for many students. A student may "take control" of this problem by budgeting time among the various questions and by answering questions in his or her own order rather than the given order. Taking control in this situation may enable the student to supply a better "solution" to the "problem" than if the test format controls the response.

Formulate a list of strategies to try on your next examination.

Exercise 18: Journal

How can the idea of taking control be applied to the effective use of your time?

Exercise 19: Invent

A persistent difficulty in problem solving is remembering relevant information. *Mnemonic* devices—aids to memory—can often help. For example, the question, "May I have a large container of coffee?" is a way to remember the often-needed value of π, correct to seven decimal places ($\pi = 3.1415926\ldots$ and 3, 1, 4, etc., are the numbers of letters in each word in the question).

Recall or invent other devices to "take control" of memory and enable you to retrieve needed information.

Note The Appendix contains partial solutions or hints for Exercises 16 and 19.

THE ROLE OF VALUES*

Mathematics provides methods for analysis and calculation, but the ultimate solution to a problem frequently involves a decision about which alternative we *value* the most.

A traveling sales representative may fly or drive from city to city. Flying costs more than driving but takes less time. The choice will be determined by the relative values of certain amounts of time and money.

One person may undergo risky surgery with the hope of being able to walk again; another may accept life in a wheelchair rather than risk surgery that might result in death. Similar circumstances lead to different decisions for different people because we each place different weights on the various outcomes.

An interesting example of the role of values in decision making follows. It is somewhat contrived, but it shows how personal values dictate the "best" solution.

*The term *values* is used here to mean "general standards," including likes and preferences. It is not restricted only to "moral values."

Example: A Hiring Decision ▶

Casper Dogood of the social welfare department needs a new secretary. Nine-to-Five Employment Agency will supply four applicants for the position. The first is to come for an interview at 10 A.M. tomorrow. Mr. Dogood knows that secretaries are in such demand that if he does not decide to hire an applicant immediately after the interview, the applicant will go on to another interview and will probably be hired elsewhere. Thus Mr. Dogood will have to make his decision on the spot, but he has no way of knowing before an interview how that applicant compares with others.

He is familiar with three methods for making hiring decisions under these circumstances:

A Hire the first applicant interviewed.

B Interview until you find an applicant who is better than the one just preceding. Hire that one. If none of the first three is hired, hire the fourth applicant.

C Interview until you find an applicant who is better than all those preceding. Hire that one. If none of the first three is hired, hire the fourth applicant.

Suppose the applicants come to Mr. Dogood in random order--with the best one equally likely to be first, second, third, or last. Which of these methods would be the best way to make the hiring decision?

If Mr. Dogood is very busy, he may choose method A, since this method requires the least time for interviewing. In fact, one brief interview is all that is needed. Methods B and C require at least two interviews and may require as many as four. If Mr. Dogood expects all candidates to be reasonably good, and—as was noted—if his time is limited, he may decide that A is the best method.

However, Mr. Dogood may be more anxious to get the best possible secretary than to save time. He observes that if he hires the first applicant scheduled for an interview, he has only one chance in four of getting the best one.

As he puzzles over what would happen if he used method B or C, Mr. Dogood begins to make a list. (His completed list is shown in Table 2-1.) He uses the quality number 1 to designate the best applicant, the quality number 2 to designate the second-best, and so on. Each group of four numbers in Table 2-1 represents one possible order in which the candidates may come for interviews. For example, the group

2 1 4 3

represents the situation in which the second-best applicant (2) appears for the first interview. The best applicant (1) is scheduled for the second interview, the worst applicant (4) comes for the third interview, and the third-best applicant (3) comes for the fourth interview.

To see what might happen if he used method B, Mr. Dogood considers, in turn, each possible order of applicants given in Table 2-1. In each group Mr. Dogood circles the quality number of the applicant that method B would select (see Table 2-2). Of course, applicants following the circled one are never actually interviewed.

When he examines Table 2-2, Mr. Dogood discovers that method B selects:

Best applicant in 10 of the 24 possible situations

TABLE 2-1
All possible orders of applicants

1 2 3 4	2 1 3 4	3 1 2 4	4 1 2 3
1 2 4 3	2 1 4 3	3 1 4 2	4 1 3 2
1 3 2 4	2 3 1 4	3 2 1 4	4 2 1 3
1 3 4 2	2 3 4 1	3 2 4 1	4 2 3 1
1 4 2 3	2 4 1 3	3 4 1 2	4 3 1 2
1 4 3 2	2 4 3 1	3 4 2 1	4 3 2 1

TABLE 2-2
Applicants selected by method B

1 2 3 ④	2 ① 3 4	3 ① 2 4	4 ① 2 3
1 2 4 ③	2 ① 4 3	3 ① 4 2	4 ① 3 2
1 3 ② 4	2 3 ① 4	3 ② 1 4	4 ② 1 3
1 3 4 ②	2 3 4 ①	3 ② 4 1	4 ② 3 1
1 4 ② 3	2 4 ① 3	3 4 ① 2	4 ③ 1 2
1 4 ③ 2	2 4 ③ 1	3 4 ② 1	4 ③ 2 1

TABLE 2-3
Applicants selected by method C

1 2 3 [4]	2 [1] 3 4	3 [1] 2 4	4 [1] 2 3
1 2 4 [3]	2 [1] 4 3	3 [1] 4 2	4 [1] 3 2
1 3 2 [4]	2 3 [1] 4	3 [2] 1 4	4 [2] 1 3
1 3 4 [2]	2 3 4 [1]	3 [2] 4 1	4 [2] 3 1
1 4 2 [3]	2 4 [1] 3	3 4 [1] 2	4 [3] 1 2
1 4 3 [2]	2 4 3 [1]	3 4 [2] 1	4 [3] 2 1

Second-best applicant in 8 of the 24 possible situations
Third-best applicant in 5 of the 24 possible situations
Worst applicant in 1 of the 24 possible situations

Comparing method B with method A, Mr. Dogood makes the observation that B selects the best applicant in almost 42 percent of the possible situations, whereas A selects the best applicant in only 25 percent of the possible situations.

Mr. Dogood next proceeds to evaluate method C. Again he considers the list of possibilities in Table 2-1. This time he marks with a square the applicant that is selected by method C. The results are given in Table 2-3.

Examination of Table 2-3 shows that method C selects:

Best applicant in 11 of the 24 possible situations
Second-best applicant in 7 of the 24 possible situations
Third-best applicant in 4 of the 24 possible situations
Worst applicant in 2 of the 24 possible situations

Thus method C selects the best applicant in almost 46 percent of the possible situations. Method C is therefore preferable to methods A and B if Mr. Dogood wants the greatest likelihood of selecting the best candidate.

TABLE 2-4
Criteria that determine Mr. Dogood's decision method

CRITERION OF GREATEST IMPORTANCE	DECISION METHOD PREFERRED
Minimize interview time	A
Select the best applicant	C
Avoid selecting the worst applicant	B

One final variation of Mr. Dogood's values is worth considering. If Mr. Dogood's primary concern in hiring is to avoid the worst applicant, the results show that method B is the preferable decision rule. Method B selects the worst applicant in only 1 of the 24 possible situations, whereas method C selects the worst in 2 situations and method A selects the worst in 6 situations.

A summary of the role of values in Mr. Dogood's decision situation is provided in Table 2-4. ◀

Mr. Dogood's hiring decision illustrates how different values can be applied to analysis of the same problem—with different solutions resulting. His problem is also useful as a model of other decision situations. Following are two analogous situations in which one might consider applying variations of decision methods A, B, and C:

Hunting for a job Vernon is a college senior and seeks a job teaching high school English and coaching football. His college placement office has provided statistics which suggest that he is likely to obtain an average of four job offers but that these offers are unlikely to come at the same time. Instead, they will be spread out so that he will need to decide whether to accept one offer before he has another one to consider. When each offer comes along, how should Vernon make his decision?

Selling a used car Catherine is moving to the city and has decided to sell her car and rely on public transportation. When she places her newspaper ad, the paper informs her that she can expect at least four offers for the car. But the newspaper also warns her that buyers are shopping around and are likely to demand an instant decision on an offer. When each offer comes along, how should Catherine make her decision?

Exercise 20

Consider Vernon's decision problem. Revise Mr. Dogood's decision methods A, B, and C so that they apply to Vernon's situation. Describe a situation in which method A would be best for Vernon to use. Do the same for method B and method C.

Exercise 21

Consider Catherine's problem. Revise Mr. Dogood's decision methods so that they apply to Catherine's situation. Describe a situation in which method A would be best for Catherine to use. Do the same for methods B and C.

Exercise 22

Complete the analysis of decision methods A, B, and C by filling in Table 2-5.

Exercise 23

If Nine-to-Five Employment Agency can send Mr. Dogood only three applicants, there are six possible orders in which the applicants can arrive. List these six possible orders.

Use your list to evaluate the effects of using decision methods A, B, and C (suitably revised to apply to this new situation) and summarize your results in Table 2-6.

TABLE 2-5
Decision methods A, B, C: Four job applicants

	PERCENT OF THE TIME THAT THE METHOD CHOOSES			
DECISION METHOD	BEST APPLICANT	SECOND-BEST APPLICANT	THIRD-BEST APPLICANT	WORST APPLICANT
A	$\frac{6}{24} = 25\%$			
B	$\frac{10}{24} = 41.7\%$			
C	$\frac{11}{24} = 45.8\%$			

TABLE 2-6
Decision methods A, B, C: Three job applicants

	PERCENT OF THE TIME THAT THE METHOD CHOOSES		
DECISION METHOD	BEST APPLICANT	SECOND-BEST APPLICANT	THIRD-BEST APPLICANT
A			
B			
C			

	PERCENT OF THE TIME THAT THE METHOD CHOOSES				
DECISION METHOD	BEST APPLICANT	SECOND-BEST APPLICANT	THIRD-BEST APPLICANT	FOURTH-BEST APPLICANT	WORST APPLICANT
A					
B					
C					

TABLE 2-7
Decision methods A, B, C:
Five job applicants

Exercise 24

If Nine-to-Five Employment Agency can send five applicants to Mr. Dogood, there are 120 possible orders in which the candidates can arrive. List these 120 possible orders.

Use your list to evaluate the effects of using decision methods A, B, and C (suitably revised to apply to this new situation) and summarize your results in Table 2-7.

Exercise 25

a Give an example of a decision for which the most important consideration is minimizing decision time. Which decision method—A, B, or C—would be preferable?

b Give an example of a decision for which the most important consideration is selecting the best alternative. Which decision method—A, B, or C—would be preferable?

c Give an example of a decision for which the most important consideration is avoiding the worst alternative. Which decision method—A, B, or C—would be preferable?

Exercise 26

What realistic factors have been ignored in the treatment of Mr. Dogood's decision problem? If Mr. Dogood can "take control" of the decision situation, what changes should he make to improve his chances of hiring a good secretary?

Exercise 27: Invent

Suppose that you have inherited a piece of property and now want to sell it to finance your college education. Formulate a detailed description of this problem—a description that is analogous to Mr. Dogood's problem and the related problems that have been discussed. Then discuss how your values would influence your choice.

Exercise 28: Journal

Describe a situation in which you would consider applying decision method A, B, or C. In what ways would your values influence your choice of method?

Exercise 29: Journal

Describe a problem or decision of yours in which the selection of a solution was dependent on values. List the various solution choices. Explain how your values influenced the solution and how different values would have led to a different solution.

Exercise 30: Test Yourself

Develop a list of new terms and key ideas presented in this chapter. Supply an example of each item on your list. (Providing your own example of an idea is a major step toward using it independently.)

Note The Appendix contains complete answers for Exercises 20 and 22. It contains partial solutions or hints for Exercises 24 and 26.

CHAPTER 3
LEARN FROM CONSTANTS

There is nothing in this world constant, but inconstancy.

—Jonathan Swift

LOOK FOR BACKGROUND CONSTANTS

Milk
(with coffee)

Coffee
(with milk)

FIGURE 3-1

Suppose that in front of you sit two cups. Neither is quite full. One contains coffee and the other milk. Carefully you transfer a tablespoonful of coffee to the cup of milk. After stirring the mixture in the second cup, you carefully transfer a tablespoonful of its contents back to the first cup. Which is the greater quantity, the coffee in the milk cup or the milk in the coffee cup (Figure 3-1)?

Solving this puzzle problem is easier if we make the following observation: *The amount of liquid in the milk cup is the same after the two transfers as it was at the start.* Thus the amount of coffee left behind in the milk cup must equal the amount of milk in the coffee cup—and the answer to the question is "Neither."

Underlying many problem-solving situations is a property similar to the one observed above here: some quantity remains constant—or unchanged—through different phases of the problem. Learning to look for constants can be a valuable problem-solving tool.

To illustrate this further, consider the following problem:

John, a senior at Schenley High School in Pittsburgh, has made a weekend trip to Philadelphia to visit the College of Pharmacy and Science, which he plans to attend next year. The trip to Philadelphia on Friday took 6 hours, with John driving at a speed of 50 miles per hour. It is now 5 P.M. on Sunday, and he is expected home at 10 P.M. How fast must he drive to make the return trip in 5 hours?

Looking beyond the information stated in this problem, we note that the distance between Philadelphia and Pittsburgh is the same for both trips. Using the fact that the distance is constant, we can solve the problem as follows:

$$\begin{aligned}
\text{Friday's distance} &= \text{Sunday's distance} \\
\text{(50 mph) (6 hours)} &= \text{(Sunday's speed) (5 hours)} \\
\text{300 miles} &= \text{(Sunday's speed) (5 hours)} \\
\text{60 mph} &= \text{Sunday's speed}
\end{aligned}$$

Exercise 1

The following questions might occur in a typical algebra textbook. For each problem, identify a quantity that is a background constant. Then use the fact that this quantity is constant to solve the problem.

a Bill has some money in his pocket. If he has enough money to buy 14 small cans of soup at 40 cents each, how many large cans costing 70 cents each can he buy?

b The total population of Coaltown has remained fixed for many years. In 1940 its inhabitants were, all together, 5000 Italian Americans and 2000 Irish Americans. In 1970 the town had only 3200 Italian Americans. How many Americans of other ethnic origins lived in Coaltown in 1970?

c Between 8 and 8:30 yesterday morning the Belmont telephone exchange processed no long-distance calls. During that time, 100 Belmont residents each placed one call and 200 Belmont residents each placed two calls (and no other Belmont residents placed calls). If all placed calls were answered and if each person who answered the phone received four calls, how many Belmont residents received calls between 8 and 8:30 yesterday?

d My eccentric aunt doesn't keep careful records of the amounts of her investments. Last year when the interest rate was 7 percent, she earned $560 from her investment account. She has asked me to determine how much income she will get this year from the same investment, now that the interest rate has risen to 9 percent. What will this year's income be?

e Ralph and Karen plan to invite eight other couples in for pizza after the basketball game. They believe that one 10-inch pizza is sufficient to serve two people; thus nine pizzas will be needed. However, when Ralph places the order, the waitress at the House of Pizza suggests that he might get a better bargain by buying 15-inch pizzas instead. How many 15-inch pizzas should he buy to equal his original order? (The measurements are the diameters of circular pizzas.)

If 10-inch pizzas cost $4.35 each and 15-inch pizzas cost $6.35, will Ralph save any money? If so, how much?

Exercise 2

As we schedule each day's activity, a background constant that limits us is the number of hours in a day: 24. To increase the amount of time we devote to one activity, we must decrease the time devoted to another. For example, a need to increase sleeping time from 6 to 8 hours per day requires a decrease of 2 hours in the time devoted to some other activity or activities. Consider the following situations that involve constants. Identify the constant in each problem and describe the effect that a change in it has on the other quantities involved.

a John has budgeted a fixed amount, $10 per week, for gasoline for his car. What effect does an increase in gasoline prices have on John's purchases? What is the effect of a decrease?

b Ellen must pay a state income tax equal to 4 percent of her earnings. What effect does an increase in her salary have on the amount of tax she must pay? What is the effect of a decrease in earnings?

c Matt spends about 5 hours per day studying for the next day's classes. If he needs extra study time to prepare for a test in one of his courses, what is the effect on his other studies?

Exercise 3

In the coffee-milk puzzle illustrated in Figure 3-1 the two cups were the same size. Reconsider the problem with the milk cup 3 times as large as the coffee cup. What is the effect of the two transfers in this case?

Exercise 4

A few toll bridges (for example, the George Washington Bridge between New Jersey and Manhattan) have reduced the effort of toll collection by collecting "double payment" from only those travelers going in one direction. What quantity must be constant for this collection method to result in the same revenue as two-way collection? What are possible advantages and disadvantages of the one-way collection method?

Exercise 5

In drawing up a budget, people can frequently predict income more easily than expenditures. How does knowledge of income allow planning for expenditures? Does the reverse situation also occur? What quantity must be constant in a balanced budget?

Exercise 6: Journal

Make a list of situations (perhaps similar to those in Exercises 1 and 2) from your own experience in which background constants control the relationship between variables. Think about whether more awareness of these constants can help you plan more wisely.

Note The Appendix contains partial solutions or hints for Exercises 1 and 4.

PERCENTS: CONFUSING CONSTANTS

Situations that involve percents can be perplexing, but often the confusion is caused simply by missing information. The following examples illustrate this point.

Constant interest percents; different interest amounts Roger was comparing his year-end statement of interest with Nick's. Both had accounts earning 7.5 percent. "You earned $60 on your account while I earned only $45," Roger remarked. "If both accounts earn the same interest rate, why didn't we get the same amount?"

Different interest percents; constant interest amounts Karen and Elaine were comparing their interest earnings for the year. Karen's savings account paid 7.5 percent while Elaine's paid only 6 percent. "We both earned the same interest on our money," observed Karen. "If you earned $60 from an account that paid only 6 percent, why didn't I earn more than $60 from my 7.5 percent account?"

These examples are almost silly because they seem deliberately to ignore part of the relevant information. In the case of Nick and Roger, since interest payments depend not only on the interest rate but also on the amount invested, Roger should not expect to earn as much interest as Nick unless he has invested as large an amount. Similarly in the case of Karen and Elaine, Karen need not expect to earn more than Elaine just because the rate is higher: as always, the size of the interest payments depends on both the rate of interest and the amount invested.

A few calculations, listed in Table 3-1, show the annual interest payments resulting from several amounts invested at 6 percent and 7.5 percent. Table 3-1

TABLE 3-1
Annual interest payments for investments

	YIELD	
AMOUNT INVESTED	6 PERCENT INTEREST	7.5 PERCENT INTEREST
$600	(0.06) × $600 = $36	(0.075) × $600 = $45
$800	(0.06) × $800 = $48	(0.075) × $800 = $60
$1000	(0.06) × $1000 = $60	(0.075) × $1000 = $75

supplies the information missing from the two examples. If Roger invested $600 and Nick invested $800 at the same rate of 7.5 percent, then Roger's interest earnings are $45 and Nick's are $60. If Elaine invested $1000 at 6 percent, and Karen invested $800 at 7.5 percent, they earned the same amount of interest, $60.

These examples direct us to an important fact about percents: *Given alone, percents are not fully useful pieces of information.*

Thus knowing that an account earns 8 percent interest does not provide us with the amount of interest unless we also know the amount invested. To find the amount of interest we must ask, "8 percent of what?"

Similarly, knowing that the number of senior citizens in Clearwater has increased by 20 percent in the last 6 years does not provide us with information on whether there are enough elderly to support a new retirement home. To calculate the number of new senior citizens we must ask, "20 percent of what?"

In each of these "of what" questions, the technical term for the number to which "what" refers is the *base* for the percent. As our examples have shown, *to fully understand the meaning of a percent we must also know the base upon which that percent has been computed.*

Example: Traffic Fatality Statistics ▶

The Morgan County Traffic Safety Committee was studying automobile accident statistics. The police chief presented an alarming pair of facts:

> Last year 60 percent of all fatal accidents (accidents that resulted in at least one death) involved a driver who was intoxicated.
>
> In the previous year only 40 percent of fatal accidents involved an intoxicated driver.

The committee, which had conducted an extensive campaign to improve highway safety—by presenting educational programs in the schools and by installing numerous roadway signs to warn drivers of potential hazards—was alarmed. The members wondered how this effort could have resulted in such failure. Fortunately, one member thought to ask the police chief for the base numbers behind the percents. The chief then provided the data shown in Table 3-2.

Examination of Table 3-2 shows a very different picture. Automobile safety did improve: there were only 100 fatalities last year, compared with 150 two years ago. The *number* of fatal accidents with intoxicated drivers remained constant, but the *percent* increased because the *bases* were different. ◀

TABLE 3-2
Automobile statistics for Morgan County

	NUMBER OF FATAL ACCIDENTS IN MORGAN COUNTY	NUMBER OF FATAL ACCIDENTS WITH INTOXICATED DRIVERS	PERCENT OF FATAL ACCIDENTS WITH INTOXICATED DRIVERS
Last year	100	60	60
2 years ago	150	60	40

Exercise 7: Sale Prices and Percents

Preliminaries A confusing aspect of percentage problems is that there is not just one type. Problems involving percents differ in the type of information offered and the quantity needing to be found. For instance, if a coat is on sale,

> We might know the original price and the percent markdown and want to find the sale price.
>
> Or we might know the sale price and the percent markdown and want to find the original price.
>
> Or we might know the sale price and the original price and want to find the percent markdown.

Possibly the easiest way to work with percentage problems is to try to fit them all into a single format. In verbal form, the format is a sentence subdivided into five word groups:

Some quantity	is	a percent	of	another quantity.

This sentence translates into the following verbal equation:

Some quantity	=	a percent, expressed as a decimal	×	another quantity

A percentage problem can be solved if information is given that allows two of the expressions in the equation to be replaced by actual numbers, so that the missing value can be calculated. Two examples follow.

- A $150 coat is marked down 25 percent. How much do we save by buying it on sale? To solve this, write the equation

  $$\text{Savings} = \text{percent of markdown} \times \text{original price}$$

  Substituting the given information, we have

  $$\text{Savings} = 0.25 \times \$150$$

  Multiplication yields

  $$\text{Savings} = \$37.50$$

- A $150 coat is marked down to $120. What is the markdown rate? Subtraction of $120 from $150 gives $30 for the amount of savings. We now know two quantities to substitute in the verbal equation:

  $$\$30 = \text{percent of markdown} \times \$150$$

> Man can learn nothing unless he proceeds from the known to the unknown.
>
> Claude Bernard

To find the percent, we use the rule, "If equals are divided by equals the results are equal," and divide both sides of the equation by \$150. This leads to

$$\frac{\$30}{\$150} = \text{percent of markdown}$$

The indicated division yields 0.80; the markdown rate is thus 80 percent.

a Choppy's Lawnmower Shop is advertising an end-of-season sale with markdowns up to 25 percent. A lawnmower regularly selling at \$369 is on sale for \$269. Find the sale price. Comment on Choppy's advertising claim.

b Larry has been shopping for a grey wool blazer. He is willing to spend up to \$90. Mercer's Men's Store is advertising wool blazers on sale at 25 percent off. At that rate, what is the original cost of the most expensive blazer that Larry can buy?

c Invent There are three types of problems for which the verbal equation "savings = percent of markdown × original price" is appropriate: one type for each possible combination of two known quantities and one unknown quantity. List the three types, telling what is known and unknown in each. Invent a verbal problem of each type.

Exercise 8

Clairton College has 1000 seniors—400 men and 600 women. A gala precommencement dance is being planned. It is traditional for all seniors to attend. An initial count shows that 30 percent of the Clairton senior men will be attending with dates who are Clairton senior women. What percent of the senior women will thus be attending with dates who are senior men?

Exercise 9*

Suppose that in a certain nation, with an adult population of 10 million, 8 million adults belong to one religious group, R_1, and 2 million adults belong to a second religious group, R_2. Exactly half of each religious group consists of men and half of women.

a If all the adult men and women marry and if, in particular, 10 percent of the men from R_1 marry women from R_2, what are the following percents?

*This exercise was adapted from an example in C. A. Lave and J. G. March, *An Introduction to Models in the Social Sciences,* Harper and Row, New York, 1975, p. 68.

(1) Percent of R_2 women involved in "mixed" marriages. (2) Percent of R_2 men involved in "mixed" marriages. (3) Percent of R_1 women involved in "mixed" marriages. (4) Percent of the total population involved in "mixed" marriages.

b For each of the following statements, tell what population (R_1 or R_2 or the entire nation) has been used as a *base* for the given percent. (1) 10 percent of individuals are involved in mixed marriages. (2) 16 percent of individuals are involved in mixed marriages. (3) 40 percent of individuals are involved in mixed marriages.

c Suppose that mixed marriages are viewed as a threat by the religious groups. Which group then has more cause for concern, R_1 or R_2? Explain.

Exercise 10

Approximately equal numbers of male and female babies are born each year,* but females in the United States tend to marry earlier and live longer than males. As a result, there are about 60 adult females (females of marriageable age) for each 50 adult males.

a If four-fifths (80 percent) of the adult males marry, what will be the corresponding fraction (percent) of married adult females?

b If four-fifths (80 percent) of the adult females marry, what will be the corresponding fraction (percent) of married males?

c The number of married males and the number of married females are equal; why, then, do the percents of married males and females differ?

Exercise 11

Accidental deaths may be grouped into two types—those resulting from motor vehicle accidents and those resulting from other causes. In 1960, motor vehicle accidents accounted for about 41 percent of all accidental deaths. In 1976, motor vehicle accidents accounted for about 47 percent of all accidental deaths.

a What percentages of accidental deaths resulted from causes other than vehicle accidents in 1960 and 1976?

b For each of the following statements, discuss whether it is a reasonable conclusion based on the given information: (1) More people died in traffic accidents in 1976 than in 1960. (2) Fewer people died in accidents not

*Actually, statistics kept by the government show that slightly over 51 percent of babies born in the United States are male.

involving motor vehicles in 1976 than in 1960. (3) Home safety educational programs have been successful because the number of accidents that occur in the home has decreased. (4) Drivers were more careless in 1976 than in 1960. (5) Automobiles were safer in 1960 than in 1976.

c What additional information is needed before we can make sensible conclusions based on the given data?

Exercise 12

The National Safety Council reported that the number of deaths resulting from bicycle accidents was around 1100 in 1977, more than double the number (about 500) in 1935.

a Was it safer to ride a bicycle in 1935 or in 1977?

b To compare bicycle safety in 1935 and 1977, what additional information is needed?

Exercise 13

Suppose that in 1935 approximately 3.6 million bicycles were in use and in 1977 the figure had risen to approximately 100 million. Combine this information with that given in Exercise 12 to make a sensible comparison of bicycle safety in the two years.

Exercise 14

Chris's father is going for a medical checkup that will include an x-ray for tuberculosis. Chris has been thinking about the possible state of his father's health. For males his father's age, tuberculosis is a rare affliction; only about 1 percent have it.

Chris is aware that x-rays are not 100 percent reliable. Consequently, he suggests that his father inquire about the chances that the x-ray will give a false reading. The medical examiner states that the x-ray is 95 percent reliable: that is, if you do not have tuberculosis, the x-ray will give you a clean bill of health 95 times out of 100; on the other hand, if you do have it, the x-ray will report that 95 times out of 100.

A few days after the checkup Chris's father receives the results of the x-ray: it indicates that he has tuberculosis.

On the basis of the information given, what are the chances that the x-ray result is correct?

Hint Suppose that there are 1 million men Chris's father's age. Only about 10,000 have tuberculosis; the rest do not. What would be the test results for these two groups?

Exercise 15

In Middle Grove, 100 people showed up at the school board meeting to protest a proposed tax increase. In nearby Lewisville, only 40 people attended the board meeting to protest taxes. If the number of taxpayers in Middle Grove is 15,000 and the number in Lewisville is 5000, in which community did a higher percent show their concern?

Exercise 16

One newspaper reported:

> The unemployment rate among black males last month rose to 15.8 percent—double the unemployment rate of 7.9 percent for white males.

A second newspaper reported:

> Bureau of Labor Statistics figures, just released, report a total of 5 million unemployed white males last month compared with only 1.25 million unemployed black males.

a Explain how it is possible that both reports can be correct.

b From the pair of news items, write a complete description of last month's unemployment situation—a description that uses both numbers and percents.

Exercise 17

a A constant annual interest rate yields an amount of interest that varies with the amount of the investment. Complete Table 3-3 by computing the amount of interest a person would earn in a year for each amount invested.

b Sometimes interest payments are made more often than once a year. For example, interest may be added monthly. A savings institution that advertised 7.5 percent interest, compounded monthly, would pay $\frac{1}{12}$ of 7.5 percent, or 0.625 percent, per month. If \$100 is invested for a year at 7.5 percent, compounded monthly, complete Table 3-4 to find the value of the investment after 1 year.

c Compare the lower right entry in Table 3-4 with the lower right entry in Table 3-3. (Both give the value of $100 after 1 year, invested at 7.5 percent interest.) Explain why they differ.

d How could column 3 of Table 3-4 be calculated directly from column 1 (without calculating column 2)?

TABLE 3-3

AMOUNT INVESTED FOR 1 YEAR	ANNUAL INTEREST RATE	INTEREST	INVESTMENT PLUS INTEREST
$10	0.075		
$25	0.075		
$50	0.075		
75	0.075		
$100	0.075		

TABLE 3-4

MONTH	VALUE OF ACCOUNT AT BEGINNING OF MONTH	INTEREST EARNED (MONTHLY RATE: 0.00625)	VALUE OF ACCOUNT AT END OF MONTH
1	$100.00	$0.62*	$100.62
2	100.62	0.63	101.25
3	101.25	0.63	101.88
4	101.88	0.64	
5			
6			
7			
8			
9			
10			
11			
12			

*Amounts are rounded to the nearest cent. When the digits dropped are 50000 . . . the last remaining digit stays the same if it is even and is increased by 1 if it is odd. Throughout this text, this same rounding rule is used. In actual practice, however, many financial institutions always round to the next *higher* digit when the digits discarded are 500 . . . instead of rounding to the nearest *even* digit.

Exercise 18

Bill, age 11, has just learned about percents in his fifth-grade arithmetic class. His textbook illustrated use of percents in calculating interest earnings. Bill wants to save money for a new bicycle and figures that he'll need about $150. From his older sister, Marty, he learns that the best local savings opportunities are: (1) an account at Eastern Savings that offers 8 percent interest, paid annually, on money invested for an entire year; and (2) an account at Central Savings that offers 7.5 percent annual interest, paid monthly, with 1⁄12 of 7.5 percent paid on money invested for the prior month. Bill asks Marty's help in deciding what to do. The questions below ask you to reproduce some of their calculations.

a If Bill wants to have $150 a year from now, how much would he need to invest now at Eastern Savings so that the amount invested plus the year's interest will total $150?

b Suppose that Bill earns money from helping with household chores and can save $12 per month this way. If he deposits these $12 amounts at Central Savings in an account earning 7.5 percent, compounded monthly, how much will he have accumulated 1 year after his initial investment? To answer this question, complete Table 3-5. (If you want to shorten your work by computing column 3 without computing column 2, it is perfectly all right to do so.)

TABLE 3-5

MONTH	VALUE OF ACCOUNT AT BEGINNING OF MONTH	AMOUNT OF INTEREST EARNED (RATE: 0.00625)	VALUE OF ACCOUNT AT END OF MONTH
1			
2			
3			
4			
5			
6			
7			
8			
9			
10			
11			
12			

c Examining the results obtained above, we see that Bill can accumulate \$150 in a year in two different ways: (1) by investing \$138.89 in an account that pays 8 percent interest compounded annually; or (2) by investing \$12 a month in an account that pays 7.5 percent interest compounded monthly. Which investment plan would you recommend for Bill? Explain why.

Exercise 19: Invent

a In each of two different mathematics classes 10 percent of the students earned A's. However, the number of A's in the one class was 3 times as great as the number of A's in the other. Invent class sizes which show that both of these statements can be true.

b Two mathematics instructors turned in grade lists containing 16 B's. However, one instructor said that 25 percent of her grades were B's while the other said that 40 percent of his grades were B's. Invent class sizes which show that both of these statements can be true.

Exercise 20

Look in current newspapers and magazines, in articles and advertisements, for statements that can make use of percents. Examine them to see whether enough information is given for the figures to be sensibly interpreted.

Exercise 21: Test Yourself

Develop a list of new terms and key ideas presented in this chapter. Supply an example of each item on your list.

Note The Appendix contains complete answers for Exercises 8, 9*b*, 10*c*, and 14. It contains partial solutions or hints for Exercises 7*a*, 17*b*, and 18*b*.

CHAPTER 4

OBSERVE WIDELY; SPECULATE FREELY; TEST CAREFULLY

> There is no harm in being wrong—especially if one is promptly found out.
>
> —John Maynard Keynes

BELIEVE IT OR NOT?

Every day, we hear—and make—statements whose truth is unconfirmed. Sometimes these statements are rules that we would like to be able to apply in problem solving. Should we believe them or not?

Let us use the term *speculations* to refer to potentially true statements. Whenever possible, speculations should be verified by deductive arguments. We can be sure that a speculation is true only if we can find an ironclad argument, based on verifiable premises, that leads logically to the speculation as an unavoidable conclusion.

But sometimes ironclad arguments are impossible—or at least impractical—to find. We don't know enough or, possibly, care enough to search for them. What seems reasonable, instead, is to find a simple means of testing whether the speculation is *probably* true. This chapter considers the testing of speculations by use of examples. Creating an example is an experiment to test whether a speculation is reasonable.

> No amount of experimentation can prove me right; a single experiment can prove me wrong.
>
> Albert Einstein

When a new and difficult problem is identified, the effort to find a solution includes these important steps:

1 Observe widely.
2 Speculate freely.
3 Test carefully.

The initial emphasis in this chapter is on *testing* speculations, since people who become skillful at testing will gradually increase their range of observations and speculations. Even wild speculations and random observations about possible solutions become useful problem-solving techniques once we have skill in testing—because we are then able to separate our good ideas from our bad ones.

TESTING BY CALCULATION: SPECULATION ABOUT POPULATION GROWTH

A continuing concern for the nations of the world is how to meet the nutritional needs of growing populations. This leads them to estimate populations 5, 10, or more years in the future.

Consider the following statement—called the *rule of 70*—for estimating how long a population will take to double:

> If the annual population growth rate is r percent, the population will double in approximately $70/r$ years.

Since no argument establishing the truth of this statement is given here, the statement is a speculation. To test it, let us invent an example:

> Consider a population of 5000 with a growth rate of 10 percent. The rule says that the population will double to 10,000 in about $70/10 = 7$ years.

(Note that when the rule of 70 is used, the percent is *not* converted to a decimal in the calculation.)

To check the rule's prediction, we can perform calculations of annual population growth. The results are given in Table 4-1. A single test cannot prove that the rule of 70 is reliable, but our calculations show at least that it is not unreasonable. Additional tests are suggested in Exercise 1.

TABLE 4-1
Effect of a 10 percent growth rate on a population of 5000

TIME	POPULATION
Present	5000
1 year hence	5500
2 years hence	6050
3 years hence	6655
4 years hence	7320
5 years hence	8052
6 years hence	8857
7 years hence	9743
8 years hence	10,717

Rousseau's Speculations*

The eighteenth-century philosopher Jean-Jacques Rousseau was interested in population problems. He made certain observations about population patterns in eighteenth-century England.

*Adapted from C. A. Lave and J. G. March, *An Introduction to Models in the Social Sciences,* Harper & Row, New York, 1975, pp. 71–73.

The birthrate in London is lower than the birthrate in rural England.

The death rate in London is higher than the death rate in rural England.

As England industrializes, more and more people are leaving the countryside and moving to London.

Rousseau reasoned that since people would continue to move to London—where the birthrate was lower and the death rate was higher—the population of England would eventually decline to zero. History has, of course, shown Rousseau's speculation to be false—but that knowledge needn't have waited for history. It is possible to invent an example that satisfies his observations and to show by calculations that his speculation was wrong.

Suppose, for example, that the information given in Table 4-2 describes the English population of Rousseau's day. We can verify readily that the rates given in Table 4-2 satisfy Rousseau's observations. Table 4-3 shows the population changes that result from these birthrates, death rates, and migration rates over a 9-year period.

In Table 4-3, for each new year the initial population of London is obtained from the preceding year's initial population by adding to it the number of births and migrations and subtracting from it the number of deaths. The initial population each year for rural England is obtained from the preceding year's initial population by adding the number of births and subtracting the number of deaths and migrations. Table 4-3 exhibits trends that show Rousseau's speculation to be incorrect: the population would not, under the given circumstances, eventually decline to zero.

TABLE 4-2
Rates of population change in eighteenth-century England

	CURRENT POPULATION	ANNUAL BIRTHRATE, PERCENT	ANNUAL DEATH RATE, PERCENT	MIGRATION TO LONDON, PERCENT
London	200,000	3	2.5	—
Rural England	1,000,000	4	2.0	1

TABLE 4-3
Effects of rates of population change over 9-year period

	INITIAL POPULATION		NUMBER OF BIRTHS		NUMBER OF DEATHS		
YEAR	LONDON	RURAL ENGLAND	LONDON	RURAL ENGLAND	LONDON	RURAL ENGLAND	NUMBER OF MIGRATIONS
1	200,000	1,000,000	6000	40,000	5000	20,000	10,000
2	211,000	1,010,000	6330	40,400	5275	20,200	10,100
3	222,155	1,020,100	6665	40,804	5554	20,402	10,201
4	233,467	1,030,301	7004	41,212	5837	20,606	10,303
5	244,937	1,040,604	7348	41,624	6123	20,812	10,406
6	256,568	1,051,010	7697	42,040	6414	21,020	10,510
7	268,361	1,061,520	8051	42,461	6709	21,230	10,615
8	280,318	1,072,136	8410	42,885	7008	21,443	10,721
9	292,441	1,082,857	8773	43,314	7311	21,657	10,829
10	304,732	1,093,685	9142	43,747	7618	21,874	10,937

Perhaps Rousseau actually tested his speculation, but he may have used examples that did not reveal its falsity. Exercise 9, below, asks you to carry out calculations for a situation in which birthrates, death rates, and migration rates are different from those of Table 4-2 and for which Rousseau's speculation does hold.

Tables 4-2 and 4-3 and, in contrast, Exercise 9 illustrate a pitfall of testing speculations by examples: different examples may lead to different results. If one example supports a speculation, we should try other examples to confirm that. Any example that contradicts a speculation shows it to be untrue. Often this result should lead us to return to the speculation and try to refine it. A contradictory example may point out a gap in our thinking; it may be the "exception that improves the rule."

TABLE 4-4
Effect of a 14 percent growth rate on a population of 5000

TIME	POPULATION
Present	5000
1 year hence	
2 years hence	
3 years hence	
4 years hence	
5 years hence	
6 years hence	

Exercise 1

a Test the rule of 70 for a population of 5000 with a growth rate of 14 percent by completing Table 4-4. How many years does the rule predict as the doubling time? Does the population actually double in that length of time?

b Test the rule of 70 for a population of 15,000 with a population growth rate of 14 percent by completing Table 4-5. How many years does the rule predict as the doubling time? Does the population actually double in that length of time?

c When we use the rule of 70, our calculations do not involve the population—they have to do with only its growth rate. Compare Tables 4-4 and 4-5 to see that the two different populations, both growing at a rate of 14 percent, double in the same length of time.

TABLE 4-5
Effect of a 14 percent growth rate on a population of 15,000

TIME	POPULATION
Present	15,000
1 year hence	
2 years hence	
3 years hence	
4 years hence	
5 years hence	
6 years hence	

Exercise 2: More Speculation

The rule of 70 applies to investments that earn interest as well as to growing populations. Specifically, if we invest money at an interest rate of r percent for a given time, then the length of time needed for the money to double is approximately $70/r$ time periods.

a At an annual interest rate of 5 percent, how long will it take to double an initial investment of $1000? First use the rule of 70 to predict an answer; then test your prediction by interest calculations.

b At an annual interest rate of 8 percent, how long will it take to double an amount invested? First use the rule of 70 to predict an answer; then test your prediction with interest calculations.

c When interest payments are accumulated semiannually, quarterly, or at the end of some period other than a year, the actual interest rate varies from the stated annual rate. To see this, consider three different savings institutions in the Land of Runaway Inflation, which advertise accounts with 20 percent interest rates.

Institution A credits 20 percent interest to accounts at the end of each year.
Institution B credits 10 percent interest to accounts at the end of each 6-month period.
Institution C credits 5 percent interest to accounts at the end of each three-month period.

Calculate values to complete Tables 4-6, 4-7, and 4-8, and observe the variations that can result from different compounding patterns with the same annual interest rate. Assume an initial investment of $1000 in each account.

d Verify that for accounts in institutions A, B, and C the rule of 70 predicts a doubling time of 3½ years. For which type of account is the prediction most accurate?

Exercise 3

a In the preceding exercises you performed lengthy calculations to test the rule of 70. Look back over your results to assess the accuracy of its predictions. Does it consistently predict doubling times that are too short? Is it more accurate where there are many growth periods at a low rate or where there are few growth periods at a high rate?

b Because the rule of 70 usually predicts a doubling time that is too short, some people prefer an alternative—the *rule of 72.* With this rule the doubling time is estimated as $72/r$. For each of the following, estimate doubling time using both the rule of 70 and the rule of 72 and compare the results with the actual doubling time to determine which rule gives a better estimate: (1) An investment grows at an annual rate of 20 percent. Refer to your calculations in Exercise 2*c.* (2) A population grows at an annual rate of 14 percent. Refer to Exercises 1*a* and *b.* (3) An investment grows at an annual rate of 8 percent. Refer to Exercise 2*b.* (4) An investment grows at an annual rate of 5 percent. Refer to Exercise 2*a.* (5) An investment grows at an annual rate of 2 percent. (6) An investment grows at an annual rate of 1 percent.

Numbers 5 and 6 require calculations not performed in earlier exercises. To save time, don't bother to round to the nearest cent at each step. Use an initial investment of $1000.

Exercise 4

It has been suggested here that examples be created as experiments to test speculations. Yet Einstein said, "No amount of experimentation can prove me right. . . ." How can these different views to be reconciled? Of what value is testing by experimentation?

TABLE 4-6
$1000 invested in A account

TIME	VALUE OF ACCOUNT, DOLLARS
Present	1000
1 year hence	1200
2 years hence	
3 years hence	
4 years hence	

TABLE 4-7
$1000 invested in B account

TIME	VALUE OF ACCOUNT, DOLLARS
Present	1000
6 months hence	1100
1 year hence	1210
1½ years hence	
2 years hence	
2½ years hence	
3 years hence	
3½ years hence	
4 years hence	

TABLE 4-8
$1000 invested in C account

TIME	VALUE OF ACCOUNT, DOLLARS
Present	1000.00
3 months hence	1050.00
6 months hence	1102.50
9 months hence	1157.62
1 year hence	1215.50
1¼ years hence	
1½ years hence	
1¾ years hence	
2 years hence	
2¼ years hence	
2½ years hence	
2¾ years hence	
3 years hence	
3¼ years hence	
3½ years hence	
3¾ years hence	
4 years hence	

Exercise 5

Speedy Construction Company has been asked to do a rush project. It has been offered a choice between a flat fee of \$20,000 and a fee computed by the formula

$$f = \$5000 + \$3000 \times (10 - d)$$

where d is the number of days it will take to complete the project. The company must choose before work starts. Should it choose the flat fee or the formula fee?

Exercise 6

Aunt Rachel has sold her house and wants to invest the money so that it will earn interest. She is planning for her retirement 10 years from now. Her investment choices yield 6 percent, 9 percent, and 12 percent, compounded annually. Because the higher-interest investments are more risky, her choice is not easy. She is considering the following pair of investment strategies:

Strategy A: Invest all the money at 9 percent.

Strategy B: Invest half the money at 6 percent and half at 12 percent.

a Which strategy, A or B, offers Aunt Rachel the greater interest?

b Which strategy would you recommend to her? (Are there questions that you would want answered before you would make a recommendation? What are they?)

Exercise 7

Bart has just learned his grade on his first chemistry examination: 69 percent. He is dismayed; chemistry is his major and he wants to earn an A. The course will include three 1-hour examinations (the one just taken and two more), each of which contribute ⅕ of the final grade. A comprehensive final examination will contribute the remaining ⅖. If Bart must have a term average of 90 percent or higher to earn an A, can he still do it? What grades does he need on future examinations to achieve his goal?

Exercise 8

Southwest School District has an overcrowded high school. The population in this suburban school district has doubled in the last 10 years, and the high school must now accommodate its students in two half-day sessions: 7 A.M. to noon and 12:30 P.M. to 5:30 P.M. The present high school can accommodate 2000 students, and there are 2500 students of high school age.

The school board is considering construction of a new building. However, construction is extremely costly and at least 5 years will pass before it can be planned and completed. Furthermore, the decision about what to do depends on which speculations about the future population of Southwest School District the board chooses to believe.

a If the high school population can be expected to grow at a rate of 7 percent per year for the next 5 years and then level off, what should the school board decide?

b If the high school population can be expected to decline at a rate of 7 percent per year for the next 5 years and then level off, what should the school board decide?

c If the high school population is expected to hold steady in the forseeable future, what should the board decide?

d What is the role of values in this decision?

Exercise 9

The calculations required by this exercise are lengthy. It's a good problem for several heads (and calculators) to work together on.

a Rousseau's conclusion that the population of England would eventually decline to zero was shown, by an example in the text, to be incorrect. Suppose, instead, that the figures in Table 4-9 describe population patterns in eighteenth-century England. Calculate values to complete Table 4-10 to observe that this example supports Rousseau's speculation.

b Compare Tables 4-2 and 4-3 with Tables 4-9 and 4-10. What appears to be the reason that the populations increase in the former situation and decrease in the latter? What relationship needs to hold between birthrates and death rates in London before Rousseau's speculation can follow?

TABLE 4-9
Rates of population change in eighteenth-century England

	CURRENT POPULATION	ANNUAL BIRTHRATE, PERCENT	ANNUAL DEATH RATE, PERCENT	MIGRATION TO LONDON, PERCENT
London	200,000	1	4	—
Rural England	1,000,000	2	2	½

YEAR	INITIAL POPULATION		NUMBER OF BIRTHS		NUMBER OF DEATHS		NUMBER OF MIGRATIONS
	LONDON	RURAL ENGLAND	LONDON	RURAL ENGLAND	LONDON	RURAL ENGLAND	
1	200,000	1,000,000	2000	20,000	8000	20,000	5000
2	199,000	995,000					
3							
4							
5							
6							
7							
8							
9							
10							

TABLE 4-10
Effect of rates of population change over 9-year period

Exercise 10

This problem is more fun to do with several people working together. Although the calculations aren't especially difficult, the problem is lengthy and it is a good idea to have someone else checking at each step.

The town of Placid has an adult population with an even distribution of ages from 21 to 70. Placid's young adults (aged 21 to 30) want to set up a center for community recreation—volleyball, dancing, bridge, etc. The town council is willing to grant use of a large room in the Placid town hall and to provide some funding, if the recreation center is open to adults of all ages. A survey of the population has shown that everybody would like to participate. However, the respondents indicate that their participation is dependent on the ages of other participants. Specifically, each person will join and stay as long as the average* age of the membership does not exceed his or her own age by more than 10 years or fall below it by more than 20 years.

Vera, a 23-year-old loan officer in the Placid bank and an active member of the group advocating the community center, has become concerned. "I'm afraid that after we get the center set up, it will gradually become a center for old people," Vera says.

a Devise an example to test the effect of the age preference on participation in the community center. (A convenient starting situation would be a community of 100 adults, of each age from 21 to 70. The 20 young adults, ages 21–30, establish a community center. The average age of this group is 25½; on the basis of this average, who else will join? Once these others join, what will happen to the average age? Will some members then drop out? Perform calculations and trace through the process of change.)

b Do your calculations in *a* support Vera's concern? What recommendations would you make to the residents of Placid about the establishment of a community center?

*The average referred to here is the (arithmetic) mean. The mean age of a group is the sum of the ages of individual members divided by the number of members.

Exercise 11: Invent

Make up a problem about investment earnings or population changes. Guess an answer. Test your guess by performing calculations.

Exercise 12: Journal

As we try to make a decision, we often speculate about what will happen if we make one choice or another. Sometimes calculations can be useful in testing our speculations. For example, Richard may speculate that if he takes a part-time job, he may not have enough time for his studies. Or Mary may speculate that if she takes a low-paying but pleasant summer job, she may not earn enough to meet her college expenses next fall. Both Richard and Mary could carry out calculations to help them decide.

What decision situations do you face in which calculations could help you? Carry out the calculations. What do they indicate are the best decisions? How are values as well as numbers involved in your decisions?

Exercise 13

a Consider the quotation from Whitehead in the box. Explain his views in your own words and develop an argument to support them.

b Would Whitehead approve of the use of shortcuts in calculations and other steps is problem solving?

c If Whitehead were alive today, what would you expect him to think about the use of calculators and computers?

> Civilization advances by extending the number of important operations which we can perform without thinking of them.
>
> Alfred North Whitehead

Note The Appendix contains complete answers for Exercises 2*c* and *d.* It contains partial solutions or hints for Exercises 1*a,* 3*b,* 5, 6, 8*a* and *b,* 9*a,* and 10*a.*

STUDYING RANDOM EVENTS BY SIMULATION

As we have seen, a speculation can sometimes be tested by calculations. Often, however, we do not know formulas or calculations to use in generating or testing speculations about a problem situation. In such cases we may be able to use simulation.

Simulation is a techique in which a problem solver sets up an artificial situation that mimics the real situation, performs experiments within the artificial situation, and uses the results to predict what will happen in the real situation. For example, tire manufacturers test their product using equipment that mimics or simulates road conditions, so that they can determine durability. Aircraft manufacturers test various designs in wind tunnels that mimic or simulate the stresses of actual flight.

This chapter will consider a particular type of simulation procedure: use of a table of random digits. Such a table (see Table 4-11) can be used as a basis for experiments to study patterns exhibited by random events. In many cases, even though the outcome of an event cannot be predicted, the results of many similar events may show a pattern.

For example, if a fair coin is tossed once, we do not know how it will fall, but if it is tossed 100 times we can estimate the outcome—about 50 heads and about 50 tails. Actual tossing of coins is easily done and thus does not require simulation. Nevertheless, the process is simulated in Example 1 (page 60) because it provides a simple introduction to the technique.

TABLE 4-11
10,000 random digits

Row 1	82698	26610	90511	08055	80364	70233	91451	34528	30357	27456
	93680	27051	67692	57437	08779	81065	50586	20621	28296	43353
	45153	17985	74725	08526	09220	89778	59814	02387	78112	16035
	65055	40547	20834	50243	23998	59708	12313	89349	25103	43682
	80863	76681	73173	48970	91202	81344	89446	60285	12653	95567
Row 6	65704	35329	80233	67505	22518	58994	63968	79316	53447	65610
	16862	82356	69963	61171	96043	56593	73637	82198	51634	71363
	76048	34462	57543	98743	80838	42517	42094	98970	07496	22223
	92003	32221	39595	99113	43596	90842	87684	80098	54888	32782
	74244	90661	80795	20305	92055	54532	99534	34660	41569	88305
Row 11	38128	35924	55245	97971	52694	92422	15875	18971	20058	78333
	33729	56998	99535	52712	21558	36734	24131	95807	80922	85010
	63971	68875	13322	07349	73991	41072	31419	29611	10297	85465
	57653	56330	22804	71402	62635	33217	85828	69039	77095	57063
	36395	30423	96224	53481	23420	44921	30883	56083	32038	63699
Row 16	90543	52660	09346	76795	89783	87944	92379	34576	18055	67418
	58133	19098	70130	16092	43843	80508	96387	42270	35335	18264
	57487	88972	50914	65331	87902	42601	85407	19867	77391	48159
	77128	23219	48346	02047	63984	66444	83317	40167	39020	00798
	13964	87042	24341	25448	30779	30472	92064	71532	47311	33061
Row 21	03114	30226	65252	72519	11706	72966	95952	93649	64857	57621
	41182	02953	20581	46556	03312	24241	54804	29809	04113	75128
	94953	59747	35056	70403	17822	04416	08601	45680	69568	35183
	26528	96679	08165	34005	90199	48983	99761	51229	31275	27314
	71479	66012	23245	49574	10116	41521	06750	29164	63007	55902
Row 26	42292	82996	86159	79513	84410	45582	38596	55311	04895	13515
	63234	72661	32908	22815	30490	01502	52419	97075	95007	03410
	95770	34807	06273	59221	42470	68812	28923	28313	16271	06813
	09463	63823	29643	62401	06537	63918	52056	83389	86422	88943
	37271	74277	85283	81867	66660	40978	80906	50846	32802	18984
Row 31	29627	48227	75458	13027	68341	24267	89088	40988	53103	79923
	64074	46280	63010	53561	12276	25624	43287	38239	56965	03913
	41630	85293	87811	97757	34504	72791	18594	21759	58785	72898
	49567	12521	70419	45853	84408	65065	47690	61921	43879	15782
	16466	91260	29140	78385	54921	16937	33072	60675	18101	77288

TABLE 4-11
Continued

Row 36	78056	96364	45121	63361	65742	03964	01998	36442	97701	85267
	53747	76678	84504	88985	58473	24216	20720	99730	44223	81412
	35832	52303	07091	97235	22488	73307	63024	42020	82151	67453
	00316	98955	28410	62443	65885	14166	61937	71899	05764	73820
	03281	21503	85071	43185	37724	27177	54710	86306	83226	05223
Row 41	88027	73621	54167	61642	15908	74027	32009	40957	00489	50941
	58412	57958	10784	91459	05057	09259	04821	48798	94313	22552
	26713	91036	37943	39567	35577	92419	27216	48996	88339	11204
	74903	84443	30195	38568	91675	53618	40088	24647	70893	99334
	63137	40699	42046	10281	57445	67771	00976	20883	79039	54049
Row 46	45053	55031	75934	73950	89878	20357	03257	14471	24981	42633
	39121	28849	82954	36481	10444	85221	55466	27512	03441	57984
	45968	12528	55047	03065	63942	45232	22368	05620	22057	13135
	62420	23116	23984	50249	42438	47611	68085	84966	08318	18250
	69928	18998	17186	08202	28286	60156	27066	95713	47429	64033
Row 51	68420	91594	18774	99086	97471	11096	83934	54694	99278	53366
	10766	65860	45180	01167	45771	87610	05272	85867	82672	68059
	59782	87239	59260	65113	59876	10642	79247	45118	65702	96858
	50136	54869	77626	25256	27837	49592	37705	01488	05843	88203
	24186	44144	99986	26937	10126	47675	36747	22790	65792	22751
Row 56	32149	32697	40420	71863	11583	02864	25198	15551	90871	10326
	71152	46507	83616	93181	68659	77281	18518	27371	68757	18253
	98480	39095	31712	53194	51924	06287	57890	12455	01641	85052
	88139	98726	64244	13517	03796	92669	46544	77797	63147	47743
	83133	03717	59230	24429	89823	69684	22210	56398	71136	22323
Row 61	53546	43190	65854	17069	11174	88317	85699	13809	67271	94418
	48162	65787	83194	80075	59176	32700	38999	41747	54312	69993
	84814	97006	58212	24471	00035	20523	67758	63351	69789	62877
	17462	44657	39043	05410	13946	13306	73265	42812	81182	47604
	69490	13754	57511	73595	60986	91695	36815	68175	46810	17198
Row 66	44817	89371	87318	64743	96118	62417	34657	53990	18410	85309
	78895	96529	26425	94164	79378	85802	35855	47916	07173	33690
	61302	09781	24426	50261	49587	09675	11506	48489	86292	32713
	37458	11722	96040	26021	43539	68552	16742	38625	30907	03649
	41800	34521	69828	47464	16216	72943	00330	93677	38492	08708
Row 71	14708	41640	22349	89030	62190	14042	13371	62037	33843	04555
	08497	13633	90824	32538	31091	81954	91588	80743	90094	31006
	19085	81561	47008	01014	23479	26661	70725	77994	95119	28515
	05856	76772	94061	87862	56015	86653	82671	06105	50992	52662
	38818	06893	03319	32736	25017	70617	49879	73150	12355	30007
Row 76	74183	52870	70880	87765	25043	05881	69958	33040	06060	99228
	71626	80724	43948	82019	56251	40368	63507	10557	74890	25340
	60240	60570	16600	16414	70969	69191	33937	47968	29374	93538
	05759	07744	12089	90706	94402	10132	75795	27739	88054	67702
	24124	49735	10951	60217	65867	16628	80069	31145	42728	72525

Continued.

TABLE 4-11
Continued

Row 81	49727	52958	52316	95660	66210	64217	20436	32849	24576	40591
	81607	30289	71071	02563	32613	66914	00753	60781	09185	76051
	60087	18737	05805	27414	42912	77982	68504	67410	87694	49195
	93765	51974	86490	26739	16100	58912	99557	29283	52530	14750
	87941	61498	16658	05112	29020	34744	25975	59405	88830	48603
Row 86	04392	42554	34738	97944	69423	22576	36792	02929	35868	45485
	77057	75328	28431	68407	96972	18792	53721	48557	94522	22621
	56462	32852	12144	00576	24336	97318	40797	73018	91053	70169
	39735	44807	82227	46221	60117	04324	05759	45700	13299	65149
	83050	41721	29138	38823	34923	93301	98190	27037	72070	56465
Row 91	62314	32142	24102	69218	20065	76827	04831	07796	34291	50754
	38950	28005	63258	67274	33980	68269	89313	38086	13712	95206
	52433	45631	05969	75331	64046	29691	13143	55478	53127	64545
	64146	82545	12934	80945	73589	33866	10603	70451	66164	44446
	09412	55663	59584	83213	69608	28923	66469	17481	71246	89910
Row 96	04503	72085	84585	52396	99506	26123	61681	38641	65181	34728
	91891	38326	29940	45907	45271	45001	97684	24776	73079	91367
	91115	19370	48461	77755	42845	87704	48785	96845	62677	84985
	34907	68127	68011	77047	78265	87344	65971	04187	40044	73874
	99856	31340	10282	20389	65561	25532	40236	48359	90606	33979
Row 101	27037	72070	56465	62314	32142	24102	69218	20065	76827	04831
	07796	34291	50754	38950	28005	63258	67274	33980	68269	89313
	38086	13712	95206	52433	45631	05969	75331	64046	29691	13143
	55478	53127	64545	64146	82545	12934	80945	73589	33866	10603
	70451	66164	44446	09412	55663	59584	83213	69608	28923	66469
Row 106	17481	71246	89910	04503	72085	84585	52396	99506	26123	61681
	38641	65181	34728	91891	38326	29940	45907	45271	45001	96784
	24776	73079	91367	91115	19370	48461	77755	42845	87704	48785
	96845	62677	84985	34907	68127	68011	77047	78265	87344	65971
	04187	40044	73874	99856	31340	10282	20389	65561	25532	40236
Row 111	48359	90606	33979	09262	40436	09883	01575	68238	27119	17924
	07467	09580	91949	40502	88651	71376	75607	04357	90371	55872
	49553	14062	00424	02124	37379	05349	75145	03491	39624	85800
	33440	18028	58970	77255	79334	17183	16797	99398	21953	62722
	94755	55821	45393	22103	24316	91264	00600	80582	09369	12410
Row 116	40878	94006	39046	35972	34043	37932	18147	85569	27222	78437
	76132	77876	78520	68648	06763	27722	33982	07731	09320	41612
	68604	38954	98664	75455	11612	39453	75065	14744	59921	71415
	50293	93191	28657	26168	37711	39518	95835	44677	36107	36507
	07572	15191	30913	31779	31882	90985	61839	49827	24802	60107
Row 121	21226	86662	23537	76993	20755	68716	04093	17531	53987	29851
	25315	35166	39491	66700	23618	10847	71758	05973	63890	23930
	62320	60272	14194	25682	22076	78290	11417	96950	64187	44749
	33095	96290	79305	94440	15733	97670	60119	33895	25756	68336
	58181	97886	19181	72545	45880	92045	45942	97354	24978	44064

TABLE 4-11
Continued

Row 126	79137	85011	25825	77765	48999	27891	41211	97539	29325	73742
	74608	11499	79360	04111	45656	74946	35111	80164	47756	71792
	73792	92958	81557	58549	41130	81741	79532	01009	95449	36689
	14460	65984	03756	10929	16602	56985	00066	82737	88829	89605
	65415	23352	22492	61876	33008	84859	39280	36179	72862	56485
Row 131	42665	91326	49901	98208	32107	71929	80351	72674	21477	92890
	73228	68631	21125	69520	54548	27554	24653	07923	80316	20491
	42600	44409	28867	46699	87567	93070	24929	47224	50096	61129
	47795	04434	64756	94677	32392	77386	10028	87136	67855	36325
	00674	55308	78280	52420	56172	35985	32502	07917	60875	32698
Row 136	96948	49982	34663	81020	08403	41764	80806	98992	33011	83328
	09165	79907	39990	25101	13110	59756	53055	38281	15223	55161
	51521	72704	16153	91634	00414	39927	18382	84077	38849	69007
	24470	48081	21733	32191	43908	90379	16210	29556	90426	65542
	79487	91238	75526	59403	49369	18605	80756	07663	91091	51813
Row 141	48699	06823	83560	81900	40815	00607	50010	78251	98775	05996
	62469	93915	04158	27301	73810	07137	80416	82445	51490	79066
	38220	15919	33125	71986	79159	12636	86029	84827	85675	18296
	67209	67163	99349	39213	22891	73833	92740	91488	28799	00213
	06540	99623	13250	90540	13108	89249	62160	44891	54297	56239
Row 146	14753	02313	03396	37114	25045	34352	90955	57786	51942	32876
	07757	10548	60790	46231	67260	34411	98285	47594	71957	54096
	23751	94093	31487	75997	91215	80803	59431	13014	69782	54283
	46205	93850	94388	07478	96634	32036	33138	70445	38034	54464
	43158	43643	57121	76627	71454	97036	72365	13356	92051	73963
Row 151	05685	67447	73286	07963	62056	60432	63421	07580	72783	51002
	11660	88785	25096	15556	68804	51431	08324	46281	59901	75352
	19106	09665	36862	59300	57953	89717	41554	32228	99023	79115
	20960	45950	54946	44967	23065	26049	10121	79641	96642	85528
	65018	57366	60878	71150	17045	63518	79555	13778	44533	72487
Row 156	96558	42015	54031	77461	44942	96934	36057	54767	77631	94379
	77942	86764	27579	66873	76015	49750	57561	52388	31914	15033
	81681	81749	30071	81901	82813	91797	71713	39843	72933	04286
	52570	35966	06913	47050	51782	83549	35859	60841	87978	46089
	84425	73503	28578	11685	16815	15978	71920	60815	53895	16900
Row 161	25535	79807	68457	14232	78851	32810	76576	39616	18218	22350
	48138	11077	43245	25517	98105	79268	41262	05899	99484	52920
	11627	66583	65057	70008	51932	73879	68870	82455	48444	10163
	31856	07704	91883	61734	18850	65907	98368	11990	52967	05709
	37440	43130	86051	46031	06637	34541	62925	29648	94378	56422
Row 166	35746	62282	17094	04250	27396	04887	56023	53245	92661	85005
	97804	87773	82534	15870	49848	80953	31171	43955	80603	27364
	91175	98163	28490	92192	06701	09367	32946	69758	00663	96301
	38302	33327	44826	09816	96198	79200	63473	66838	34831	50893
	09180	07928	29973	81575	15131	51129	60741	16445	36119	49166

Continued.

TABLE 4-11
Continued

Row 171	55666	50999	13101	18800	83267	68427	99189	89986	40883	22136
	29038	91236	46064	39305	25733	73713	58019	61522	70278	45161
	84577	95714	88417	56123	14788	03531	11364	15741	54252	80109
	13885	72669	93446	02807	17864	56620	29762	51934	65284	97767
	31558	13888	11120	13444	25443	62756	99367	65893	82758	71926
Row 176	51899	52677	07831	47908	40581	22500	91302	52878	90316	06545
	20509	72146	63389	23334	13954	80998	83254	22812	38984	72714
	03217	22731	94040	21712	76040	63934	88640	12379	25720	75254
	88124	29703	13149	86345	03022	36374	74372	16250	40762	29498
	41194	15611	78675	81355	85500	01372	06289	87351	32462	15277
Row 181	20934	10776	24867	85067	30259	57928	13688	40697	12574	80184
	44809	12889	67228	84743	49216	38306	81830	34922	71881	37172
	62731	23281	74601	51960	90468	28746	80548	07593	71242	10987
	54498	35814	84711	86001	27335	75586	48296	10307	33053	45522
	20236	85252	25616	86696	28140	76975	56010	17620	31577	78843
Row 186	50128	86377	67646	45256	80934	14582	93088	82511	68214	30086
	18022	72895	58151	83733	21216	80519	92430	82367	29961	00058
	25245	68730	00615	18502	99241	78775	52898	96479	47632	23135
	64830	21037	12282	53755	34260	95694	58986	05376	69326	73447
	18707	81752	39539	50986	24542	48826	40438	01398	30363	17431
Row 191	82453	18982	80155	17210	62812	68135	25603	88137	58265	34246
	00981	58914	69021	10403	47073	57405	83520	62114	19764	14452
	08392	82666	30928	69790	21875	57349	73750	41290	32488	49351
	51013	70767	27662	54188	27703	95973	97846	46031	44783	25761
	37469	09296	35058	19965	47830	17051	53582	78327	05699	66906
Row 196	32261	50791	29185	29158	35977	65910	96837	80985	95785	62837
	96226	15685	44491	50286	36609	17567	46168	90818	04408	41019
	25690	99568	88380	91418	43852	99198	98576	42972	33829	27422
	00504	88071	48494	14313	22705	83812	13743	98413	34617	32793
	89449	00747	33760	18193	19696	68067	26374	98229	38250	12026

Source: *UMAP Journal,* vol. 1, no. 2, 1980, pp. 82–85; reprinted with permission.

Example 1 ▶

A pair of coins is tossed many times. What portion of the tosses can be expected to result in a pair of heads? A pair of tails? A pair showing one head and one tail?

To solve this problem, we will use a table of random digits, Table 4-11. A *table of random digits* is a table, generated by a computer, in which any of the digits 0, 1, 2, 3, 4, 5, 6, 7, 8, 9 is equally likely to occur in any position. Since, in such a table, half of the digits would be expected to be even and half odd, there are these correspondences between digits of the table and results of a toss of a fair coin:

Even digits (0, 2, 4, 6, 8) designate heads.

Odd digits (1, 3, 5, 7, 9) designate tails.

For example, the first 10 digits of row 106 of Table 4-11 are:

17481 71246

They are interpreted as:

TTHHT TTHHH

Starting in row 106 of Table 4-11, we can read 100 pairs of random digits and interpret them as 100 tosses of a pair of coins. The details of doing this are shown in Table 4-12.

A summary of the simulation results, obtained by counting the number of outcomes of each type in Table 4-12, is given in Table 4-13, which suggests answers to the questions at the beginning of this example: both coins will turn up heads about one-fourth of the time; both will turn up tails about one-fourth of the time; one will turn up heads and the other tails about half of the time. Exercise 14 will ask you to test this speculation by further simulation. ◀

TABLE 4-12
Simulation of 100 tosses of a pair of coins

17481 71246 89910 04503 72085 84585 52396 99506 26123 61681
TTHHT TTHHH HTTTH HHTHT THHHT HHTHT THTTH TTTHH HHTHT HTHHT

38641 65181 34728 91891 38326 29940 45907 45271 45001 96784
THHHT HTTHT THTHH TTHTT THTHH HTTHH HTTHT HTHTT HTHHT THTHH

24776 73079 91367 91115 19370 48461 77755 42845 87704 48785
HHTTH TTHTT TTTHT TTTTT TTTTH HHHHT TTTTT HHHHT HTTHH HHTHT

96845 62677 84985 34907 68127 68011 77047 78265 87344 65971
THHHT HHHTT HHTHT THTHT HHTHT HHHTT TTHHT THHHT HTTHH HTTTT

TABLE 4-13
100 simulated tosses of a pair of coins: Summary of results

OUTCOME	NUMBER OF EXPERIMENTS WITH GIVEN OUTCOME	PROPORTION OF EXPERIMENTS WITH GIVEN OUTCOME
Both heads	24	24/100 or 24%
Both tails	27	27/100 or 27%
One head, one tail	49	49/100 or 49%

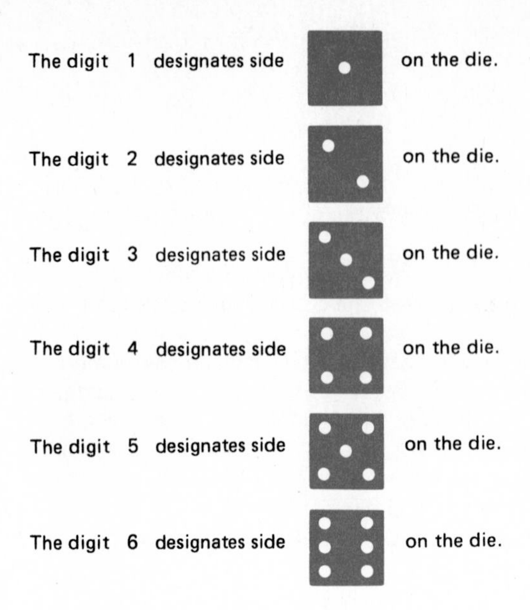

FIGURE 4-1 The digits 7, 8, 9, 0 are ignored.

Example 2 ▶

A slightly different way of using the random-digit table would be suitable if we wanted to simulate tossing a balanced six-sided die. The interpretation of digits shown in Figure 4-1 would be easy to use. When the digits 7, 8, 9, and 0 are ignored, the remaining digits are still randomly distributed.

It is reasonable to speculate that in a large number of tosses of a die, the proportion of times that each outcome occurs will be about $\frac{1}{6}$. Exercise 16 will ask for a test of this speculation by simulation. ◀

Example 3 ▶

We now turn to a realistic—but more complicated—application of simulation.

Faye and Rick were married recently and are beginning to plan for the time when they will have a family. They both want a boy and a girl. As they look ahead, economic considerations seem important. They wonder, "How many children will we need to have in order to get our boy and our girl? Will we be able to afford that many children?" Friends advise them just to wait and see what happens. But Faye and Rick want to know what is likely to happen.

Tentatively, they have decided on the following rule for planning their family:

> Stop having children when we have at least one boy and at least one girl, or whenever we have four children—whichever comes first.

Although the rule seems reasonable, Faye and Rick remain curious about what will probably happen: "How likely are we to achieve our goal of at least one boy and one girl?"

To answer Faye and Rick's questions about their future family, let us turn to simulation. To simplify matters, suppose that when children are born to Faye and Rick, each child has an equal chance* of being a boy or a girl.

The random-digit table can be used to simulate possible family situations in the following way:

> Even digits (0, 2, 4, 6, 8) designate girls.
> Odd digits (1, 3, 5, 7, 9) designate boys.

Let us use row 43 as a starting point in Table 4-11. Row 43 begins with these digits:

26713 91036

They represent the outcomes

GGBBB BBGBG

The first three digits represent the family GGB, which is satisfactory according to Faye and Rick's family-planning rule. The first simulation experiment has thus yielded a 3-child family with 2 girls and 1 boy.

For the second experiment, continue reading random digits. The next four digits are odd—designating four boys. Despite the fact that there is no girl, Faye and Rick will stop with four boys, since that is what their rule stipulates.

We go on with this procedure for 40 simulation experiments—as shown in Table 4-14. Table 4-15 summarizes the results of the simulation experiments in Table 4-14. These results suggest that Faye and Rick have a good chance (45 percent) of having one child of each sex with only two children. But there also seems to be a significant chance (27½ percent) that they will have four children of the same sex. This possibility may cause them to think twice about considering a large family while they try to achieve one child of each sex. Exercise 17 will ask you to continue this simulation experiment to see if these patterns continue to hold. ◀

---

*Actually, statistics kept by the government reveal that slightly over 51 percent of the babies born in the United States are male. In some population subgroups the deviation from 50 percent is even greater. However, our examples all will assume the (approximately true) distribution of births as 50 percent male and 50 percent female.

TABLE 4-14
40 simulation experiments testing Faye and Rick's family-planning rule

Random digits	26713 91036 37943 39567 35577 92419 27216
	GGBBB BBGBG BBBGB BBBGB BBBBB BGGBB GBGBG
Experiment	1 2 3 4 5 6 7 8 9 10 11 12 13
Family size	3 4B 2 2 3 4B 2 4B 3 2 2 2 2
Random digits	48996 88339 11204 74903 84443 30195 38568
	GGBBG GGBBB BBGGG BGBGB GGGGB BGBBB BGBGG
Experiment	14 15 16 17 18 19 20 21 22 23 24 25
Family size	3 2 3 4B 4 2 2 4G 3 4B 2 3
Random digits	91675 53618 40088 24647 70893 99334 63137
	BBGBB BBGBG GGGGG GGGGB BGGBB BBBBG GBBBB
Experiment	26 27 28 29 30 31 32 33 34 35 36 37
Family size	2 4B 2 4G 4G 3 2 2 4B 2 2 4
Random digits	40699 42046 10281 57445 67771 00976 20883
	GGGBB GGGGG
Experiment	38 39 40
Family size	3 2 4G

TABLE 4-15
Summary of 40 simulation experiments testing Faye and Rick's family-planning rule

EXPERIMENTAL RESULTS	NUMBER OF EXPERIMENTS WITH GIVEN RESULT		PROPORTION OF EXPERIMENTS WITH GIVEN RESULT	
2 children	18		18/40 or 45%	
3 children	9		9/40 or 22½%	
4 children (mixed sexes)	2	13	2/40 or 5%	32½%
4 children (same sex)	11		11/40 or 27½%	

Simulation is a problem-solving technique that enables us to create an experimental model of a problem situation. Using the model, we can

- Observe what happens when the experiment is repeated many times.
- Speculate about patterns that we observe in the results of repeated experiments.
- Test our speculations by continued experimentation.

Simulation is a technique to use when we don't know how to solve a problem but do know how to mimic it with a model. As the examples and the exercises below indicate, performing the large number of experiments needed for accurate results is tedious. When a computer is available, however, much of the tedium is avoided. Most computers offer random-digit generators, and simple programs can be written to perform experiments and tabulate the results.

How to Use a Table of Random Digits for Simulation

1 Describe the problem to be simulated.

2 List the random events in the problem situation.

3 Decide on a scheme for associating the random digits of the table with random events.

4 Describe a method for interpreting consecutive random digits read from the table as experiments that mimic the problem situation.

5 Decide on the number of experiments you will perform. Remember: The more repetitions, the more accurate the results.

6 Pick a random starting point in the table.* Read enough digits to complete one instance of the experiment. From that point on, continue reading the digits consecutively, starting each new experiment where the last one ended, until you have completed as many repetitions as desired.

7 Tabulate the results. Examine them. Do you observe any patterns? Formulate tentative conclusions to test with further simulation.

Illustration: Following Steps 1–7 for Example 1

1 Two coins are tossed. What is the likelihood of each of the outcomes: two heads, two tails, one of each?

2 Each of the coins can turn up either H or T.

3 $\{0,2,4,6,8\} \longleftrightarrow$ H
$\{1,3,5,7,9\} \longleftrightarrow$ T

4 Read pairs of digits. Interpret each pair as the result of a two-coin toss.

5 Our example included 100 repetitions.

6 Our random starting point was row 106.

7 Table 4-13 summarizes the results. These experiments suggest that when two coins are tossed, both will be heads about one-fourth of the time, both will be tails about one-fourth of the time, and there will be one of each about one-half of the time. Exercise 14 asks for further simulation to test this tentative conclusion.

* This is often done by the "blind stab" method: without looking at the table, place your finger down on it, and start reading the table at the number to which your finger points.

Caution Accurate completion of the exercises that follow will require a great deal of care and organization. Check your results as you go along. Working with another person and checking your results can enhance speed, accuracy, understanding, and enjoyment. (It will be easier to do the exercises involving Table 4-11 if you make one or two photocopies of the table. Then, on your copies, you can mark digits and carry out simulation—so that you will not need to hand-copy any parts of the table.)

TABLE 4-16
200 simulations of two-coin-toss experiment: Summary of results

OUTCOME	NUMBER OF EXPERIMENTS WITH GIVEN OUTCOME	PROPORTION OF EXPERIMENTS WITH GIVEN OUTCOME
Both heads		
Both tails		
One head, one tail		

Exercise 14

This exercise is a continuation of Example 1. Use the first digit of row 161 in Table 4-11 as your random starting point. Perform 200 repetitions of the experiment of tossing two coins and tabulate the results in Table 4-16. Do your results support the speculation that one-fourth of the time such tosses will result in two heads, one-fourth of the time they will result in two tails, and half the time they will result in one of each?

FIGURE 4-2

1 ⟷ die turns up

2 ⟷ die turns up

3 ⟷ die turns up

4 ⟷ die turns up

5 ⟷ die turns up

6 ⟷ die turns up

Digits 7, 8, 9, 0 are ignored.

Exercise 15

In Example 1 we used as a basis for our simulation the correspondence

$$\{0,2,4,6,8\} \longleftrightarrow \text{H}$$
$$\{1,3,5,7,9\} \longleftrightarrow \text{T}$$

Would the following correspondence have served as well?

$$\{1,2,3,4,5\} \longleftrightarrow \text{H}$$
$$\{6,7,8,9,0\} \longleftrightarrow \text{T}$$

Why or why not? What other correspondences might be used?

Exercise 16

Example 2 sets up the correspondence shown in Figure 4-2 for use of random digits to simulate the tossing of a die. Using the thirty-first digit of row 132 as a random starting point, simulate 300 tosses of a balanced die using Table 4-11. Before you begin the simulation, guess how many times each number will turn up. Compare your experimental results with your guesses. Does your experi-

TABLE 4-17
300 repetitions of die-tossing experiment: Summary of results

OUTCOME: NUMBER OF SPOTS SHOWING ON DIE	GUESS: HOW MANY TIMES WILL THIS OUTCOME OCCUR?	ACTUAL NUMBER OF EXPERIMENTS WITH THIS OUTCOME	PROPORTION OF EXPERIMENTS WITH THIS OUTCOME
1			
2			
3			
4			
5			
6			

mentation confirm your guesses as reasonable? What would you expect if you had performed the experiment 3000 times? Table 4-17 is provided for a summary of your simulation results.

Exercise 17

This problem is a continuation of Example 3. Our simulation, which involved 40 repetitions of Faye and Rick's future family situation, ended with the fifteenth digit of row 45 in Table 4-11. Continue with this simulation—repeating the experiment 60 more times—to obtain a total of 100 repetitions. Start with the sixteenth digit of row 45. Summarize your results in Table 4-18. Use the simulation results as a basis for answering the following questions:

a What percent of the time would a family expect to have (at least) one child of each sex when following Faye and Rick's family-planning rule?

b What percent of the time would a family expect to have all children of the same sex when following Faye and Rick's family-planning rule?

TABLE 4-18
Summary of 100 simulation experiments testing Faye's and Rick's family-planning rule

	NUMBER OF EXPERIMENTS WITH GIVEN RESULT			
EXPERIMENTAL RESULTS	EXAMPLE 3: 40 REPETITIONS	EXERCISE 17: 60 REPETITIONS	TOTAL	PROPORTION OF EXPERIMENTS WITH GIVEN RESULT
2 children	18			
3 children	9			
4 children (mixed sexes)	2			
4 children (same sex)	11			

TABLE 4-19
George's predictions

OUTCOME	EXPECTED PROPORTION
2 children	½ or 50%
3 children	¼ or 25%
4 children (mixed sexes)	⅛ or 12½%
4 children (same sex)	⅛ or 12½%

Exercise 18

George, a friend of Faye and Rick who has studied probability theory, told the couple that the proportions given in Table 4-19 describe the likelihood of each possible outcome of their family-planning rule. Do your experimental results (Exercise 17) agree with George's expected proportions?

Exercise 19

In some societies, male children have been favored, for various reasons (for example, they "carry on the family name"). Suppose that in one such society all families have agreed on the following family-planning rule:

> Have children until the first boy is born or until the number of children is five; then stop.

Devise a simulation experiment, using Table 4-11, to discover the results of such a family planning policy. Use the correspondence

$$\{0,2,4,6,8\} \longleftrightarrow G$$
$$\{1,3,5,7,9\} \longleftrightarrow B$$

to interpret the digits of the table. Repeat the experiment 100 times starting with the eleventh digit of row 87. Summarize your results in Table 4-20. Look for patterns in your simulated data. Speculate about the probable consequences of the given family-planning rule.

TABLE 4-20

OUTCOME: NUMBER AND SEXES OF CHILDREN		NUMBER OF EXPERIMENTS WITH THIS RESULT	PROPORTION OF EXPERIMENTS WITH THIS RESULT
1	B		
2	GB		
3	GGB		
4	GGGB		
5	GGGGB		
	GGGGG		

Total number of boys born in 100 simulated families: ________
Total number of girls born in 100 simulated families: ________

TABLE 4-21

NUMBER AND SEXES OF CHILDREN IN FAMILY		PROPORTION OF FAMILIES IN THIS CATEGORY
1	B	½ or 50%
2	GB	¼ or 25%
3	GGB	⅛ or 12½%
4	GGGB	1⁄16 or 6¼%
5	GGGGB	1⁄32 or 3⅛%
	GGGGG	1⁄32 or 3⅛%

Exercise 20

If all the families in Exercise 19 are financially and biologically able to have as many as five children, and if each birth has equal chances of producing a boy or a girl, then one might reason as follows about the outcomes of the given family planning rule:

Half the families will have a boy as their first child and will then stop having children.

Half the remaining families will have a boy as their second child and will then stop.

Half the remaining families will have a boy as their third child and will then stop.

Half the remaining families will have a boy as their fourth child and then stop.

Half the remaining families will have a boy as their fifth child, and *all* families will stop.

This reasoning suggests the proportions of families in each group shown in Table 4-21. Compare these proportions with the simulation results of Table 4-20. How well do they agree? Can either list of proportions be used as predictions of what will actually happen in a group of families that follow the given family-planning rule?

Exercise 21

a In your experimental results in Exercise 19, what percent of the families achieved their goal of having a boy?

b What percent of the families had at least one girl?

c Is it surprising that the answers to *a* and *b* differ substantially? Explain why they do.

d What percent of the total number of children born to the 100 simulated families in Exercise 19 are boys?

e Is it surprising that the answers to *a* and *d* differ substantially? Explain why they do.

Exercise 22

Testing by simulation enables us to consider a more realistic version of Mr. Dogood's problem of hiring a secretary (Chapter 2). Suppose now that Nine-to-Five Employment Agency promises to keep sending applicants until Mr. Dogood finds one who is satisfactory. Mr. Dogood must, as before, make a hiring decision on the spot. On the basis of his previous hiring experience, Mr. Dogood thinks that he would like to use the following decision rule:

> Keep interviewing until you find an applicant of top quality or one whose quality is higher than that of the preceding applicant. Hire that one.

Mr. Dogood will rate each candidate using a scale from 1 (super) to 5 (ugh).

The day before interviewing is scheduled to start, Mr. Dogood decides to test his decision rule. He uses a table of random digits and interprets digits as shown in Figure 4-3. Starting at a random point in the table, Mr. Dogood reads digits until he comes to a 1 or to a digit smaller than the one just preceding. This ends one experiment, and the final digit read gives the quality of the applicant hired. The next experiment begins with the next digit of the table.

Perform simulations like those used by Mr. Dogood. Pick column 16 of row 112 as a random starting point in Table 4-11. Repeat the hiring experiment 100 times; keep track of both the quality level of each applicant selected and the number of interviews needed. Summarize the results in Table 4-23. Your simulation will begin as shown in Table 4-22. On the basis of the simulation results you have summarized in Table 4-23, would you recommend to Mr. Dogood that he use the proposed hiring decision rule? Explain why or why not.

FIGURE 4-3

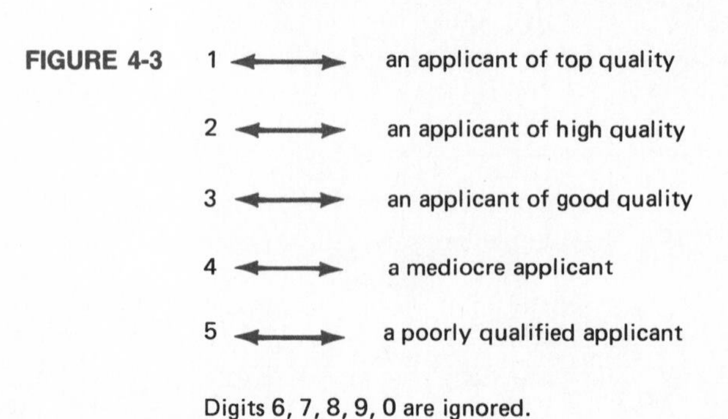

TABLE 4-22

Random digits	40̸50̸2	8̸8̸651	7̸1	37̸6̸ 7̸56̸0̸7̸ 0̸4	357̸ 9̸0̸3	7̸1	558̸7̸2
Experiment	1	2	3	4	5	6	7
Quality of selected applicant	2	1	1	4	3	1	2
Number of interviews needed	3	2	1	3	3	1	3

TABLE 4-23
Testing Mr. Dogood's hiring decision rule: Summary of results from 100 simulation experiments

QUALITY OF SELECTED APPLICANT	NUMBER OF EXPERIMENTS WITH THIS RESULT	PROPORTION OF EXPERIMENTS WITH THIS RESULT
1		
2		
3		
4		
5		
Average number (mean) of interviews needed: ________		

Exercise 23

For each of the following problems, *describe* a simulation that could be used to study it.

a Carol and Fred want four children—two boys and two girls. They reason that since half of children born are boys and half are girls, then they have a 50 percent chance of having two of each sex. Are they correct?

b Dr. Kidney is a surgeon who specializes in a risky operation on critically ill patients. The success rate for this operation is 50 percent—that is, 50 percent of patients who have this operation survive. Two such operations are scheduled for next week. Dr. Kidney reasons that there is a 50 percent chance that both patients will survive. Is he correct? Three such operations are scheduled for next month. What is the likelihood that all three of these patients will survive? If the success rate changes—say, to 70 percent—how would you modify your simulation?

c A certain lottery sells tickets which, when a covering substance is rubbed away, reveal a hidden letter. The letter may be N, O or W. Of the tickets, 60 percent contain the letter N, 30 percent contain the letter O, and 10 percent contain the letter W. You win if you hold three cards spelling NOW. What proportion of the time will you win with only 3 cards? With only 4 cards? What is the average number of cards needed to win?

Exercise 24: Invent

Make up a problem involving random events that can be studied by simulation. Describe how you would set up a simulation experiment and how you would use the table of random digits. Make a guess about how the experiments will turn out.

Exercise 25: Journal

It is difficult to find applications of random-digit simulations that apply to personal problems. In spite of this, the general process of simulation can be quite useful. For example, you might simulate taking a test in physics by selecting a varied group of problems from relevant chapters and trying to solve them in a fixed time. Or you might try to prepare for a speech by delivering your words to a mirror or to a friend. These preparations mimic or simulate the actual experience that is yet to come and help you to foresee possible difficulties.

Stop and think about ways that you can use simulation as a personal problem-solving tool; note these ideas in your journal.

Exercise 26: Test Yourself

Develop a list of new terms and key ideas presented in this chapter. Supply an example of each item on your list.

Note The Appendix contains complete answers for Exercises 16 and 17. It contains partial solutions or hints for Exercises 14 and 22.

PART TWO

APPLYING MEASUREMENT AND ARITHMETIC TO PROBLEM SOLVING

PART CONTENTS

CHAPTER 5

USING NUMERICAL MEASURES TO PROMOTE PRECISE THINKING

The Value of Measurement Scales
Types of Measurement Scales
Utility: Measuring Strengths of Preferences

CHAPTER 6

WHEN IS ADDITION APPROPRIATE?

What Mathematical Process Suits a Given Problem?
When Should We Add?

CHAPTER 7

USING MEASUREMENT IN DECISION MAKING: MATRICES AND WEIGHTED AVERAGES

Forming Matrices
Assigning Weights and Computing Weighted Averages

CHAPTER 8

MAXIMIZING SATISFACTION; MINIMIZING COST

Introduction
A Problematic Situation
A Utility Matrix
Getting the Most from a Fixed Budget
More on Marginal Utility
Shortcuts
Summary

CHAPTER 5

USING NUMERICAL MEASURES TO PROMOTE PRECISE THINKING

> When you can measure what you are speaking about, and express it in numbers, you know something about it; but when you cannot measure it, when you cannot express it in numbers, your knowledge is of a meager and unsatisfactory kind: it may be the beginning of knowledge, but you have scarcely, in your thoughts, advanced to the stage of science.
>
> —William Thomson, Lord Kelvin

THE VALUE OF MEASUREMENT SCALES

Lynn had just returned to the dormitory from her first date with Paul. Her friends gathered around. Questions came quickly from all directions: "Did you have fun?" "How did you like him?" "Will you go out with him again?"

Lynn collected her thoughts for a moment and then replied, "On a scale from 1 to 10, Paul rates a 6." Some of the questioners were satisfied; others kept on: "What did you like about him?" "What didn't you like?" Did you like him better than Ben?"

Measurement scales, such as the one invented by Lynn to answer the questions fired at her, can be useful in solving problems. The effort to assign numbers to objects is an attempt at precise thinking. Increased knowledge about something and increased ability to measure its attributes generally go together.

Portions of this chapter first appeared in: UMAP Instructional Module 546, Consortium for Mathematics and Its Applications, Lexington, Mass., 1981.

The concurrence of increased knowledge and increased ability to measure is evident in science. Our knowledge of the universe has increased as we have become better able to measure distance, radiation, and gravitational attraction. Knowledge of the human body has increased along with our ability to measure body temperature, blood pressure, heart rate, and glandular activity.

Because measuring an object's attributes enables us to know more about it, measurement is a useful problem-solving device. We can use established measurement scales—such as those in the sciences—or we can devise our own scales (as Lynn did) suitable for describing new situations and solving new problems.

The last section of this chapter ("Utility: Measuring Strengths of Preferences") illustrates possibilities for devising new measurement scales. First, however, some background material will be considered: common types of measurement scales and their uses and limitations.

TYPES OF MEASUREMENT SCALES

Exactly what does it mean to measure? First, when we measure, we do not describe an object in its entirety; we describe only some chosen attribute of it. For example, if the object is a pencil, we might measure

Its length
Its weight
Its circumference
The hardness of its lead

Second, the result of a measurement is always a number. To *measure* an attribute means to assign numbers to entities as a way of making distinctions related to that attribute.

Any prescribed assignment of numbers to entities is called a *measurement scale.* Following are some examples of measurement scales:

The coding scheme 0 ↔ male, 1 ↔ female used in a computerized employment file to designate the sex of each employee.
House numbers that label dwellings on a street.
Year numbers that tell how much time has passed since the birth of Christ.
Numbers of centimeters that tell the heights of different people.

Measurement scales differ in the amounts of information that they provide. A measurement scale is called *nominal* if the numbers tell only the categories into which the measured entities fall. The coding scheme

$$0 \longleftrightarrow \text{male}$$
$$1 \longleftrightarrow \text{female}$$

is an example of a nominal scale. Postal zip codes, telephone numbers, and numbers on football players' jerseys are also nominal measures.

A measurement scale is called *ordinal* if the numbers tell whether one entity has more of the measured attribute than another. Odd (or even) house numbers provide an example of an ordinal scale. Knowing that one house number is larger than another tells us that the house bearing the one number is farther along the street than the house bearing the other number. A ranking of choices—first, second, third, and so on—is another example of an ordinal scale. The numbers tell whether one ranked item is preferred over another.

A measurement scale is called an *interval* scale if the numbers are expressed in terms of a fixed unit. When an interval scale is used, subtraction of measure numbers leads to sensible answers. For example:

> When we use calendar year numbers to locate events in time, we are using an interval scale. Using year numbers enables us to put events in order (as with an ordinal scale) and also to calculate differences in measures.
>
> If we know that Kenneth graduated from high school in 1975 and graduated from college in 1981, we not only know that his college graduation occurred later than his high school graduation but also know how much later. Subtraction of the year numbers gives $1981 - 1975 = 6$ (years).

But, in contrast,

> For house numbers, subtraction does not give precise information. Although a house at 1981 Lincoln Avenue is farther along that street than the house at 1975 Lincoln Avenue, even if we subtract ($1981 - 1975 = 6$) we do not know how much farther, because we have no unit of measure.

Measurement of temperature, in degrees Celsius or Fahrenheit, supplies another example of an interval scale.

A measurement scale is called a *ratio* scale if it is based on a fixed unit of measure and zero designates the minimum possible measure. Most scientific measurements—such as length, mass, temperature (on the Kelvin scale), and elapsed time (length of time needed for an event to take place)—are measured using ratio scales. When a ratio scale is used, ratios of measure numbers lead to sensible answers. But ratios are not sensible for an interval scale, as is shown by the comparison below between spans of bridges and spans of time.

When we use feet to measure lengths of objects, we have a ratio scale. The foot is the fixed unit of measure; 0 feet designates the minimum possible length. If we know that the George Washington Bridge to New York City is 3500 feet in length and that the Benjamin Franklin Bridge to Philadelphia is 1750 feet in length, not only do we know that the George Washington Bridge is longer than the Ben Franklin Bridge and the difference in length between the two bridges is 1750 feet; we also know that the George Washington Bridge is *twice as long* as the Ben Franklin Bridge ($3500/1750 = 2$).

In contrast, if we know that in the year 611 A.D. Mohammed received a revelation from Allah that led to the development of Islam, and in the year 1833 A.D Oberlin College, the first coeducational institution of higher education in the United States, was founded, even though $1833/611 = 3$, it is not sensible to say that Oberlin's founding occurred three times as late in time as Mohammed's revelation.

The ratio of bridge lengths is sensible because the measure numbers describe entire spans, 0 feet being the minimum possible measure. On the other hand, 611 A.D. and 1833 A.D. are not measures of the entire spans of time leading up to Mohammed and Oberlin; they measure only the time elapsed since the birth of Christ. Because 0 A.D. does not designate the beginning of time, ratios of times measured from that point are not meaningful.

An example involving heights may clarify the distinction between interval and ratio scales. Suppose that 1 year ago Chester and his brother Barry were the same height but that in the past year Chester has grown 6 inches and Barry has grown 2 inches. If we use last year's height as a starting point (Figure 5-1), we have an interval scale for measuring height. The 2-inch and 6-inch measurements do indeed measure the boys' heights. Chester is 6 inches taller than he was last year; Barry is 2 inches taller than he was last year. It makes sense to subtract these measurements and find that Chester is 4 inches taller than Barry. It does *not* make sense to form a ratio ($6/2 = 3$) of the measures and say that Chester is three times as tall as Barry.

On the other hand, we may use 0 as a starting point and measure growth instead of height. For the attribute of growth, the measurements form a ratio scale. It does make sense to say that Chester has grown three times as much in the past year as Barry has grown.

Why should we be concerned about what type of measurement scale we are using? The answer to this question follows from the definitions: *Certain arithmetic operations are sensible only for certain types of measurement scales.*

- For nominal scales, no arithmetic is sensible.
- For ordinal scales, only comparisons are sensible. (Is one measure greater than, less than, or the same as another?)
- For interval scales, comparisons, subtractions, and averages* are sensible operations on measure numbers.
- For ratio scales, one may compare, add, subtract, and average measure numbers; one may also take multiples and ratios of measure numbers—all with sensible results.

Types of measurement scales can be ranked according to the amount of information they carry and the number of calculations that are sensible. The ordering of the scales from most to least versatile is:

Ratio
Interval
Ordinal
Nominal

* Throughout this chapter, *average* refers to the arithmetic mean, which is obtained by dividing the sum of a set of terms by the number of terms. (In this discussion, the terms would be measurements.)

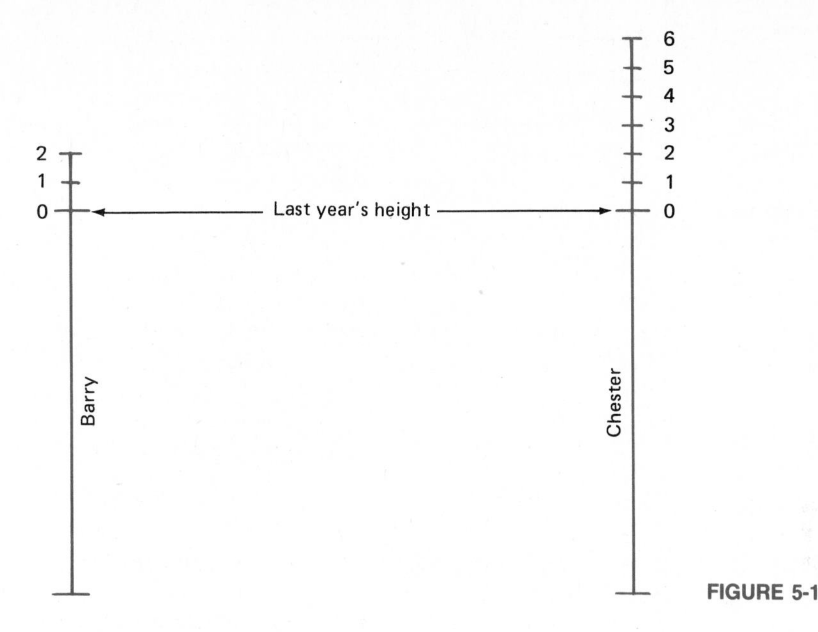

FIGURE 5-1

Here is a final example: The list in Table 5-1 of daily high and low temperatures illustrates some distinctions between sensible and nonsensible calculations for an interval scale. Temperature measures, expressed in degrees Celsius, form an *interval* scale.

The *difference*

$$24°\mathrm{C} - 10°\mathrm{C} = 14°\mathrm{C}$$

which gives the range between last week's high and last week's low, is a meaningful value. The *average* high temperature last week,

$$\frac{18°\mathrm{C} + 20°\mathrm{C} + 24°\mathrm{C} + 23°\mathrm{C} + 18°\mathrm{C} + 17°\mathrm{C} + 20°\mathrm{C}}{7} = 20°\mathrm{C}$$

is also a meaningful value. However, the *sum* of Monday's and Tuesday's high temperatures

$$20°\mathrm{C} + 24°\mathrm{C}$$

has no meaningful interpretation. Nor is it meaningful to say that Monday's high was twice as warm as Sunday's low.

TABLE 5-1
Temperatures in degrees Celsius

DAY	HIGH	LOW
Sunday	18	10
Monday	20	16
Tuesday	24	18
Wednesday	23	16
Thursday	18	14
Friday	17	12
Saturday	20	14

Exercise 1

Identify the type of measurement scale—nominal, ordinal, interval, or ratio—described in each of the following uses of numbers. (If more than one scale is possible, indicate that.) For interval and ratio scales, give the unit of measure.

a Class standings (first, second, third, etc.) for the 1984 graduates of a certain high school.

b Social security numbers of American taxpayers.

c Numbers of milligrams per cup of sodium in dry breakfast cereals.

d Students' scores on a recent sociology test.

e Lengths of time required for runners to complete the 1985 Boston marathon.

f Positions of runners who completed the 1985 Boston marathon (first, second, third, etc.).

g Distances by air from Washington, D.C., to other cities.

h End-of-season standings of National League, Eastern Division, baseball teams.

i Classification of eggs as peewee, small, medium, large, or extra large.

j Relative hardness of minerals as measured by Mohs' scale, arranged in 10 ascending degrees: 1—talc, 2—gypsum, 3—calcite, . . . , 10—diamond. (One mineral is considered harder than another if the first can be used to scratch the second.)

k The air-quality scale used in certain cities: 1—unhealthy air, 2—unsatisfactory air, 3—acceptable air, 4—good air, 5—excellent air.

Exercise 2

Multiples of measurements are meaningful whenever a ratio scale is used—that is, whenever measurements are expressed in terms of a fixed unit, based at zero. Use this criterion to decide which of the following statements are meaningful; for each meaningful statement give the unit of measure.

a I like you twice as much as I like Carl.

b I spent twice as much on your Christmas gift as I spent on Carl's gift.

c The Eagles are twice as good a football team as the Steelers.

d The Eagles won twice as many games last season as the Steelers.

e Ellen knows twice as much mathematics as Jennifer knows.

f Ellen's score on the last mathematics test was twice Jennifer's score.

g Betty is twice as good a swimmer as Laura.

h Betty can swim 100 meters in half the time that it takes Laura to swim the same distance.

Exercise 3

Explain why:

a Zip code assignments do not yield an ordinal scale.

b Air-quality ratings do not constitute an interval scale.

c Celsius or Fahrenheit temperature measurements do not constitute a ratio scale, but Kelvin temperature measurements do.

Exercise 4

Milk is scalded by heating it to a temperature of about 180°F. When it is taken from the refrigerator, its temperature is about 45°F. Explain why, although $4 \times 45 = 180$, scalded milk is not four times as warm as milk taken from the refrigerator.

Exercise 5

When final grades are assigned in college courses, units called *quality points* are used. Usually an A is worth 4 points, a B is worth 3 points, a C is worth 2 points, a D is worth 1 point, and an E is worth 0. What type of scale is this? Discuss.

Exercise 6

In primary elections in which there are several candidates for a given office, and none of them is especially favored, it is not unusual for the candidate whose name is listed first on the ballot to get the most votes. Might voters be treating ballot position as an ordinal scale that ranks the candidates? Discuss.

Exercise 7

A recent edition of *The People's Almanac* reported a survey of thirty-five American sportswriters who were asked to name the 15 greatest male athletes since 1900. The votes were counted by awarding 10 points to a writer's first choice, 9 points to the second choice, and so on, with no points for rankings below 10.

The overall results were:

1: Jim Thorpe
2: Babe Ruth
3: Muhammad Ali
4: Jack Dempsey
5: Jack Nicklaus
6: Ty Cobb
7: Bobby Jones
8: Joe Louis
9: Jesse Owens
10: Red Grange
11: O. J. Simpson
12: Jackie Robinson
13: Hank Aaron
14: Arnold Palmer
15: Mark Spitz

Examining the results, one might conclude that Muhammad Ali is "near the top." However this is not an interval scale with a fixed unit of measure to use in comparing differences between rankings. In fact, a further examination of the voting results reveals that Jim Thorpe received 287 points, Muhammad Ali received 157 points, and Mark Spitz received 40 points. Thus, in terms of points received, Ali was nearer to Spitz than to Thorpe.

a It is tempting to interpret ordinal scales as if the differences between rankings had a uniform meaning. Using the example above, why it is not sensible to do that.

b Look for illustrations in newspapers and magazines in which it is tempting to deduce too much information from a measurement scale.

Exercise 8: Making Meaningful Comparisons

Comparisons may lead to arguments because people are using different measurement scales. For example, it is hard to agree on an answer to the question, "Which is more—an apple or an orange?" unless we first agree on a basis for comparison. We might settle on cost as a basis for comparison, and use cents as the unit of measure. Then, at a given supermarket on a given day, we can check prices and agree on an answer. Or we might compare on the basis of caloric content (a medium-size apple has more calories than a medium-size orange), or on the basis of vitamin C content (a medium-size orange contains considerably more vitamin C than a medium-size apple). The point is this: *Sensible comparison of entities requires that both be measured on the same scale.*

What would be appropriate measurement scales to use to make meaningful comparisons between the following pairs of objects?

a John's intelligence and Mary's intelligence.

b John's knowledge of mathematics and Mary's knowledge of mathematics.

c My liking for chocolate ice cream and my liking for vanilla ice cream.

d My liking for chocolate ice cream and your liking for chocolate ice cream.

e John's size and Fred's size.

f The popularity of two presidential candidates.

g Caroline's culinary ability and Dave's culinary ability.

h A job offer in Pittsburgh and a job offer in Philadelphia.

i The value of an astronomy course and the value of a sociology course.

j The difficulty of getting an A in general physics and the difficulty of getting an A in French Literature.

Exercise 9

A mother, administering a punishment to her child, was heard to say, "This hurts me more than it hurts you." Is the mother's statement meaningful? Discuss.

Exercise 10

Average per capita income in Florida was $3700 in 1970 and $6700 in 1977. Is it valid to conclude that the average Floridian was better off in 1977? Discuss.

Exercise 11

Consider a man who is entertaining two guests. He wants to serve either coffee or tea—whichever the guests prefer—but not both. It occurs to him that if one guest prefers coffee and the other tea, he will not know what to serve. He devises the following scheme: he will ask both guests to rank seven beverages and, on the basis of their rankings, choose between tea and coffee. The seven beverages are: coffee, milk, lemonade, tea, cola, hot chocolate, orange drink. When the guests arrive, they are willing to indulge the eccentricities of their host and supply him with the following lists of beverages in order of preference:

Guest A	Guest B
Tea	Coffee
Coffee	Cola
Hot chocolate	Lemonade
Milk	Orange drink
Orange drink	Milk
Lemonade	Hot chocolate
Cola	Tea

a When the host examined the lists, he decided to serve coffee. On what basis did he make that decision? Is it a reasonable one?

b Suppose that the guests change their minds and revise their preference lists as follows:

Guest A (revised list)	Guest B (revised list)
Tea	Coffee
Hot chocolate	Tea
Milk	Cola
Orange drink	Lemonade
Lemonade	Orange drink
Cola	Milk
Coffee	Hot chocolate

Now the host decides to serve tea. How did he decide this? Is it reasonable?

c Notice that guest A has consistently preferred tea to coffee and Guest B has consistently preferred coffee to tea. Does it make sense that the host comes to different decisions in the two cases?

d What is the problem with the host's method? How can he discover which beverage—coffee or tea—will give the greatest total satisfaction to his guests?

Exercise 12

Explain, in your own words, the characteristics of each of the four types of measurement scales. Supply examples that illustrate their differences.

Exercise 13: Journal

At the beginning of this chapter it was stated that increased knowledge of something and increased ability to measure its attributes generally go together. This exercise follows up on that assertion; its steps are designed to help you gain increased knowledge about a decision situation through measurement.

Consider a difficult decision that you now face (or have faced or may someday face) and examine it by following steps 1 to 4 below. You could consider these:

Which courses to take next semester
Whether to find a summer job or go to summer school
What to major in
Whether to pledge a fraternity or sorority
What career to pursue

Steps to follow:

Step 1: List the decision choices that are available to you.

Step 2: List the criteria that you consider important for assessing the choices. (For example, if you are selecting a course, your criteria might include difficulty, relevance to your major, and interest. If you are selecting a summer activity, your criteria might include income, location, and opportunities for fun.)

Step 3: Using each criterion as a heading, rank the decision choices in a list beneath it, in order of preference.

Step 4: Examine your rankings. Do they reveal some choices as obviously better than others? Is it clear which is the best alternative? Do you need additional information to help you decide? What type of information?

If steps 1 to 4, developing and analyzing rankings, have revealed the best decision, hurrah! If not, don't be dismayed. This systematic use of measurement scales is not an automatic decision-making process. Rather, it is a process for organizing decision-related information. The organized information serves as a base for future deliberations about the decision. Additional information may be gathered and steps 1 to 4 repeated.

Eventually, either the decision will become clear or you will tire of the effort—and will make a "let's get it over with" choice by picking an alternative at random. If careful thought has preceded your fatigue, even this ending can be a good one.

Exercise 14: Journal

The nineteenth-century Italian economist Vilfredo Pareto stated a principle that is often called the *80-20 rule.* According to this rule, 80 percent of the value of a group of items is generally concentrated in only 20 percent of the items. Applying the 80-20 rule to ranked lists suggests that the top 20 percent of the listed items account for 80 percent of the total value of the list.

Test the 80-20 rule by making and examining several lists, with items in each ranked in order of decreasing importance. (Your lists might be things to get done today, people you care about, or goals for the future.) Do your lists support Pareto's rule? In what way can you make use of the 80-20 rule in combination with measurement scales?

Note The Appendix contains complete answers for Exercises 1 and 3*b*. It contains partial solutions or hints for Exercises 5, 8, and 11.

UTILITY: MEASURING STRENGTHS OF PREFERENCES

When we are able to measure an attribute using a ratio scale, there is a satisfying sense of really knowing the extent to which that attribute is present. Height, weight, distance, and a great variety of scientific attributes can be measured by ratio scales. Since a ratio scale satisfies all the properties of an interval scale as well as an additional property, it is common to think of a ratio scale as "better than" or "of a higher order than" an interval scale. Likewise, an interval scale is of a higher order than an ordinal scale, and an ordinal scale is of a higher order than a nominal scale. Scientific study of a particular attribute involves measuring with a scale of as high an order as possible.

One of the most commonly discussed and most controversial attributes that people are interested in measuring is personal preference. The measurement of preference ("utility theory") is an important topic in economics, and you are likely to encounter it elsewhere as well. Only a brief introduction is given here, to illustrate the use of measurement scales.

FIGURE 5-2
Lynn's ranking of five dates

1	Abe
2	(Tie) Ben, Sam
3	Paul
4	Ted

Consider again Lynn, who was introduced at the beginning of the chapter. Lynn has recently had dates with five young men and has ranked them as shown in Figure 5-2. The ordinal scale in Figure 5-2 does not tell Lynn's strength of preference for one man over another. In fact, when we recall that she had earlier rated Paul as a 6 on a scale from 1 to 10, it is a surprise to find him next to last on this list.

We gain more information from Lynn if she rates all the men on a scale of 1 to 10—as shown in Figure 5-3. Because the scale in Figure 5-3 conveys more information than the ordinal scale in Figure 5-2, we might ask whether Figure 5-3 is an interval scale. With this question in mind, let us look for a unit of measure.

FIGURE 5-3
Lynn's rating of five dates on a scale of 1 to 10

Abe:	9
Ben:	7
Sam:	7
Paul:	6
Ted:	3

Although there is no "standard unit" for measuring pleasure, we might suppose that, on the basis of her past dating experience, Lynn has subdivided the 1–10 range into approximately equal intervals of "liking." *If* this is the case, then the common size of these intervals (see Figure 5-4) serves as a unit of measure.

The formal name for the study of preference measurement is *utility theory.* The numerical value assigned to the satisfaction we get from a given alternative is called its *utility,* and a unit of measure on a utility scale is called a *util.* A util is a subjective unit of measure, the meaning of which varies from person to person and, for a given person, from situation to situation.

Lynn's scale, shown in Figure 5-4, is based on a personal definition—for this type of situation only—of a util. From her scale we learn the relative amounts of pleasure that she receives from certain dates, but we cannot deduce how much pleasure another woman would gain from dates with the same men. Indeed, if Lynn were to rate movies she has recently seen on a scale of 1 to 10, we would expect her to base her ratings on a different meaning of *util.*

FIGURE 5-4
Lynn's date-rating interval scale

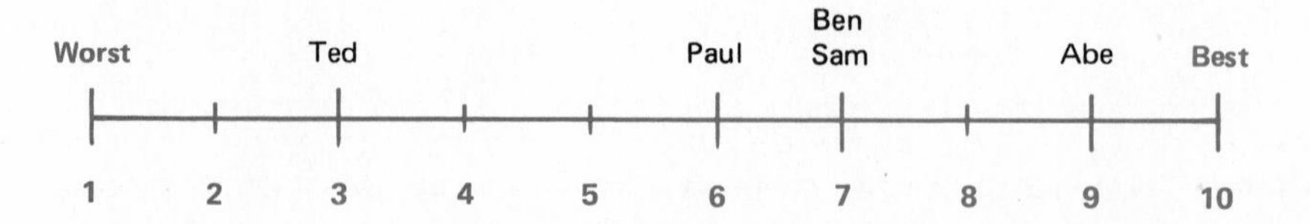

The advantage to be gained from development of an interval (utility) scale, instead of an ordinal (ranking) scale, to describe preferences is this: with the interval (utility) scale, differences and averages of measures may be computed and compared. (Chapters 7 and 12 will investigate the usefulness of computing averages of measures.)

The following example further illustrates the variety of possible utility scales.

Example ▶

On Tuesday morning Mark and Marilyn hear a weather report predicting a 50 percent chance of rain. Each wonders, "Should I carry an umbrella?" Each considers the following four possible outcomes for the day:

Carry an umbrella; it rains.
Carry an umbrella; it doesn't rain.
Don't carry an umbrella; it rains.
Don't carry an umbrella; it doesn't rain.

Suppose that Mark and Marilyn assess these outcomes on a utility scale. Each wants to decide correctly. Mark is planning to wear a new sport coat and is anxious not to get it wet. Marilyn will be shopping after work and would rather risk getting wet than be encumbered with an umbrella. After some experimentation with numbers, Mark and Marilyn come up with the utility assignments shown in Table 5-2. Their numerical assignments indicate that they have envisioned different scales on which to evaluate the day's events (see Figure 5-5).

TABLE 5-2

	UTILITY ASSIGNMENTS	
POSSIBLE OUTCOMES	MARK	MARILYN
Carry an umbrella; it rains.	+5	+3
Carry an umbrella; it doesn't rain.	−2	−3
Don't carry an umbrella; it rains.	−5	−1
Don't carry an umbrella; it doesn't rain.	+5	+3

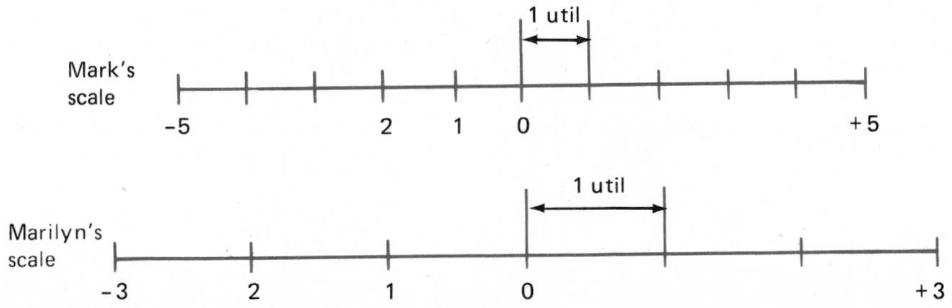

FIGURE 5-5
Utility assignments for weather situations

TABLE 5-3

POSSIBLE OUTCOMES	YOUR UTILITY ASSIGNMENTS
Carry umbrella; it rains.	_____
Carry umbrella; it doesn't rain.	_____
Don't carry umbrella; it rains.	_____
Don't carry umbrella; it doesn't rain.	_____

Both Mark and Marilyn have used negative numbers to show dissatisfaction with an outcome and positive numbers to show satisfaction. We cannot deduce that Mark gains more satisfaction than Marilyn from a correct decision, even though he assigns it a larger number. Their scales are personal, and each has his or her own private meaning for a util. ◀

If you have never tried to assign numerical values to your preferences, you may be wondering where Mark and Marilyn got their utility values. These values were invented, just as you might invent a scale to describe your pleasure from a date (as Lynn did), or just as you might invent a scale to describe your reaction to a bad joke. ("That joke rates 2 on a scale from 1 to 10.") Take a moment to experiment: invent your own utility values for the possible weather outcomes for a given day; use the form given in Table 5-3.

Usually it is easiest to assign utility values by first picking a (high) number to assign to the best outcome, then picking a (low) number for the worst outcome, and finally selecting in-between numbers for the in-between possibilities.

The example of Mark and Marilyn makes a lot of a situation that may be unimportant to you. But it illustrates an idea that can be valuable in a variety of situations: *Assignment of utility values to a list of alternatives provides a way of thinking clearly and precisely about our attitudes toward these alternatives.* It is thus a useful technique in decision making.

Exercise 15

Return to the example of Mark and Marilyn.

a Which of the two will carry an umbrella even though there is only a 50-50 chance of rain? Explain how the given utility values reveal this.

b Which of the two will not carry an umbrella even though there is a 50-50 chance of rain? Explain how the given utility values reveal this.

Exercise 16

Exercise 11 described an eccentric host who is trying to decide whether to serve coffee or tea to two guests. Suppose that the host asked his guests to rate coffee and tea on a utility scale with values ranging from 1 to 10. If the guests responded as follows, how should the host interpret the results? Discuss.

Guest A		Guest B	
Coffee:	7	Coffee:	7
Tea:	9	Tea:	1

Exercise 17

The examples in the text included utility scales developed by Lynn, Mark, and Marilyn. Explain why these scales are not ratio measurement scales.

Exercise 18: Invent

Suppose that you are planning for next summer and wonder whether to try to get a job or to attend summer school. Using a scale from 1 to 20, assign a utility value to each of the following possible outcomes:

Possible outcomes	Utility assignments
Find interesting job with high pay	___
Find interesting job with low pay	___
Find boring job with high pay	___
Find boring job with low pay	___
Attend summer school	___

Use your utility assignments as a basis for describing your plans for next summer; complete the following sentence: I will go to summer school next summer if ____

____.

Exercise 19: Invent

Suppose that you are selecting courses for next semester. After scheduling required courses, you have room for one elective. As an aid in selecting the final course for your schedule, develop a utility scale and use it to rate each of the following types of courses:

Type of course	Utility value
Interesting, useful, convenient time	____
Interesting, useful, inconvenient time	____
Interesting, not useful, convenient time	____
Interesting, not useful, inconvenient time	____
Uninteresting, useful, convenient time	____
Uninteresting, useful, inconvenient time	____
Uninteresting, not useful, convenient time	____
Uninteresting, not useful, inconvenient time	____

If you can't find an elective that is interesting, useful and offered at a convenient time, what type of course will you look for?

Compare your utility values with those of other students. In what ways are they similar? In what ways do they differ?

Exercise 20: Journal

Consider a decision that you face now (or have faced or will face). (You may reconsider the decision examined in Exercise 13.)

a List the choices available to you.

b List the criteria you consider important for judging these choices.

c Considering each criterion separately, devise a utility scale on which to evaluate your choices; rate the choices using the scale.

As an example, consider Ben, who is trying to decide whether to take swimming or archery as his physical education course next semester. His criteria for judging the course choice are usefulness and fun. Using scales of 1–10 in both cases, Ben comes up with two sets of ratings:

Usefulness		Fun	
Archery:	2	Archery:	7
Swimming:	9	Swimming:	3

After you have assigned your utility values, examine the results. Has the use of utility scales helped you to think carefully about the decision situation? In what ways can you make effective use of utility scales in personal decision making?

Exercise 21

Use of numbers to measure preferences is sometimes criticized. Numbers, it is said, "give an aura of precision in a context in which precision is impossible." Do you agree that numbers have this effect when used in preference measurement? Explain why or why not.

Exercise 22: Test Yourself

Develop a list of new terms and key ideas presented in this chapter. Supply an example of each item on your list. (Providing your own example of an idea is a major step toward your own independent use of that idea.)

Note The Appendix contains partial solutions or hints for Exercises 16, 17, 18, and 20.

CHAPTER 6

WHEN IS ADDITION APPROPRIATE?

Life is painting a picture, not doing a sum.

—Oliver Wendell Holmes, Jr.

WHAT MATHEMATICAL PROCESS SUITS A GIVEN PROBLEM?

As we learn more and more mathematics, the question, "When will I ever use this?" keeps surfacing. We already seem to know more mathematics than we can make sensible use of—and the gap keeps widening.

How can we make more use of the mathematics we know? One way to broaden the use of mathematics is to experiment. When we encounter a new problem, we can ask, "Is there any mathematics that I know that I can use here?" Instead of being cautious about trying mathematics, we can defer caution until later—when we examine the solution to see whether it is reasonable.

This chapter focuses on experiments with the mathematical process of addition. We will try using addition to solve problems for which it may or may not be the appropriate process; then we will check to see whether the answer obtained is sensible.

This chapter was inspired by Philip J. Davis, "Mathematics by Fiat," *Two-Year College Mathematics Journal*, vol. 11, no. 4, September 1980, pp. 255–263. A later version appeared in *The Mathematical Experience*, Birkhauser, Boston, 1981, pp. 68–76.

As was stated in Chapter 4, three steps are important in problem solving:

- Observe widely.
- Speculate freely.
- Test carefully.

Past mathematics courses can be considered the equivalent of observation. The usefulness of mathematics can be increased by putting emphasis on the other two steps: speculation and testing.

When you encounter a new problem, sift through the knowledge that you have stored. Do you find a mathematical process that might fit the problem? *Speculate* that it does; try it; and then *test* to see if the answer makes sense. If so, fine. If not, repeat the steps.

WHEN SHOULD WE ADD?

Addition is the most familiar of the mathematical processes. We learn it early, and we use it more often than any other process; but sometimes we use it incorrectly.

Because of its familiarity, addition is a good process to use as a basis for experimentation. Let us now look at a variety of problems and ask, "Can they be solved by addition?"

Consider the following problems:

a Marilyn is making pickles. In a kettle, she already has 2 cups of vinegar. She pours in 4 more cups of vinegar. What is the total volume of the mixture?

b Jeremy is making candy. In a pan he already has 2 cups of water. He pours in 6 cups of sugar. What is the total volume of the mixture?

c Marty and Karen are planning a party. Marty has a list of 15 people to invite and Karen has a list of 18 people to invite. How many guests do they want to invite all together?

These problems have similar formats. For each, we might speculate that addition is the process to use to solve it:

a $2 + 4 = 6$
b $2 + 6 = 8$
c $15 + 18 = 33$

Experience suggests that these tentative solutions can be tested and evaluated as follows:

a Vinegar undergoes no change when combined with more vinegar; thus the total volume will be the sum of the separate volumes. The solution is correct.

b Sugar dissolves in water. The total volume of the mixture is less than the sum of the separate volumes. The answer is therefore too large.

c If Marty's and Karen's lists have some names in common, then the combined lists will total fewer than 33 names. For example, if two names—Joan and

It is impossible to offer foolproof rules that prescribe when a mathematical process should be used. Although mathematics itself is objective, its use is subjective. Each use must pass the test of reasonableness: does this process lead to a sensible solution in this case?

Jeffrey—appear on both lists, then the combined list contains 31 names. The problem as stated gives too little information to be solved.

You may find it disturbing to try a process that turns out to be incorrect; obviously, when we encounter a problem we want to solve it correctly. However, trying a method that doesn't work offers at least one advantage over doing nothing at all: it gets you started. You will detect the mistake when you test the solution—and the false start will serve as a basis for improvement.

Experimentation with mathematical processes is not as difficult as it seems at first—but it does require practice. The following exercises give opportunities to experiment. In particular, they offer:

A variety of problem situations with which to experiment using addition. Is addition appropriate in each situation? (Exercises 1, 2, 3, 4, and 8.)

Opportunities to consider generalizations of observations about addition to other mathematical processes. (Exercises 5 and 6.)

Questions that investigate attitudes toward the use of mathematics in problem solving. (Exercises 7, 9, and 10.)

Exercise 1*

For each of the following problems, first assume that addition is the appropriate way to solve it, then find a sum using values given in the problem, and finally consider whether that sum is likely to be a good answer—look for reasons why it might not be reasonable or situations in which it would not be reasonable. Are there questions that you need to have answered before you can decide on a correct solution method? If addition does not yield a good solution, how would you solve the problem?

a Individual cans of chicken noodle soup cost 35 cents. How much does a case of 24 cans cost?

b Searching through junk in the garage, Rhonda finds first one half and then the other half of an automobile tire. How many whole tires has she found?

c A cup of boiling water (100°C) is mixed with a cup of lukewarm water (40°C). What is the temperature of the mixture?

d Jay's earnings statements from last year showed $12,345 in after-tax income from his teaching job and $4,321 in after-tax income from his summer job as a construction worker. What was Jay's total after-tax income?

e Clyde can paint a large house in 6 days. Darren, who works more slowly, can paint a house of the same size in 10 days. How long would it take them both, working together, to paint one large house?

* Some of these exercises first appeared in JoAnne Simpson Growney, "Contrariness: A First Step in Applying Mathematics" *UMAP Journal,* vol. 4, no. 4, 1983, pp. 381–388.

f A $100,000 advertising campaign gains 50,000 new customers for Crust toothpaste. How many new customers will result from a $200,000 advertising campaign?

g The label on a bottle of diet pills states that taking 1 pill per day will result in a weight loss of 5 pounds per week. How much weight will be lost by taking 2 pills per day?

h Philip gets 50 utils of satisfaction from completing his mathematics homework and 25 utils of satisfaction from jogging 3 miles. How much satisfaction will he get from doing both?

i Julie gets 10 utils of pleasure from having a candy bar for desert after dinner and 15 utils of pleasure from an ice cream sundae. How much pleasure will she get from having both? (Compare this situation with *h,* above. How do they differ?)

j A 10-inch apple pie requires 45 minutes baking time (at 425°F); a 10-inch walnut pie requires 55 minutes (at the same temperature). How much time is required to bake both pies together?

k The college store observed the purchases of 100 students who attended its sale of records and T shirts last Tuesday. The manager reported that 75 students bought records and 65 bought T shirts. How many students bought at least one of the sale items?

l Edward recently completed a mathematics course in which he took three tests. His scores were 80, 60 and 85 (each time out of 100 points). What score represents Edward's total achievement in the course?

m Marge is trying to decide on a college major. During her first semester she is taking five courses in five different fields that are possible majors for her. Using a utility scale from 1 to 10, Marge has just finished evaluating her courses using three criteria; her rankings are given in Table 6-1. What overall ratings should Marge assign to these five subjects?

n Steve has an estimated 3 hours of homework to get done this evening. He also needs to eat (which generally takes 1 hour) and to do his laundry (which takes 1½ hours). He also wants to jog (another 1 hour gone) and to spend at least 1 hour relaxing with friends. It's 6 P.M.; he needs to be in bed by midnight. Can he do it all?

o Veronica has a 1-hour job to do. Dennis has four 15-minute jobs to do. If both start at the same time, who will finish first?

TABLE 6-1

	CRITERIA		
SUBJECTS	Interesting	Job opportunities	I learn the subject easily
Biology	6	8	5
Economics	7	3	5
History	9	2	5
Philosophy	7	2	6
Psychology	5	6	8

Exercise 2

Consider the following answer to the question, "When may two quantities be added?"

> Two quantities may be added if both are expressed in terms of the same unit of measure.
>
> Do you agree with this answer? Supply examples to support your opinion.

Exercise 3: Invent

Describe three problem situations in which addition seems at first to be the correct process but testing the answer shows a need for modification.

Exercise 4

Consider the question, "When should addition be used to solve a problem?" and the response, "Addition should be used to solve the problem if the answer obtained by addition is reasonable." What does the response imply about the need to check answers obtained by using addition?

Exercise 5: Beyond Addition

Consider the question, "What mathematical process should I use to solve this problem?" and the response, "Use any mathematical process that leads to a reasonable solution." Is the response satisfactory? What does the response imply about how to use mathematics in problem solving?

In the game of life it's a good idea to have a few early losses, which relieves you of the pressure of trying to maintain an undefeated season.

Bill Vaughan

Exercise 6

Joyce worries about doing mathematics problems. "I'm always afraid I'll do the wrong thing," she says. "I can get the right answer if I know what method to use—but it's hard for me to decide on a method." Develop an argument to convince Joyce that her caution should come later—that she should be daring at first and speculate about possible methods, and be cautious later, testing whether a given solution is sensible.

Exercise 7

Discuss the truth of the following statements:

a In problem solving, 95 percent of success is due to persistence and only 5 percent is due to creative ability.

b In problem solving, 95 percent of failure is due to unwillingness to work hard and only 5 percent is due to lack of ability.

Exercise 8

For each of the following problems, explain why addition fails as a method of solution:

a Six young men who room together in the same house decide to cook a special dinner for six young women. They have subdivided this project into five activities, with time estimates as follows:

Clean house	1 hour	Prepare and cook food	1½ hours
Decide on menu	15 minutes	Set table	15 minutes
Purchase food	1 hour		

The sum of the times listed is 4 hours. Must completing all these tasks take 4 hours? Explain. If the hosts have invited their guests for 6 P.M. next Tuesday, what is the latest time on Tuesday that their preparations can begin if they are to finish by 6?

b Twentieth Century Realty Associates publicized the following information about the prices of homes it sold in New Jersey communities in August:

30 homes sold at prices under $30,000
100 homes sold at prices under $50,000
150 homes sold at prices under $75,000
175 homes sold at prices under $100,000
5 homes sold at prices of $100,000 or more

The sum of the numbers of homes in the various categories is 460. Why is 460 not the correct number of New Jersey homes that the agency sold in August? What is the correct number?

c Laura is trying out for the high school field hockey team and also for the cheerleading squad. She analyzes her chances this way: "I have a 50 percent chance of being selected for the hockey team and a 50 percent chance of being selected as a cheerleader. Thus my chance of being selected for one or the other are 50 percent + 50 percent, or 100 percent." Convince Laura that she's wrong.

Exercise 9: The Story Behind Success

Other problem solvers make mistakes too. A common misconception that people have in mathematics classes is that everyone else knows exactly what to do and solves problems correctly the first time. In fact, teachers, texts, and fellow students often share only their correct solutions; they are not candid about their mistakes and false starts.

Identify two or three people who are successful at mathematical problem solving or have other skills you admire. Investigate the story behind their success. What role did mistakes and failures play? Was persistence a major factor?

Exercise 10: Journal—Personal Qualities of a Good Problem Solver

An enterprising person who wants to be an effective problem solver needs to work continually to develop certain personal characteristics:

Knowledge. The effective problem solver learns a variety of mathematical procedures to draw on.

Daring. Good problem solvers are willing to try procedures in new situations.

Caution. Effective problem solvers check to see if the results their procedures produce are sensible.

Persistence. If at first they don't succeed, good problem solvers learn from their mistakes and try again.

Evaluate these characteristics; are there others you would add? Which ones do you consider most important? Assess yourself: to what degree do you possess these characteristics? Which do you need most to develop? Set goals.

Exercise 11: Test Yourself

a Make three lists of problem situations: (1) those for which addition can be expected to give a correct answer; (2) those for which addition seems to be the correct process but gives an answer that is too small; (3) those for which addition seems to be the correct process but gives an answer that is too large.

b Have you developed attitudes toward addition that can be useful in working with other mathematical methods? List these views; briefly speculate on how they can be useful.

Note The Appendix contains partial solutions or hints for Exercises 1, 7, 8*a, b,* and *c,* and 10.

CHAPTER 7

USING MEASUREMENT IN DECISION MAKING: MATRICES AND WEIGHTED AVERAGES

Guess if you can; choose if you dare.

—Pierre Corneille

FORMING MATRICES

A rectangular array of numbers—a table—is called a *matrix.* Matrices can be used in a variety of situations, but this chapter will focus on a particular type of application: the organized display of information relevant to a decision-making situation.*

In Chapter 5, Exercises 13 and 20, you were asked to describe a puzzling decision situation and evaluate the possible choices using several criteria. Let us now consider several examples of this type. For each decision, we will use a matrix to organize and display the information; we will then go a step further and ask, "On the basis of the given information, what is the best decision?"

*Here, matrices will be used solely for organizing and displaying information. However, their actual importance extends far beyond this use; the algebra of matrices has many applications.

Example 1 ▶

Gayle is trying to decide which college to attend next year; she has been accepted at all three colleges to which she applied. She (and her parents) consider three criteria important:

Cost
Distance from home
Quality of programs

Table 7-1 shows a matrix that Gayle has drawn up to display her decision alternatives and criteria. For each criterion, Gayle devises a utility scale with values ranging from 1 (worst) to 10 (best). She first evaluates each school on the basis of cost and fills in the first column of her matrix. She next evaluates each school on the basis of distance, and then on the basis of quality. Her completed matrix is shown in Table 7-2.

When we examine Gayle's completed matrix, we observe the following:

1 Cost is not a useful factor in the decision, since all three schools received the same cost rating.

2 Each entry in the second row is at least as large as the corresponding entries in the other rows.

The second observation shows how easily Gayle's decision can be made. All ratings for Southwestern are as high as or higher than those for the other choices. Thus Southwestern is, on the basis of these ratings, obviously the "best" decision for Gayle.

The phenomenon exhibited by the numbers in Gayle's choice matrix is called *dominance;* one decision choice dominates the others. Specifically, one row of a matrix is said to *dominate* another row if each entry in that row is at least as large as the corresponding (same-column) entry in the other row. When matrix rows represent choices, any row that is dominated by another row may be eliminated from further consideration. ◀

TABLE 7-1
Gayle's choice matrix

	CRITERIA		
CHOICES	Cost	Distance	Quality
Central State College			
Southwestern University			
Mountainview College			

TABLE 7-2
Gayle's completed choice matrix

	CRITERIA		
CHOICES	Cost	Distance	Quality
Central State College	5	4	5
Southwestern University	5	7	7
Mountainview College	5	3	6

ASSIGNING WEIGHTS AND COMPUTING WEIGHTED AVERAGES

As Gayle's situation illustrates, decisions in which one choice dominates the others are easy to make. We now turn to a situation where the choice is less easy.

Example 2 ▶

Carl is a college freshman who must decide on a major and, with it, a career. His own abilities and interests, and the expectations that others have for him, suggest these possibilities as most reasonable: accounting, biology, computer science, and journalism. Carl's father is an accountant, and Carl knows that he could earn a good income as an accountant—and also please his father by following in his footsteps. On the other hand, the job opportunities in computer science look even more promising, and Carl is fascinated with the idea of being part of that seemingly elite group that understand computers. Of all his studies, however, Carl likes biology best—especially the laboratories—although he is aware that a graduate degree will be needed for the type of laboratory job he'd enjoy. Finally, an idea that has stuck with Carl since the days of the Watergate scandal is to major in journalism, in preparation for a career in investigative reporting.

Six criteria* are important to Carl, and he will use them as a basis for judging these possible majors:

1 Enjoyment—both of the preparatory studies and of the career itself
2 Potential for obtaining a good job—one with an above-average salary and opportunities for advancement
3 Opportunities to gain the respect of others
4 Opportunities for personal satisfaction
5 Free time to devote to family and personal interests
6 Cost—in time and money—of preparation for the career

Because this decision is a significant one for Carl, he is willing to devote considerable time and thought to it.

Carl begins his analysis by constructing a choice matrix on which the row heads are choices and the column heads are criteria. He next goes through the laborious process of assigning utility values. The result is shown in Table 7-3. To develop this choice matrix, Carl has used utility scales with values ranging from 1 to 10. But because his criteria differ in their importance, a util on one scale does not measure the same amount of satisfaction as a util on another scale.

Carl examines his choice matrix for dominance but finds none. Each major ranks high on some criteria and not so high on others. He wonders what do do.

---

* Developing a list of criteria is one of the most important parts of making a decision. A list of criteria should be sensible and complete, but as short as possible, and the criteria should not be redundant.

TABLE 7-3
Carl's completed choice matrix

	CRITERIA					
CHOICES	Fun	Good job	Respect	Personal satisfaction	Free time	Low cost
Accounting	4	8	7	4	5	6
Biology	9	6	5	7	5	3
Computer science	6	9	7	4	3	5
Journalism	8	3	5	9	4	6

"What about finding an average rating for each major?" Carl thinks, but he quickly discards that idea. Averages are meaningful only when measurements are made using a scale with a fixed unit of measure, and he has not used the same unit of measure for all six criteria. A util on his "fun" scale, for example, represents a different amount of satisfaction from a util on his "good job" scale.

While mulling over what to do, Carl decides to discuss the decision with Scott, his roommate. "You have too many criteria, Carl," Scott suggests. "For me the choice would be a lot simpler. A good job is the criterion I consider most important. I'd ignore all those other criteria and simply ask which choice offers the best job potential. In your case that would narrow your decision down to accounting or computer science. For now you could take further courses toward both majors and make a final decision later."

Although he does not agree fully with Scott's assessment, Carl does find in it an idea to work with: the differences in importance of his criteria. The six criteria definitely do not all have the same importance to him. Thinking further about this, Carl decides that he can assign a percentage value to each criterion, indicating its importance for his decision. The technical term for Carl's "importance values" is *weights;* the process of identifying the relative importance of each criterion is called *assigning weights.*

The weights that Carl assigns are given below:

Criterion	Weight, percent
Fun	30
Good job	10
Respect	10
Personal satisfaction	30
Free time	10
Low cost	10

Carl next converts these weights to decimal values and adds a new row to his choice matrix to include them. (See Table 7-4.) He stares at the matrix for a while; what shall he do next?

Suddenly Carl recalls a situation that seems analogous to this. In high school last year, in a writing course, his final grade was based on three reports and a term paper. His grades on these items had been 75, 80, 85, and 98 points, respectively; but the items did not all have the same importance. Their weights, assigned by the

TABLE 7-4
Carl's choice matrix and weights

	CRITERIA					
	Fun	Good job	Respect	Personal satisfaction	Free time	Low cost
Accounting	4	8	7	4	5	6
Biology	9	6	5	7	5	3
Computer science	6	9	7	4	3	5
Journalism	8	3	5	9	4	6
WEIGHTS	0.30	0.10	0.10	0.30	0.10	0.10

teacher, were 15 percent, 20 percent, and 40 percent. His semester grade of 88 was a "weighted average" and had been calculated in the following way:

$$\frac{(15)(75) + (20)(80) + (25)(85) + (40)(98)}{100} \approx 88$$

"The points assigned to one paper weren't worth the same as the points assigned to another," Carl observes. "My teacher compensated for the differences in value of a point by computing a 'weighted average.' I think I might do the same with my utility values."

Carl decides to discuss this possibility with Scott, who by now is quite interested in the problem. Scott thinks that computing weighted averages is a good idea—and he points out an alternative format for Carl's calculation:

$$(0.15)(75) + (0.20)(80) + (0.25)(85) + (0.40)(98) \approx 88$$

"When you compute an ordinary average, you must divide by the total number of items being averaged," Scott reminds Carl. "When you compute a weighted average, you must divide by the total of all the weights. But if your sum of weights turns out to be 1, then you avoid any division."

Carl is now anxious to apply the concept of a weighted average to the values in his choice matrix. "I think," he repeats, "that the use of weights compensates for the differences in my utility scales. I'm going to compute a weighted-average rating for each major." Table 7-5 shows the results of Carl's calculations. From now on, as in Table 7-5, a matrix that incorporates weights and weighted averages as will be called a *decision matrix.*

TABLE 7-5
Carl's decision matrix

	CRITERIA						
CHOICES	Fun	Good job	Respect	Personal satisfaction	Free time	Low cost	WEIGHTED AVERAGES
Accounting	4	8	7	4	5	6	5.0
Biology	9	6	5	7	5	3	6.7
Computer science	6	9	7	4	3	5	5.4
Journalism	8	3	5	9	4	6	6.9
WEIGHTS	0.30	0.10	0.10	0.30	0.10	0.10	

Carl's Use of Weighted Averages—Is It Sensible?

In Example 2, Carl speculates that the weighted average can be used to combine the measures on his different utility scales. Let us examine the reasonableness of this process.

Recall that Carl gave his criteria the following weights:

Criterion	Weight, percent
Fun	30
Good job	10
Respect	10
Personal satisfaction	30
Free time	10
Low cost	10

Carl's use of these ratings in computing a weighted average assumes the following interpretation: Each util on the scales for measuring good job, respect, free time, and low cost designates the same amount of satisfaction. Furthermore, each util on the scales for measuring fun and personal satisfaction designates three times as much satisfaction as a util on each of the other four scales.

If Carl agrees with this interpretation, his computation of weighted averages makes sense.

Here is a sample of Carl's calculations: the weighted average for accounting is

$$(0.30)(4) + (0.10)(8) + (0.10)(7) + (0.30)(4) + (0.10)(5) + (0.10)(6) = 5.0$$

When he examines Table 7-5, Carl is rather disappointed: there is still no clear-cut decision. Journalism has obtained a slightly higher weighted average than biology, but the difference seems to Carl to be hardly conclusive.

Wisely, Carl decides to set the decision aside for a while. Perhaps the passage of time will reveal some new information that will help him decide. Using the framework he has established for organizing information relevant to his decision, Carl can continue to ask himself questions such as these:

Is my list of criteria satisfactory?

Are my weights honest reflections of what is really important to me?

What sources can I use to learn more about future possibilities in these fields—so that my utility estimates can be based on more information? ◀

A third example will illustrate how different people can use the same background information to arrive at different decisions.

Example 3 ▶

Janet shops in fashionable stores for new styles as soon as they are available.

Lisa likes to dress fashionably but waits for sales before making her purchases.

Terri buys her clothing at secondhand thrift shops.

Which woman is getting the most for her money?

These three shoppers have different buying habits. Let us suppose, however, that each has initially considered the same three alternatives:

1 Buy clothes early, at the beginning of the fashion season, at high prices.
2 Buy clothes on sale, trading some variety of selection and fashion in exchange for savings.
3 Buy secondhand clothing, trading fashion and choice for large savings.

Shopping decisions are generally based on numerous criteria; but for simplicity, let us suppose that Janet, Lisa, and Terri have considered just two:

1 Dress well.
2 Save money.

Now let us suppose that these three women have constructed a choice matrix—shown in Table 7-6—which includes utility values for each choice criterion combination. (The values are measures on utility scales with a range of 0 to 10.)

It may seem remarkable that the three women—who exhibit different buying habits—can agree on a choice matrix. However, as we know from Example 2, decisions depend not only on how decision makers rate choices but also on what importance they attach to various criteria. To illustrate this point further, consider Table 7-7, which includes each woman's weights and the resultant weighted averages, incorporated in decision matrices.

Finally, let us return to our initial question, "Which woman is getting the most for her money?" *If* "getting the most for her money" means "deriving the greatest satisfaction from her purchases," then we cannot answer the question, since utility scales are personal and subjective rather than public and standardized.

TABLE 7-6
Shoppers' choice matrix

	CRITERIA	
CHOICES	Dress well	Save money
Buy early	9	0
Buy on sale	6	4
Buy secondhand	0	8

TABLE 7-7

(a) Janet's decision matrix

CHOICES	CRITERIA: Dress well	CRITERIA: Save money	WEIGHTED AVERAGES
Buy early	9	0	8.1
Buy on sale	6	4	5.8
Buy secondhand	0	8	0.8
WEIGHTS	0.90	0.10	

(b) Lisa's decision matrix

CHOICES	CRITERIA: Dress well	CRITERIA: Save money	WEIGHTED AVERAGES
Buy early	9	0	4.5
Buy on sale	6	4	5.0
Buy secondhand	0	8	4.0
WEIGHTS	0.50	0.50	

(c) Terri's decision matrix

CHOICES	CRITERIA: Dress well	CRITERIA: Save money	WEIGHTED AVERAGES
Buy early	9	0	0
Buy on sale	6	4	4
Buy secondhand	0	8	8
WEIGHTS	0	1	

However, on the basis of our analysis, we can provide the following partial answers:

> Janet gets the most satisfaction for her money by shopping early for fashionable clothes.
>
> Lisa gets the most for her money by waiting to buy fashionable clothes on sale.
>
> Terri gets the most for her money by buying secondhand clothes. ◀

In summary, a matrix of utility values provides a format for an organized display of choices and criteria pertinent to a decision problem. With a matrix display, comparisons and evaluations are readily made. When dominance is evident, some choices can be eliminated. When dominance fails to pinpoint a choice, assignment of weights and calculation of weighted-average utility ratings may help the decision maker to select a best choice.

How to Compute a Weighted Average of Measures

Suppose that, on the basis of a list of criteria, a choice has these utility ratings:

$$r_1, r_2, \ldots, r_n$$

Suppose also that the criteria have been given the following weights:

$$w_1, w_2, \ldots, w_n$$

Then the *weighted average* rating is computed thus:

$$\text{Weighted average} = \frac{w_1r_1 + w_2r_2 + \cdots + w_nr_n}{w_1 + w_2 + \cdots + w_n}$$

For example, suppose that on the criteria of plot, quality of acting, and special effects a new movie receives (on scales from 1 to 10) ratings of

$$r_1 = 5$$
$$r_2 = 6$$
$$r_3 = 9$$

A moviegoer who considers plot more important than quality of acting and acting more important than special effects might assign to these criteria the following weights:

$$w_1 = 3$$
$$w_2 = 2$$
$$w_3 = 1$$

This moviegoer would then compute the following weighted average rating:

$$\text{Weighted average} = \frac{w_1r_1 + w_2r_2 + w_3r_3}{w_1 + w_2 + w_3}$$

$$= \frac{(3)(5) + (2)(6) + (1)(9)}{6}$$

$$= 6$$

Exercise 1

Each of the choice matrices in Table 7-8 gives ratings of three job applicants on three different criteria. All ratings are utility values measured on scales with a range of 0 (worst) to 5 (best). In each case, tell which applicant you would select. Begin by eliminating dominated choices; if more than one choice remains after this step, discuss how you would decide among those remaining.

TABLE 7-8

Matrix 1

	CRITERIA		
CHOICES	Educational background	Recommendation of previous employer	Health
Applicant A	4	2	4
Applicant B	3	3	2
Applicant C	4	3	4

Matrix 2

	CRITERIA		
CHOICES	Educational background	Experience	Legibility of handwriting
Applicant A	5	4	1
Applicant B	3	4	3
Applicant C	4	2	1

Matrix 3

	CRITERIA		
CHOICES	Educational background	Experience	Pleasant disposition
Applicant A	3	5	2
Applicant B	4	4	3
Applicant C	4	5	2

Matrix 4

	CRITERIA		
CHOICES	Educational background	Experience	Pleasant disposition
Applicant A	3	4	3
Applicant B	5	2	4
Applicant C	2	5	3

TABLE 7-9

	CRITERIA						
CHOICES	Fun	Good job	Respect	Personal satisfaction	Free time	Low cost	WEIGHTED AVERAGES
Accounting							
Biology							
Computer science							
Journalism							
WEIGHTS	0.30	0.10	0.10	0.30	0.10	0.10	

Exercise 2

Suppose that Carl, in Example 2, had used the following weights to indicate the relative importance of his criteria:

Criterion	Weight
Fun	3
Good job	1
Respect	1
Personal satisfaction	3
Free time	1
Low cost	1

a Using these weights, compute the weighted-average rating for each of Carl's choices.

b Why are the weighted averages that you computed in *a* the same as those in Table 7-5?

c List three different sets of weights for Carl's criteria that will lead to the same weighted averages as those given in Table 7-5.

Exercise 3

Suppose that Carl shows you his choice matrix (Table 7-4) and asks for your advice. On the basis of your knowledge and experience, what weights would you assign to the six criteria? Using your weights, compute weighted averages. Does the choice with the highest weighted average seem to be the best choice? Discuss.

Exercise 4

The discussion of Example 2 concluded with the comment that perhaps the passage of time will reveal new information that will help Carl decide. Suppose that Carl chooses you as a source of new information. On the basis of your knowledge of the four fields he is considering, suggest changes in his utility ratings. Insert your utility ratings in the matrix in Table 7-9 and compute weighted averages. Does the choice with the highest weighted average seem to be the best choice? Discuss.

Exercise 5

a Experiment with matrix 1 in Exercise 1 by trying out at least three different sets of weights for the criteria and computing weighted averages for applicants A, B, and C. Which applicant do your weighted averages consistently choose?

b Explain why it will *always* happen that when a choice matrix has a row that dominates all the others, this row will be selected by every weighted average as well. Does it make sense to assign weights and compute weighted averages in such a case?

Exercise 6

Betsy is about to graduate from college. She has obtained a job in another town, and before she moves, she wants to buy a new car. She has spent considerable time reviewing consumer reports and shopping around. She has narrowed her choice to five models. Three criteria are important to her: initial cost, gasoline economy, and body style and comfort. On the basis of what she has learned from her reading and shopping, Betsy has formulated a matrix (Table 7-10). Under each criterion she has ranked the cars in order using 1 (best), 2, 3, 4, and 5.

Betsy examines her matrix for dominance and finds none. She wonders about computing an average ranking for each car. As she considers this, a warning from a mathematics class comes to mind:

> Averages of measure numbers are meaningful only if the measures are obtained using an interval scale or a ratio scale; a single unit of measure must be used.

Heeding this warning, Betsy proceeds to translate her rankings into utility measures. For each criterion she rates the cars using a scale with a range of 1 (lowest) to 10. The matrix shown in Table 7-11 results.

a Suppose that Betsy gives the following weights to her criteria:

Cost	30 percent
Gasoline economy	50 percent
Style and comfort	20 percent

Examine Table 7-11 and guess which brands of automobiles will have the highest and lowest weighted averages. Check your guesses by calculating the weighted-average ratings for all five choices. What type of car should Betsy buy?

b What are additional criteria that you think should be used in deciding what type of car to purchase?

TABLE 7-10
Betsy's initial matrix of rankings

	CRITERIA		
CHOICES	Cost	Gasoline economy	Style and comfort
Ford	3	3	3
Chevrolet	1	5	4
Renault	2	1	5
Plymouth	4	2	2
Volkswagen	5	4	1

TABLE 7-11
Betsy's choice matrix

	CRITERIA		
CHOICES	Cost	Gasoline economy	Style and comfort
Ford	5	5	6
Chevrolet	8	3	4
Renault	7	9	1
Plymouth	4	8	7
Volkswagen	2	4	8

Exercise 7: Invent

Raymond is a senior at Euphoria College, majoring in accounting. He is in the throes of deciding on a job. A long list of applications and several interviews have resulted in four job offers that he is seriously considering. The criteria that he considers important are location, salary, and opportunity for advancement. The list below shows his rankings; each job is identified by the city in which it is located.

	Location	Salary	Opportunity for advancement
Best	Chicago	Atlanta	Birmingham
Second best	Atlanta	Birmingham	Dallas
Third best	Birmingham	Dallas	Chicago
Fourth best	Dallas	Chicago	Atlanta

a Give Raymond some advice. What steps can he follow in making his decision?

b Invent whatever new information is needed to carry out the steps you have recommended in *a*.

Exercise 8

Interpret the quotation from Chang Tzu in your own words and illustrate it with an example from your own experience.

All men know the utility of useful things; but they do not know the utility of futility.

Chang Tzu, 369–286 B.C.

TABLE 7-12

	GEORGIA FRIED CHICKEN'S CHOICES		
BURGER QUEEN'S CHOICES	Lower prices	Raise prices	Increase advertising
Lower prices	−3	5	−5
Raise prices	−10	5	−5
Increase advertising	5	10	−3

Exercise 9

Matrices can be used to display the outcomes of competitive decision-making situations. (The formal name for the study of such situations is *game theory.*) To illustrate this type of application, let us consider a hypothetical struggle for customers between Burger Queen and Georgia Fried Chicken. In a certain college town these are the only two competitors for shares of the fast-food market; together they have 100 percent of the market. At present Burger Queen has 45 percent of the customers and Georgia Fried Chicken has 55 percent. Burger Queen's manager, Sid McQueen, is trying to decide what to do. On the basis of results of past competitive campaigns, McQueen is considering three strategies, and he expects his competitor to do likewise. The estimated effects—in percentage points of increase or decrease of Burger Queen's market share—of the combinations of strategies are shown in Table 7-12. McQueen must make his decision without knowing what his competitor will do. If the estimates provided in the matrix are reliable, what is his best choice? Why?

Exercise 10

The French philosopher and mathematician Blaise Pascal made a lifelong attempt to reconcile his religious faith with the rationalism of mathematics and science. He gave an argument for believing in God that is summarized by the matrix in Table 7-13. In the matrix, m denotes the largest imaginable positive number and $-m$ denotes the smallest imaginable negative number.

Pascal believed that a life of adherence to religious rules and doctrines was every bit as worthwhile as a life that pursued other pleasures; the equal utility values of 100 in the second column of the matrix show this. The entries m and $-m$ express Pascal's belief that if God existed, an eternity of happiness would be in store for the believer and an eternity of suffering for the nonbeliever.

a Pascal's choice matrix exhibits dominance. What is the dominant choice?

b Many people disagree with the values in Pascal's matrix. For some, the matrix in Table 7-14 is a better representation. Explain why the decision whether or not to believe in God is more difficult as formulated in Table 7-14 than as formulated in Table 7-13.

Belief is a wise wager. Granted that faith cannot be proved, what harm will come to you if you gamble on its truth and it proves false? . . . If you gain, you gain all; if you lose, you lose nothing. Wager, then, without hesitation, that He exists.

Blaise Pascal

TABLE 7-13
Choice matrix illustrating Pascal's view of whether to believe

CHOICES	POSSIBLE SITUATIONS	
	God exists (believers are rewarded and nonbelievers are punished)	God doesn't exist (this earthly life is all there is)
Believe (and live a religious life)	m	+100
Don't believe (and pursue whatever pleasures suit you)	$-m$	+100

TABLE 7-14
A view other than Pascal's

CHOICES	POSSIBLE SITUATIONS	
	God exists	God doesn't exist
Believe	m	−100
Don't believe	$-m$	+100

Exercise 11

Some athletic performances—for example, in diving, gymnastics, and figure skating—are evaluated by judges who each assign up to 10 points. The points from the different judges are averaged to determine a single score.

a Compare this method of determining the best athletic performance with decision making using weighted averages as described in the examples and exercises in this chapter. What similarities do you find?

b Controversies sometimes arise over judges' assessments of an athletic performance. The same judge may evaluate similar performances as if they were quite different, and different judges may evaluate the same performance very differently; bias and inconsistency seem to slip into the decision process. Discuss these difficulties; why do you suppose that judges continue to be used despite the problems?

c Bias and inconsistency can also be found in our own endeavors to assign points (or utils) as a means of evaluating alternatives in making a decision. Carefully consider this fact and any other difficulties (list them) that you find with decision making using weighted averages. From these difficulties, develop some cautionary guidelines that can help you to use the method wisely.

Exercise 12: Journal

Scrutinize some decision. (You may want to reconsider a decision analyzed in Exercises 13 and 20 in Chapter 5.) Begin by listing the choices and criteria. Examine the criteria: do they fulfill the conditions stipulated in the footnote to Example 2? Examine the choices: is the list complete? Using utility scales, assess the extent to which each choice meets each criterion and organize the results in a choice matrix. Examine your matrix for dominance: can some choices be eliminated? Assign weights; compute weighted averages; and examine the results—do they pinpoint a decision? Is more information needed?

Exercise 13: Journal

Make a list of decision situations in which construction of a choice matrix and computation of weighted averages could be useful to you.

The examples in the text focused on decisions that were important enough to justify considerable decision-making effort. However, some apparently minor decisions are repeated over and over again, so that their cumulative effect may make them also worthy of analysis. (For example, you may face the daily decision whether to get up early to beat rush-hour traffic on your way to school and find a parking space easily, or to sleep later and spend more time driving and finding a parking space.) Add to your list some minor decisions which you make repeatedly and for which decision analysis could prove beneficial.

Exercise 14: Test Yourself

Develop a list of new terms and key ideas presented in this chapter. Supply an example of each item on your list.

Note The Appendix contains complete answers for Exercises 1*a* and *c*, 2*c*, 5*b*, and 6*a*. It contains partial solutions or hints for Exercises 6*b*, 7, 9, and 10*b*.

CHAPTER 8

MAXIMIZING SATISFACTION; MINIMIZING COST

> When I get a little money, I buy books; and if any is left, I buy food and clothes.
>
> —Desiderius Erasmus

INTRODUCTION

Some people have a lot of money but get very little satisfaction from spending it. Others have less money but are well pleased with what they can purchase. The difference may be due primarily to self-knowledge: people in the second group have thought carefully about what they want and have directed their purchases toward that end.

This chapter presents a process for deciding how to spend a limited resource—like money or time—to obtain the greatest satisfaction. Most of the ideas will be introduced by means of a story about an imaginary student named Patricia. The story is somewhat stilted and unrealistic, but its artificiality may help make it memorable and the methods are, in any event, easily extended to more realistic examples.

A PROBLEMATIC SITUATION*

Last year, when she was a freshman at Learnwell University, Patti lived it up. She went to all the parties she could squeeze in, played bridge every afternoon at the student union, and spent many an evening in the television room. Despite all this, she managed to avoid probation, and she has now returned to Learnwell for her sophomore year.

But this year Patti's attitude has changed. She wants to be a serious student and to get the maximum educational benefit from her college years. Her new outlook is even reflected in a name change: Patti establishes her seriousness by asking to be called Patricia.

In addition to her increased emphasis on doing well in her studies, Patricia has decided to increase her learning outside classes by spending whatever she can on books for her personal library. There are limitations to how many books Patricia can buy. One, of course, is her budget; but another is her craving for hamburgers. Patricia has a meal ticket at the college dining hall; but the meals there are sometimes dull, and although she has to pay for hamburgers at a grill off campus, they are often just the thing to boost her morale and her energy.

Sometimes Patricia finds that she has spent so much of her budget on books that she cannot afford a hamburger; at other times she gives in to her craving for hamburgers only to wish later that she had the money to spend on books. "How can I learn to spend my money on what I really want?" she wonders. Suppose that Patricia has come to us for financial advice. We can give it if she will provide the information we need.

A UTILITY MATRIX

We first ask Patricia to devise a utility scale to measure the satisfaction she gets from her purchases. As a starting point for her measures we suggest that she assign 0 utils as the satisfaction value of no purchases. After some discussion with Patricia we learn that she values one book more than one hamburger, but not twice as much. Eventually she settles on the following initial utility assignments:

1 hamburger: 20 utils

1 book: 30 utils

In this situation the utility scale will be "open-ended" rather than restricted to a fixed range (such as 1 to 10). The minimum possible utility value is 0, but we do not fix an upper limit. This will allow for increasing utility values associated with the acquisition of more hamburgers or books.

The use of 0 for minimum satisfaction makes our measurement scale a ratio scale; the unit of measure, or util, is $\frac{1}{20}$ of the satisfaction gained from the purchase of an initial hamburger (or, equivalently, $\frac{1}{30}$ of the satisfaction from the purchase of a book).

---

*The adventures of Patricia related in this chapter were inspired by the adventures of Al in C. A. Lave and J. G. March, *An Introduction to Models in the Social Sciences,* Harper and Row, New York, 1975, chap. 5, pp. 160–170.

TABLE 8-1
A matrix for Patricia's utility values

NUMBER OF HAMBURGERS (PER MONTH)	NUMBER OF BOOKS (PER MONTH)										
	0	1	2	3	4	5	6	7	8	9	10
0	**0**	**30**									
1	**20**										
2											
3											
4											
5											
6											
7											
8											
9											
10											

TABLE 8-2
Patricia's initial attempt at utility assignments

NUMBER OF HAMBURGERS (PER MONTH)	NUMBER OF BOOKS (PER MONTH)										
	0	1	2	3	4	5	6	7	8	9	10
0	**0**	**30**	**60**	**90**	**120**	**150**	**180**	**210**	**240**	**270**	**300**
1	**20**										
2	**40**										
3	**60**										
4	**80**										
5	**100**										
6	**120**										
7	**140**										
8	**160**										
9	**180**										
10	**200**										

We next ask Patricia to enter values in the utility matrix shown in Table 8-1. Each entry should give the total utility to Patricia from the combined purchase of hamburgers and books indicated in the row and column headings.

Patricia starts by filling in the first row and column of the utility matrix; see Table 8-2. At this point we stop Patricia with a few comments and a question. We observe that her utility values exhibit the pattern "more is better." That is, the more hamburgers Patricia buys, the more satisfaction she has, and likewise with books. We agree with this pattern: if having a particular commodity is desirable, then the more one has of it, the more satisfaction one experiences. But we also point out to Patricia the phenomenon of *saturation.* "After you have purchased five books in a month," we ask, "do you gain as much pleasure from buying a sixth book as you did from the purchase of your first book?" Patricia admits that she does not. Her capacity for gaining pleasure from buying books is decreasing. The technical term for this "saturation" that occurs with increased acquisition of a commodity is

TABLE 8-3
Patricia's revised utility assignments

NUMBER OF HAMBURGERS (PER MONTH)	NUMBER OF BOOKS (PER MONTH)										
	0	1	2	3	4	5	6	7	8	9	10
0	0	30	50	70	85	100	110	120	125	128	130
1	20										
2	35										
3	50										
4	65										
5	75										
6	85										
7	95										
8	100										
9	105										
10	110										

declining marginal utility. The *marginal utility* of a hamburger or a book or anything else is the increase in utility that results from acquiring it. When Patricia has no books, the marginal utility of a book is 30 utils. But her response to our question tells us that when she has already purchased five books in a month then the marginal utility of one more book is less than 30 utils.

For anyone, as acquisition of any commodity increases, the marginal utility of each additional unit eventually declines. Even though the measurement scale is a ratio scale, saturation or declining marginal utility means that addition of utilities is not always sensible. Patricia agrees with the notion of declining marginal utility and revises her entries in the first row and first column of the utility matrix to those shown in Table 8-3. But she is becoming impatient with the laboriousness of the process. "How will I ever come up with numbers to fill that entire matrix?" she says.

We are pleased to be able to tell her that finishing the matrix is easy. The remaining utility values for combinations of hamburgers and books may be found by adding utilities already given. For example, the total utility for a combination of 2 hamburgers and 3 books is 105 utils, since the utility for 2 hamburgers is 35 utils, the utility for 3 books is 70 utils, and

$$35 + 70 = 105$$

We can justify the addition of utilities in this case because purchases of hamburgers do not contribute to saturation of the desire for books, nor is the reverse true (books do not contribute to saturation of the desire for hamburger).

Patricia's completed utility matrix is shown in Table 8-4. It summarizes her preferences. (Later in this chapter, in "Getting the Most from a Fixed Budget," we will examine the equally important and competing notion of what she can actually afford.)

TABLE 8-4
Patricia's complete utility matrix for monthly purchases of hamburgers and books

NUMBER OF HAMBURGERS (PER MONTH)	NUMBER OF BOOKS (PER MONTH)										
	0	1	2	3	4	5	6	7	8	9	10
0	0	30	50	70	85	100	110	120	125	128	130
1	20	50	70	90	105	120	130	140	145	148	150
2	35	65	85	105	120	135	145	155	160	163	165
3	50	80	100	120	135	150	160	170	175	178	180
4	65	95	115	135	150	165	175	185	190	193	195
5	75	105	125	145	160	175	185	195	200	203	205
6	85	115	135	155	170	185	195	205	210	213	215
7	95	125	145	165	180	195	205	215	220	223	225
8	100	130	150	170	185	200	210	220	225	228	230
9	105	135	155	175	190	205	215	225	230	233	235
10	110	140	160	180	195	210	220	230	235	238	240

Exercise 1

a How much satisfaction will Patricia gain from purchasing 8 hamburgers? 6 hamburgers and 2 books? 4 hamburgers and 4 books? 2 hamburgers and 6 books? 8 books? Use Table 8-4 to answer.

b If a friend offers to treat Patricia to one of the purchases listed in *a*, which will she choose? Explain why. (Again, use Table 8-4.)

Exercise 2

A nutritionist is studying the eating habits of Sam, whose special cravings are for marshmallow sundaes and candy bars. Sam provides the utility values that form the first row and the first column of the utility matrix in Table 8-5.

a Can Sam's utility matrix be completed by adding pairs of values already present? Discuss.

b Do Sam's utility values show declining marginal utility? Give examples.

TABLE 8-5
Sam's utility matrix

NUMBER OF MARSHMALLOW SUNDAES	NUMBER OF CANDY BARS					
	0	1	2	3	4	5
0	0	150	250	325	375	400
1	200					
2	350					
3	450					
4	500					
5	525					

TABLE 8-6
Jeff's utility matrix

HOURS OF STUDY	HOURS OF LEISURE								
	0	1	2	3	4	5	6	7	8
0	0	10	15	19	22	24	25	25	25
1	20								
2	35								
3	45								
4	50								
5	54								
6	57								
7	59								
8	60								

Exercise 3

On Mondays Jeff has 5 hours of classes; eating requires an additional 3 hours, and he needs 8 hours of sleep. These necessities leave 8 hours for him to apportion between studies and leisure.

a Suppose that Jeff has supplied the utility values in Table 8-6 to describe the satisfaction he receives from hours of study and hours of leisure. Complete Jeff's utility matrix by adding pairs of given values. (Why is it reasonable to add in this case?)

b Does it make sense that the last three entries in the first row of Jeff's utility matrix are equal? How do you interpret this?

c As long as Jeff's schedulable time is fixed at 8 hours, only the diagonal entries from lower left to upper right are useful. Focus on these entries by circling them; then examine them to see how Jeff should allocate his 8 hours of time to gain the greatest satisfaction. How many hours of leisure and how many hours of study should he schedule to gain the greatest satisfaction from his Mondays?

d If Jeff joins the tennis team, which practices for 2 hours on Mondays, how much time does he have to divide between studies and leisure? Circle the matrix entries that correspond to this new situation. Now how should Jeff allocate his time to obtain the greatest satisfaction?

Exercise 4

When development of Patricia's utility matrix was begun, it was suggested that Patricia assign 0 utils as the satisfaction value of no purchases. Need 0 have been used? What about the satisfaction of saving money for more purchases later?

Note The Appendix contains complete answers for Exercises 1*b*, 2*a*, and 3*c*. It contains partial solutions or hints for Exercises 3*a* and *b*.

GETTING THE MOST FROM A FIXED BUDGET

Patricia has limited funds and watches her expenditures carefully. Let us suppose that each month she allows herself $16 for the purchase of hamburgers and books. Let's also suppose, for the sake of our calculations, that the cost of a hamburger (with a Coke) is $2 and an average paperback book costs $4. (These costs are, obviously, as imaginary as the rest of Patricia's story.) Therefore her possible monthly purchases include these:

8 hamburgers, no books
6 hamburgers, 1 book
4 hamburgers, 2 books
2 hamburgers, 3 books
No hamburgers, 4 books

We suggest to Patricia that she circle the entries in her utility matrix (Table 8-4) that correspond to these possible monthly purchases. She does so; the result is shown in Table 8-7. Patricia is quick to see what to do with the circled values. "The circled numbers measure the satisfaction for the purchases that I can afford. My best purchase is the one whose circled utility is highest."

Patricia gets maximum utility from her $16 monthly budget if she spends it on

6 hamburgers and 1 book

or on

4 hamburgers and 2 books

since these combinations offer a total utility of 115 utils—higher than those offered by any other purchase that fits her budget.

Now that she understands the process, Patricia would just as soon stop thinking about it. But we urge her to spend some more time with the problem. Since she has devoted so much effort to developing her utility matrix, we'd like her to see how it can be used to solve a variety of problems.

TABLE 8-7
Patricia's utility matrix;
$16-budget purchases are circled

NUMBER OF HAMBURGERS (PER MONTH)	NUMBER OF BOOKS (PER MONTH)										
	0	1	2	3	4	5	6	7	8	9	10
0	0	30	50	70	(85)	100	110	120	125	128	130
1	20	50	70	90	105	120	130	140	145	148	150
2	35	65	85	(105)	120	135	145	155	160	163	165
3	50	80	100	120	135	150	160	170	175	178	180
4	65	95	(115)	135	150	165	175	185	190	193	195
5	75	105	125	145	160	175	185	195	200	203	205
6	85	(115)	135	155	170	185	195	205	210	213	215
7	95	125	145	165	180	195	205	215	220	223	225
8	(100)	130	150	170	185	200	210	220	225	228	230
9	105	135	155	175	190	205	215	225	230	233	235
10	110	140	160	180	195	210	220	230	235	238	240

Example 1: If the Total Budget Changes ▶

Patricia's sister's birthday is in March. Suppose that, in order to purchase a gift, Patricia takes $6 from her budget. What combination of hamburgers and books should she purchase in March to gain maximum satisfaction from her remaining $10?

To solve this problem, refer to the matrix in Table 8-7. The pertinent information from that matrix is listed below:

Possible purchases	Total utility
5 hamburgers	75 utils
3 hamburgers, 1 book	80 utils
1 hamburger, 2 books	70 utils

In this case, Patricia should purchase 3 hamburgers and 1 book, to obtain a maximum total satisfaction of 80 utils. ◀

Example 2: If Prices Change ▶

Patricia's bookstore is having a January clearance sale—with many books that Patricia wants reduced to $2. How many $2 hamburgers and $2 books should she purchase in January to gain maximum satisfaction from her $16 budget?

To solve this problem, turn again to Patricia's utility matrix (Table 8-7). The matrix supplies the following information pertinent to this case:

Possible purchases	Total utility
8 hamburgers	100 utils
7 hamburgers, 1 book	125 utils
6 hamburgers, 2 books	135 utils
5 hamburgers, 3 books	145 utils
4 hamburgers, 4 books	150 utils
3 hamburgers, 5 books	150 utils
2 hamburgers, 6 books	145 utils
1 hamburger, 7 books	140 utils
8 books	125 utils

In this situation Patricia's most satisfying purchases are 4 hamburgers and 4 books or 3 hamburgers and 5 books, for a total satisfaction of 150 utils.

Example 3: If the Goal Is a Certain Level of Satisfaction at Minimum Cost ▶

Suppose that Patricia is tired of worrying about money and begins to think instead about trying to achieve a certain amount of satisfaction. We hasten to remind her that no matter how much she *wants*, her expenditures are limited by her budget.

Still Patricia pursues her idea. "Suppose I feel that I deserve 105 utils of satisfaction," she says. "What are the different ways I can achieve it?"

From her utility matrix (Tables 8-4 and 8-7) Patricia finds the following list of combinations that yield 105 utils of satisfaction:

9 hamburgers (cost: $18)
5 hamburgers, 1 book (cost: $14)
2 hamburgers, 3 books (cost: $16)
1 hamburger, 4 books (cost: $18)

Patricia is pleased with the discovery she has made: she can gain 105 utils of satisfaction at a cost of only $14. ◀

When several alternatives provide us with the same satisfaction, we say that we are *indifferent* about which alternative is the best choice. Stated differently, *indifference* is the attitude or feeling that we have toward choices with equal utility values. When faced with choices that offer the same satisfaction, we might let someone else choose, or pick the least expensive choice, or even flip a coin.

MORE ON MARGINAL UTILITY

Sometimes the preferences that people express seem to be contradicted by the way they spend money.

- When asked which he valued more, spending for education or for highways, the President replied, "Education," and then proposed a budget that included slight increases in spending for education and huge increases in spending for highways.
- When asked which they valued more, an adequate supply of good drinking water or new tennis courts, the members of a rural community unanimously responded that they placed a higher value on drinking water—but last year they spent $100,000 on new tennis courts while investing nothing to improve the community's water-supply system.
- Carolyn Vitamin is an outspoken proponent of healthful eating habits, yet she is seen twice weekly purchasing and eating candy bars.

However, such apparent contradictions between preferences and expenditures may have a reasonable explanation in terms of declining marginal utility. For example, the President may indeed have placed a high priority on education, but he may also have believed that the expenditures for education were nearly adequate to begin with and that the greatest gain from additional spending would occur if it went for highways. In other words, the marginal utility of additional expenditures for education was less than the marginal utility of additional expenditures for highways.

Similarly, the citizens of the community might indeed have placed a high value on a good water supply—but if this need was already satisfied they would gain more satisfaction from spending for recreation. That is, *more* good water had a lower marginal utility than *more* recreation.

Finally, Carolyn may be very conscious of meeting her body's nutritional needs—but once these have been met she may legitimately gain more satisfaction from a candy bar than from another cup of yogurt. Although the total utility that Carolyn gains from nutritious foods is high, after a great deal of nutritious eating, saturation drives its marginal utility down.

To explore further the concept of marginal utility, let us return to Patricia's problem. We will consider three similar situations to see how marginal utility affects the outcomes.

Situation 1

Suppose that in a certain month Patricia has acquired 5 hamburgers and 2 books and a friend offers her a choice—at no cost to her—of one more hamburger or one more book. Which will she choose?

Analysis

Here Patricia's current acquisitions have yielded a utility of 125. An additional hamburger would raise her total utility to 135 (see Table 8-7) and an additional book would raise her total utility to 145. Because of her saturation with hamburgers, the marginal utility of a book is higher than the marginal utility of a hamburger, and Patricia will choose the gift of a book.

Situation 2

Suppose that in a certain month Patricia has acquired 2 hamburgers and 5 books and a friend offers her a choice of one more hamburger or one more book. Which will she choose?

Analysis

In this situation, Patricia's current acquisitions have yielded a total utility of 135. An additional hamburger would raise her utility total to 150 whereas an additional book would yield a total utility of only 145. In this case, Patricia is more saturated with books than with hamburgers. The marginal utility of a hamburger is higher than the marginal utility of a book, and Patricia will choose the gift of a hamburger.

Situation 3

Suppose that in a certain month Patricia has acquired 3 hamburgers and 4 books and a friend offers her a choice of one more hamburger or one more book. Which will she choose?

Analysis

In this case, Patricia's current acquisitions have yielded a total utility of 135. Either one more hamburger or one more book will raise the total utility to 150. Since the marginal utilities of a hamburger and a book are the same, Patricia will be indifferent about the choice offered to her. If her friend is cost-conscious, we might suppose that he or she will choose to give the less expensive gift.

SHORTCUTS

The analysis of Patricia's financial situation has been long and slow, largely because of the time devoted to constructing her utility matrix. Fortunately, once we understand how budget constraints and preferences work together, we can often solve specific problems with only a small amount of effort. To illustrate this, let us consider two more situations that involve Patricia.

Situation A: Obtaining Maximum Satisfaction from a Fixed Budget

This month, Patricia has a $20 budget to spend on hamburgers and books—both of which, as we know, she likes a great deal. If hamburgers cost $2 each and books $4 each, how many of each should Patricia buy to get the most satisfaction from her monthly budget?

Solution

Step 1: List the possible combinations of purchases that fit the budget.

Step 2: Estimate the amount of satisfaction that will be gained from each combination.

Step 3: Select the combination that offers the greatest satisfaction.

The utility values needed to solve the problem are obtained from Table 8-7. Following steps 1 to 3, we have:

Possible purchase combinations	Total utility	
10 hamburgers	110	
8 hamburgers, 1 book	130	
6 hamburgers, 2 books	135	Best purchases
4 hamburgers, 3 books	135	
2 hamburgers, 4 books	130	
5 books	110	

Note In some situations of this type, it may be possible to omit step 2. When the possible purchases have been listed, the decision maker may be able to tell with ease which is the most desirable—*without* assigning specific utility values.

Situation B: Obtaining a Certain Satisfaction at Minimum Cost

Patricia's parents are proud of her recent achievements at college and want to reward her—within their limited budget. They offer to buy her five books. Patricia, knowing that her parents' resources are limited, wonders, "Could my parents give me the same amount of satisfaction at less cost?"

Solution

Step 1: Determine the satisfaction level to be met.

Step 2: List possible purchases that provide the desired level of satisfaction; for each of these find the cost.

Step 3: Select the purchase with the least total cost.

Following steps 1 to 3, we determine that 5 books will give Patricia a satisfaction of 100 utils (see Table 8-7). The combinations of purchases that achieve this satisfaction level are

5 books (cost: $20)
3 hamburgers, 2 books (cost: $14) } Least cost
8 hamburgers (cost: $16)

Note It may not be necessary to give a numerical utility value in step 1; one need only list purchases that yield the same satisfaction. For example, suppose that you and a date are considering a variety of possible activities for your time together next Saturday: picnicking and hiking in a state park, going to the beach, and going to an amusement park. It is possible for you to determine—without ever assigning numerical values—that you are indifferent among these choices; that is, that each has the same utility value. In such a case it makes sense to select the alternative with the lowest cost.

SUMMARY

Allocation of limited resources—such as money and time—to obtain the greatest possible satisfaction is a process that can require considerable thought. As input information for our spending decisions we need to know:

Budget: How much we can afford
Preferences: How much satisfaction different purchases will provide
Prices: How much the desired items cost

The examples involving Patricia have illustrated steps that may be taken to decide what to purchase. The exercises that follow will help you extend these steps to other situations having to do with spending time or money.

Exercise 5

Use the utility matrix in Table 8-7 for information to decide what Patricia's best purchase would be in each of the following situations.

a Patricia has $12 a month to spend on hamburgers and books.

b Patricia has been losing weight and decides that a trip to the grill off campus should result in a hamburger dinner (including fries and a salad) that costs $4 rather than just a hamburger and a Coke. Books still cost $4, and her budget is $16 a month.

c Last May, Patricia splurged and purchased 10 books and 1 hamburger. Now she looks back on that combination as a foolish one. "I could have gotten the same satisfaction (150 utils) for a lot less money," she said. At what minimum cost could Patricia have gained the same amount of satisfaction?

Exercise 6: Journal

Think of a situation in which a pair of commodities compete for your money or your time. Describe the situation in detail and decide on the best purchase plan under the circumstances. (You may want to pattern your analysis on the steps in Situation A or B.) Examples of commodities that you may find competing include: studies versus sleep, arcade games versus pool, and class attendance versus sunbathing.

Exercise 7

(This exercise is a variation of Exercise 6 in Chapter 2.) Right now Terry has D averages in physics and French, but he intends to take positive action. He estimates that for each additional hour per week that he devotes to French he can raise his grade one letter. For each additional 3 hours a week that he can spend on physics he can raise that grade one letter. The matrix in Table 8-8 describes the utility to Terry of each letter-grade combination. Notice that in Terry's utility matrix the other entries are not the sum of the leftmost and

TABLE 8-8
Terry's utility matrix

FRENCH GRADE	PHYSICS GRADE			
	D	C	B	A
D	0	30	50	60
C	20	45	60	65
B	30	50	70	75
A	35	55	70	80

topmost entries. This reflects the fact that raising his grade in one subject reduces his desire to raise his grade in the other—that is, his desire for achievement becomes saturated.

a If Terry can find 6 additional hours to devote to his studies, how should he allocate these hours to receive maximum satisfaction? What grades will he earn?

b If Terry can find 7 additional hours to devote to his studies, how should he allocate these hours to receive maximum satisfaction? What grades will he earn?

c Compare the grades earned in *a* and *b*. Are the results surprising? What explanation can you provide?

d **Invent** Make up variations of Terry's situation. For each new set of conditions that you give, find the best decision.

Exercise 8: Journal

On the basis of ideas gained from thinking about Terry's situation in Exercise 7, assess the different amounts of satisfaction you gain from the courses that compete for your study time. Find an allocation of time that provides you with the greatest satisfaction.

Exercise 9

Donna wants to go to medical school. She knows that the grades she earns in college are important. Specifically, she needs A's in her biology and chemistry courses and a B average in her other courses. Right now she has courses in art history, sociology, and statistics (in addition to biology and chemistry). To earn a D in art history she must spend at least 1 out-of-class hour per week on the subject; 1 additional hour will be required for each letter grade higher than D. To earn D in sociology requires 1½ hours per week and an additional 1½ hours for each letter grade higher than a D. To earn a D in statistics requires 2 hours, and each increase of one letter requires 2 more hours. List the various grade combinations that will yield a B average and give the time cost of each.

For example, the grades

Art history: C
Sociology: B
Statistics: A

yield a B average; the time cost per week is 2 + 4½ + 8 = 14½ hours. What allocation of weekly study hours will yield a B average in the minimum time?

Exercise 10

What are the merits and the weaknesses of Donna's approach to her studies in Exercise 9?

Exercise 11: Invent; Journal

Make up a problem, perhaps similar to Donna's in Exercise 9, in which the object is to achieve a certain goal in a minimum amount of time. Analyze and solve your problem.

Exercise 12

Provide explanations, in terms of marginal utility, for the following inconsistencies between what people say and what they do.

a Wayne, a student at State University, was asked, "Which is more important to you—obtaining an education or owning a car?" Wayne replied that an education was much more important to him. However, shortly thereafter Wayne received a $900 gift from his grandfather and, without hesitation, purchased a used car.

b The citizens of Quiettown are outspoken about their desire for crime control. Yet in the last election 90 percent of them voted "no" to a proposed tax increase that would have funded an annex to the county jail.

c Students surveyed at the college dining hall almost unanimously chose a salad bar as an important option for the food service organization to offer. But in its first few weeks of existence, the salad bar has been used by fewer than 20 percent of the students at each lunchtime meal.

d "Before we married," Dave complained to Rosa, "You used to like to go fishing with me. Now you hardly ever want to go."

Exercise 13

Spending patterns are affected by preferences (and the degree to which they have already been satisfied), by budget constraints, and by prices. With these ideas in mind, supply at least two plausible explanations for each of the following situations.

a Students at Middletown High School tend to spend their daily lunch allowances for candy and cigarettes at the shop across the street from the school instead of buying lunch at the school cafeteria.

b Bob works long hours as a systems analyst for a computer company and then spends additional evening hours in a computer store selling microcomputers and teaching people how to use them.

c Lila spends a maximum of 3 hours a day studying and then stops even if she has not finished her assignments. Linda sticks with every assignment until it is completed, no matter how long it takes.

d Last year the Clark family spent $5000 on home maintenance and repairs; this year they spent only $2000.

Exercise 14

Some people speak of the difficulty of comparing time and money. (But all of us can do it to an extent: we trade some of our hours on a job in exchange for pay.) Other things that are difficult to compare include happiness and money and pleasure today and the possibility of pleasure a year from now. Do utility scales provide a way of "comparing the incomparable?"

Exercise 15

A skeptic says, "This notion of utility might be all right for little decisions, like whether to buy a red balloon or a blue one. But big things—like marriage or a career—are just too important for numbers." Do you agree? Discuss.

One can drink too much, but one never drinks enough.

Gotthold Ephraim Lessing

Exercise 16

How may the cryptic quotation from Lessing be related to utility determinations?

Exercise 17: Journal—The Cost of Thinking

When faced with a task, some people first spend a lot of time thinking about how best to do it. Others quickly start the task; they are more concerned about getting it completed than about doing it the best way. In such situations there are two competing costs—the cost of thinking and the cost of doing the task in some way other than the best way.

For example, suppose that Kurt faces the task of writing a 500-word essay giving his opinions on the subject "The Trend Away from Excellence in College Studies." Kurt may decide to invest ½ hour or so in organizing his thoughts—deciding what is important to say and in what order to say it—and then write an essay that earns a B. Or he may simply start to write, producing an essay that earns a C.

In Kurt's case organizing costs ½ hour and gains a letter grade. Or, to state it differently, not organizing saves ½ hour and costs a letter grade.

We could offer Kurt this advice: If a higher grade gives more satisfaction than other ways of spending ½ hour of his time, then the extra time for thinking is a good investment.

Examine your own circumstances for situations similar to Kurt's—situations in which you face tasks and must decide whether to invest time in thinking in order to improve your performance on the task. In each case, assess your preferences to decide whether the investment of time for thinking will be worthwhile. Will it produce a gain that is greater than the cost of thinking?

> Thinking in its lower grades is comparable to paper money, and in its higher forms it is a form of poetry.
>
> Havelock Ellis

Exercise 18: Test Yourself

Develop a list of new terms and key ideas presented in this chapter. Supply an example of each item on your list.

Note The Appendix contains complete answers for Exercises 5*a*, 7*c*, 12*b*, and 13*b*. It contains partial solutions or hints for Exercises 5*c*, 7*d*, 10, and 17.

PART THREE

ORGANIZING COMPLEX INFORMATION

PART CONTENTS

CHAPTER 9

ORGANIZING INFORMATION IN MUTUALLY EXCLUSIVE CATEGORIES

Addition and Mutually Exclusive Groups
Organizing Information with Venn Diagrams

CHAPTER 10

CALCULATING NUMBERS OF POSSIBILITIES: A MULTIPLICATION RULE

How Many License Plates?
What Is the Multiplication Rule?
A Solution to the License Bureau's Problem
Use of the Multiplication Rule in Solving Problems
Binary Coding of Numbers

CHAPTER 11

VISUALIZING THE STRUCTURE OF INFORMATION WITH A TREE DIAGRAM

An Introduction to Tree Diagrams
Numbered Tree Diagrams
Summary

CHAPTER 12

SUMMARIZING INFORMATION USING A FEW NUMBERS

Locating the "Center" of a Collection of Data Items
Measuring Variation from the "Center"

CHAPTER 9

ORGANIZING INFORMATION IN MUTUALLY EXCLUSIVE CATEGORIES

A picture shows me at a glance what it takes dozens of pages of a book to expound.

—Ivan Sergeyevich Turgenev

ADDITION AND MUTUALLY EXCLUSIVE GROUPS

In Chapter 6 this addition problem was considered:

> Mary and Karen are planning a party. Marty has a list of 15 people she'd like to invite, and Karen has a list of 18 people to invite. How many guests do they want to invite all together?

To obtain a correct answer, we must know how many names appear on both lists. If we know that this number is 5, then we also know that the number of names on Marty's list alone is 10, and the number of names on Karen's list alone is 13. We then can correctly deduce that the total number of names is

$$10 + 13 + 5 = 28$$

Calculating the total number of objects or items resulting from the combination of two or more groups is an easy process only if the groups we combine are *mutually exclusive*. That is, the groups should not overlap; each object is in exactly one of the groups that are combined. When groups are mutually exclusive, the total number in a group formed by combining several smaller groups may be found by *addition*.

ORGANIZING INFORMATION WITH VENN DIAGRAMS

A Venn diagram is a pictorial device consisting of a rectangle with ovals inside. It provides a useful way of sorting overlapping information into mutually exclusive categories. First let us examine how Venn diagrams can be used to organize information. Our goal, however, goes beyond such diagrams. The ultimate aim is to organize information carefully—into mutually exclusive categories—so that Venn diagrams will not be needed to make sense out of overlapping information.

The confusion that can result when information is given in overlapping categories is illustrated by the following example. (This problem appeared earlier as Exercise 8*b* in Chapter 6.)

Example 1 ▶

Twentieth Century Realty Associates publicized the following information about the prices of homes it sold in New Jersey communities in August:

30 homes sold at prices under $30,000
100 homes sold at prices under $50,000
150 homes sold at prices under $75,000
175 homes sold at prices under $100,000
5 homes sold at prices of $100,000 or more.

How many New Jersey homes did the agency sell in August?

At first reading, it is tempting to add the given numbers and to say that the agency sold 460 homes. However, examination of the information given shows that there are only two mutually exclusive categories:

1 Homes that sold at prices under $100,000
2 Homes that sold at prices of $100,000 or more

Category 1 includes the three other categories listed.

A Venn diagram that illustrates categories 1 and 2 is given in Figure 9-1. The rectangular region represents the total collection of homes sold. The oval curve labeled A and its interior represent homes sold at prices under $100,000. The region outside the oval represents the homes sold at prices of $100,000 or more. Each region is labeled with the number of homes in its category. (Note that the Venn diagram is not drawn to scale.)

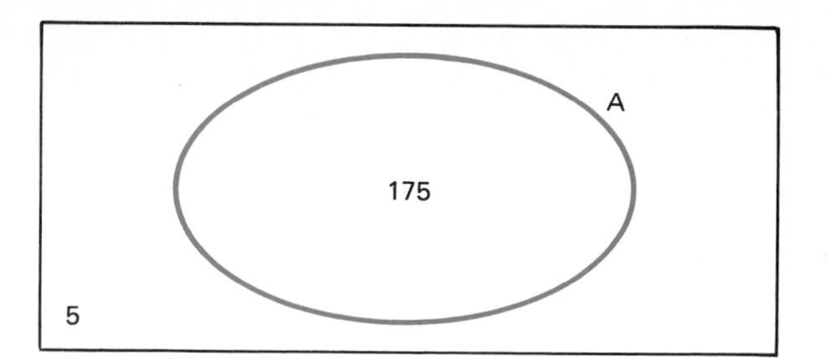

FIGURE 9-1
Partial Venn diagram for August home sales

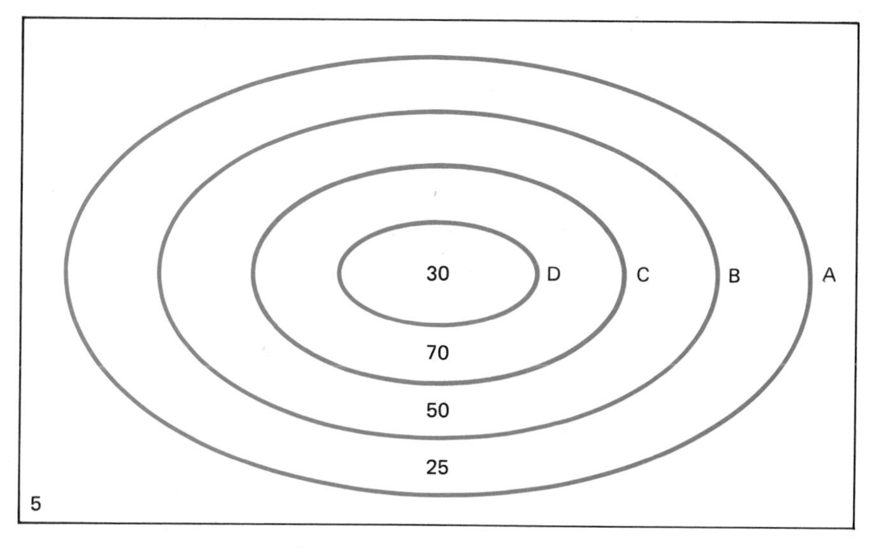

FIGURE 9-2
Complete Venn diagram for August home sales

Key to symbols:

Inside oval A: prices under $100,000
Inside oval B: prices under $75,000
Inside oval C: prices under $50,000
Inside oval D: prices under $30,000

Additional information noted above on the number of homes sold may be incorporated into the Venn diagram. The result is shown in Figure 9-2. Here, an oval region is used for each price category. Each of the doughnut-shaped regions is labeled with a number that tells how many homes are included in its outer oval but not in its inner oval. The sum of the numbers inside each oval gives the total number of homes in that category.

If Twentieth Century Realty Associates wants to present a clear picture of its August sales, it should use the numbers given in the Venn diagram and organize the information as in Table 9-1—in mutually exclusive categories. ◀

TABLE 9-1
Price breakdown for 180 homes sold in August by Twentieth Century Realty Associates

PRICE RANGE	NUMBER OF HOMES SOLD
0–$29,999	30
$30,000–49,999	70
$50,000–74,999	50
$75,000–99,999	25
$100,000–	5

TABLE 9-2
City Hospital blood bank data

ANTIGENS PRESENT	NUMBER OF PINTS AVAILABLE
A	50
B	53
Rh	85
A and B	15
A and Rh	42
B and Rh	45
A, B, and Rh	12

Example 2 ▶

This example further illustrates the difficulties caused by failure to insist on mutually exclusive categories. When blood is supplied to a patient, care must be taken to find the right "blood type" so that no allergic reaction will take place. Three types of antigens ("foreign agents") are especially important in determining whether a person will accept donated blood. These antigens are designated by the symbols A, B, and Rh. A person will react against any blood containing antigens that his or her own blood does not contain.

City Hospital stores 100 pints of blood for emergencies. When Dr. Clearhead is assigned to emergency room duty, she requests records to find out how many pints of each type were available. You can imagine her confusion and irritation when she is given the information shown in Table 9-2.

"This information is useless to me," Dr. Clearhead says. "It's obvious that some well-meaning aide has done a lot of counting, but some pints have been counted more than once. These data make our 100-pint supply appear to be 302 pints."

At this point we step into Dr. Clearhead's problem with the goal of reorganizing the data into mutually exclusive categories. We will use a Venn diagram.

We start with a rectangle to represent the total blood supply of 100 pints. Pints of blood containing the A antigen are represented by an oval region. Similarly, we use ovals to represent pints containing the B and Rh antigens (Figure 9-3).

Since blood which contains one antigen may or may not contain others, the regions in Figure 9-3 are not nested (as in Figure 9-2). Instead, the ovals overlap but do not coincide. The information from Table 9-3 enables us to label the central region: there are 12 pints that contain all three antigens—A, B and Rh.

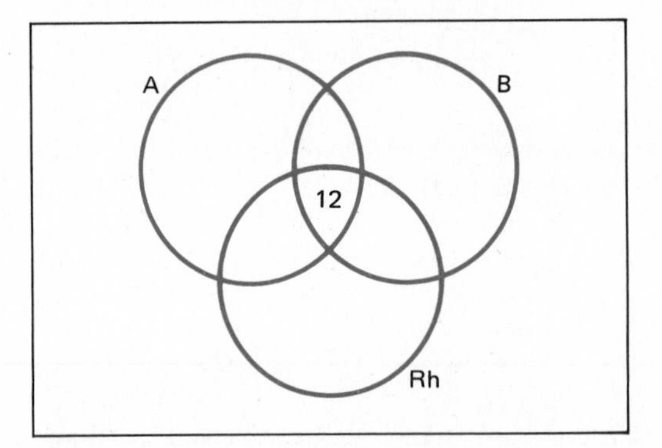

FIGURE 9-3
Initial Venn diagram for City Hospital blood supply

TABLE 9-3
City Hospital blood bank data listed in mutually exclusive categories

ANTIGENS PRESENT	BLOOD TYPE	NUMBER OF PINTS AVAILABLE
A only	A−	5
B only	B−	5
Rh only	O+	10
A and B only	AB−	3
A and Rh only	A+	30
B and Rh only	B+	33
A and B and Rh	AB+	12
None of A, B, or Rh	O−	2

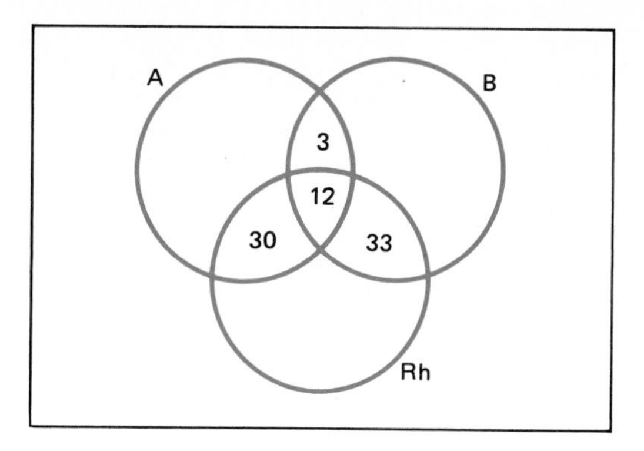

FIGURE 9-4
Partially labeled Venn diagram for City Hospital blood supply

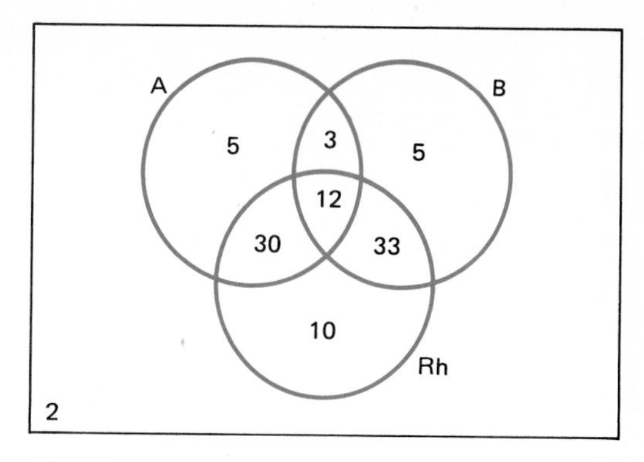

FIGURE 9-5
Completely labeled Venn diagram for City Hospital blood supply

Subtraction, to eliminate the effect of counting pints more than once, is required to continue labeling the Venn diagram. Table 9-2 tells us that 45 pints contain the B and Rh antigens. We have already considered 12 of these, which also contain the A antigen. Thus there must be

$$45 - 12 = 33 \text{ pints}$$

that contain only the B and Rh antigens. Continuing in this manner, we obtain the labels for Figure 9-4. Carefully compare Figure 9-4 and Table 9-2. The labels in Figure 9-4 are obtained from information in Table 9-2 by subtraction. Also, data in Table 9-2 can be obtained by adding the appropriate values in the Venn diagram.

The information that 50 pints contain the A antigen tells us that the upper left region within oval A must have the label 5. The information that 53 pints contain the B antigen and 85 pints contain the Rh antigen tell us to label the right-hand region of oval B with a 5 and the bottom region of oval Rh with a 10. To account for 100 pints, the region outside the ovals must be labeled with a 2, since the total inside the ovals is 98. We put this all together in Figure 9-5.

To conclude our consideration of this problem, we put the information from Figure 9-5 into the list shown in Table 9-3 and give it to Dr. Clearhead to replace Table 9-2. After checking our total and finding that it is indeed 100 pints, she gratefully accepts our list. However, she suggests a change in our terminology: she supplies the list of blood types given as the middle column in Table 9-3. ◀

The exercises that follow present some confusing mixtures of information which need to be organized into mutually exclusive categories. You may find Venn diagrams useful. But in each case a diagram is only a tool, not the goal. The goal is to classify the information into a complete list of nonoverlapping categories; here and elsewhere, organize information with that goal in mind.

Exercise 1

Alfred and Donna are planning a party. They decide to limit the number of guests to 30. Each has come up with a list of people to invite. Donna's list has 20 names, and Alfred's list has 20 names. However, when they combine their lists there is a total of only 25 names. On the basis of the information given, how many items are in each of the following mutually exclusive categories?

Names on Donna's list only	___
Names on Alfred's list only	___
Names on both lists	___
Additional people who can be invited	___
Total =	30

Exercise 2

Chapter 6 focused on the question, "In what situations is addition the appropriate problem-solving operation?" Provide a partial answer to this question for the case of problems that require finding the total number of objects in a group resulting from the combination of several other groups.

Exercise 3

A survey of 100 magazine subscribers yielded the following information:

70 subscribe to *TV Guide.*
22 subscribe to *TV Guide* and *Newsweek* but not to *Reader's Digest.*
25 subscribe to *TV Guide* and *Reader's Digest* but not to *Newsweek.*
10 subscribe to all three magazines.
45 subscribe to *Newsweek.*
10 subscribe only to *Newsweek.*
10 subscribe to none of these magazines.

How many subscribers are there in each of the following mutually exclusive categories?

Subscribers to all three magazines	___
Subscribers to only *TV Guide* and *Newsweek*	___
Subscribers to only *TV Guide* and *Reader's Digest*	___

Subscribers to only *Newsweek* and *Reader's Digest*	_____
Subscribers to only *TV Guide*	_____
Subscribers to only *Reader's Digest*	_____
Subscribers to only *Newsweek*	_____
Subscribers to none of these magazines	_____
Total =	100

Exercise 4

A mathematician friend of mine likes to classify the mathematics problems that she tries according to the amount of effort she must invest to obtain a solution. She calls a problem "simple" if she has little trouble figuring out what to do to solve it. (A "simple" problem might take her a long time to complete, however, and might involve lengthy steps or computations.) On the other hand, she calls a problem "easy" if not much time or effort is required to work out the solution. Not all "easy" problems are "simple." Sometimes a problem requires little effort to do but a great deal of effort to figure out what to do. (If you have ever struggled for a long time over an exercise, only to attend class the next day and see the instructor present a three-line solution that you readily understood but hadn't thought of yourself, then you have an example of a problem that is "easy" but not "simple.")

a Describe in your own words the types of mathematical problems in each of the following mutually exclusive categories resulting from the mathematician's designations "simple" and "easy": (1) problems that are "simple" and "easy"; (2) problems that are "simple" but *not* "easy"; (3) problems that are *not* "simple" but are "easy"; (4) problems that are neither "simple" nor "easy."

b As you work on this set of exercises, place each problem in one of the four categories in *a*.

> Intelligence is proved not by ease of learning but by understanding what we learn.
>
> **Joseph Whitney**

Exercise 5

A tax assessor examines 172 private dwellings and determines that

27 of them have values over $80,000.
85 of them have values over $60,000.
125 of them have values over $40,000.
The rest have values of $40,000 or less.

Reorganize this information to give the number of dwellings in each of the following mutually exclusive categories:

Dwellings valued at $40,000 or less	___
Dwellings valued over $40,000 but not over $60,000	___
Dwellings valued over $60,000 but not over $80,000	___
Dwellings valued over $80,000	___

Exercise 6

The college placement office has issued the following report on last year's graduates:

5 percent of the graduates obtained jobs with salaries over $30,000.
20 percent of the graduates obtained jobs with salaries over $25,000.
50 percent of the graduates obtained jobs with salaries over $20,000.
85 percent of the graduates obtained jobs with salaries over $15,000.

Reorganize this information into mutually exclusive categories and give the percent of graduates in each category.

Exercise 7

Clarissa provides the following information about 13 friends who live with her in the same dormitory suite: 6 of them are intelligent, 8 are entertaining, and 8 have high moral standards. Furthermore, 4 are both intelligent and entertaining, 6 are entertaining and have high moral standards, 3 are intelligent and have high moral standards, and 2 have all three virtues. List eight mutually exclusive categories for Clarissa's friends and tell how many belong in each category.

Exercise 8

While searching for the cause of a mysterious disease which has stricken 200 people, researchers have focused their attention on three activities occurring in a certain hotel. They have surveyed the 200 victims to determine how many attended a lunch, a banquet, or a party given there. Unfortunately, one page of the survey results has been destroyed.

Only the following information remains:

16 people attended the lunch, the banquet, and the party.
20 people had nothing to do with the hotel.
24 people attended the party and the lunch.
32 people attended the lunch and the banquet.
28 people attended the banquet and the party.
80 people attended the lunch.
28 people attended the banquet but had no other contact with the hotel.

Organize this information into mutually exclusive categories. How many of the 200 people attended the party?

Exercise 9

There are six children at a party. Three of them are Walkers, and three are Smiths. The Walker children all have the same color hair and the Smith children all have the same color eyes.

Using the following descriptions, determine each child's last name.

Tom: brown eyes, brown hair
Jill: blue eyes, blond hair
Bob: brown eyes, brown hair
Kent: brown eyes, blond hair
Mary: brown eyes, brown hair
Sue: brown eyes, blond hair

Exercise 10: Invent

During the evening rush hour at South Station, 300 people were surveyed about their use of mass transit. Each was asked whether he or she used buses, subways, or trains. The following results were recorded:

Number who ride buses:	180
Number who ride subways:	250
Number who ride trains:	145

Much of the information that could have been gained from the survey was lost because the results were not listed in mutually exclusive categories. We don't know, for example, how many people ride both the bus and the subway but not the train.

Suppose that each of the 300 people surveyed used at least one form of mass transit. List the seven mutually exclusive categories of users.

Invent survey data for each of your seven categories that are consistent with the numbers given above.

Exercise 11

A survey of 100 people was taken to determine which of the various media they use regularly to obtain news of state, national, and world events.* The following information was reported:

13 regularly use all three of radio, television, and newspapers to learn the news.

65 regularly read a newspaper for their news; 25 of these use only the newspaper.

37 regularly listen to the radio for their news; 13 of them use only the radio.

88 regularly watch television for their news; 45 of these use only television for news.

What sense can you make of these results?

Exercise 12

Edgar is a college freshman and is considering a double major in political science and economics. When he examines the requirements he discovers that

Economics requires 35 semester hours of course work.

Political science requires 36 semester hours of course work.

A bachelor's degree in any field requires 60 semester hours of "general education" courses distributed over a variety of fields.

Noting that $35 + 36 + 60 = 131$, which is a total considerably greater than the 120 semester hours needed to graduate, Edgar is dismayed.

Advise Edgar. Might there be a way he can accomplish his goal?

*This survey is fictitious. However, see *Time,* Dec. 6, 1982, p. 88, for results of an actual survey of sources of news.

Exercise 13: Invent; Journal

Look for examples in your reading or your own experiences in which reorganization of information into mutually exclusive categories is helpful for understanding and using the information.

Exercise 14: Test Yourself

Develop a list of new terms and key ideas presented in this chapter. Supply an example of each item on your list.

Note The Appendix contains complete answers for Exercises 1 and 2. It contains partial solutions or hints for Exercises 6, 7, 10, and 11.

CHAPTER 10

CALCULATING NUMBERS OF POSSIBILITIES: A MULTIPLICATION RULE

How do I love thee? Let me count the ways.
I love thee to the depth and breadth and height
My soul can reach. . . .

—Elizabeth Barrett Browning

HOW MANY LICENSE PLATES?

The license bureau of a state department of motor vehicles has been trying to arrange for the production of new automobile license plates. The number of plates needed is estimated at 1 million. The bureau wants each plate to begin with a letter of the alphabet, followed by four digits. It considers this organized approach preferable to the random mixing of letters and digits used in the past. However, as contracts are being drawn up with manufacturers of license plates, it is discovered that this design format will not allow for 1 million plates.

Several questions face the license bureau. Immediately, it must answer the question, "Is there a systematic arrangement of letters and digits that will yield at least 1 million different plates?" Moreover, as it solves this problem, it would like to learn how to avoid the embarrassment of choosing an inadequate design in the future. Thus it is also asking, "Is there a method we can use to predict the number of plates that can be made with a given design pattern?"

To answer the license bureau's questions, let us first consider a simplified version of the problem:

> Suppose that we wish to count the number of license plates that can be designed to display two symbols—a letter followed by a single digit.

The complete list of possible two-symbol license plates may be displayed using a matrix; this is done in Table 10-1. The matrix of Table 10-1 contains (if completed) 26 rows, with 10 entries in each row—a total of $26 \times 10 = 260$ entries—each one designating a possible license plate.

We next consider a slightly more difficult question:

> How many license plates can be designed with three symbols—a letter followed by two digits?

To answer this question, we can build on what we have already done. If we select B-6 from our two-symbol plates, we can convert it to a three-symbol design in ten different ways: B-6-0, B-6-1, B-6-2, B-6-3, B-6-4, B-6-5, B-6-6, B-6-7, B-6-8, and B-6-9. Each two-symbol plate design likewise leads to ten three-symbol arrangements. The number of possible three-symbol plates is thus 10 times the number of two-symbol plates:

$$26 \times 10 \times 10 = 260 \times 10 = 2600$$

In the two-symbol and three-symbol cases considered, the number of plates is found by multiplying the number of possibilities for each symbol. We can generalize this procedure to find the number of five-symbol plates with one letter followed by four digits; this number is

$$26 \times 10 \times 10 \times 10 \times 10 = 260{,}000$$

This is far short of the 1 million plates needed by the license bureau.

Later, we will return to the license bureau's problem and try to find a design that does yield 1 million plates; but first we will develop a formal statement of the multiplication rule that can be used in many, many situations for calculating the number of possibilities.

TABLE 10-1
A matrix display of possibilities for two-symbol license plates

	0	1	2	3	4	5	6	7	8	9
A	A-0	A-1	A-2	A-3	A-4	A-5	A-6	A-7	A-8	A-9
B	B-0	B-1	B-2	B-3	B-4	B-5	B-6	B-7	B-8	B-9
C	C-0	C-1	C-2	C-3	C-4	C-5	C-6	C-7	C-8	C-9
D	D-0	D-1	D-2	D-3	D-4	D-5	D-6	D-7	D-8	D-9
.										
.										
.										
Z	Z-0	Z-1	Z-2	Z-3	Z-4	Z-5	Z-6	Z-7	Z-8	Z-9

WHAT IS THE MULTIPLICATION RULE?

Consider the following situation:

> We have a certain task to complete and we want to find the number of possible ways that it can be done.

Suppose that the overall task can be broken down into k successive steps:

> Step 1, step 2, . . . step k

Suppose also that we know the numbers of possible ways to complete each of steps 1 to k. We denote these by $n_1, n_2, \ldots, n_k$. That is,

> Step 1 can be completed in n_1 ways.
> Step 2 can be completed in n_2 ways. . . .
> Step k can be completed in n_k ways.

In terms of this notation, the *multiplication rule* is this: The number N of possible ways to complete the overall task is

$$N = n_1 \times n_2 \times \cdots \times n_k$$

That is, the number of ways of completing the overall task is the product of the numbers of ways of completing each step.

In designing five-symbol license plates, $k = 5$. The five steps and the corresponding values $n_1, n_2, \ldots, n_5$ are:

> Step 1: Choose the first symbol, a letter; $n_1 = 26$.
> Step 2: Choose the second symbol, a digit; $n_2 = 10$.
> Step 3: Choose the third symbol, a digit; $n_3 = 10$.
> Step 4: Choose the fourth symbol, a digit; $n_4 = 10$.
> Step 5: Choose the fifth symbol, a digit; $n_5 = 10$.

$$\begin{aligned} N &= n_1 \times n_2 \times n_3 \times n_4 \times n_5 \\ &= 26 \times 10 \times 10 \times 10 \times 10 \\ &= 260{,}000 \end{aligned}$$

A SOLUTION TO THE LICENSE BUREAU'S PROBLEM

The multiplication rule is a tool that the License Bureau can use to check whether a design format will allow a sufficient number of plates. Let us consider three possibilities.

Design 1: One Letter Followed by Five Digits

The multiplication rule applies with $k = 6$. We have

Step 1: Choose the first symbol, a letter; $n_1 = 26$.
Step 2: Choose the second symbol, a digit; $n_2 = 10$.
Step 3: Choose the third symbol, a digit; $n_3 = 10$.
Step 4: Choose the fourth symbol, a digit; $n_4 = 10$.
Step 5: Choose the fifth symbol, a digit; $n_5 = 10$.
Step 6: Choose the sixth symbol, a digit; $n_6 = 10$.

$$\begin{aligned} N &= n_1 \times n_2 \times n_3 \times n_4 \times n_5 \times n_6 \\ &= 26 \times 10 \times 10 \times 10 \times 10 \times 10 \\ &= 2{,}600{,}000 \end{aligned}$$

This design generates more than 1 million plates and thus is adequate.

Design 2: Two Letters Followed by Three Digits

Because this problem is similar to the preceding one, the steps are not written out. In this case,

$$\begin{aligned} N &= 26 \times 26 \times 10 \times 10 \times 10 \\ &= 676{,}000 \end{aligned}$$

This design generates an insufficient number of plates.

Design 3: Four Letters

In this case we have

$$\begin{aligned} N &= 26 \times 26 \times 26 \times 26 \\ &= 456{,}976 \end{aligned}$$

This design also generates an insufficient number of plates to meet the license bureau's needs.*

USE OF THE MULTIPLICATION RULE IN SOLVING PROBLEMS

The multiplication rule often serves only indirectly as a problem-solving tool. Typically, it provides background information useful in organizing attempts at a solution rather than actually producing a solution.

*If design 3 had generated a sufficient number of plates, the bureau would have had another problem: what to do about "objectionable" four-letter combinations.

Consider, for example, Mr. Dogood's problem (first posed in Chapter 2 and reconsidered in Chapter 4) of hiring a secretary. If there are four applicants with different qualifications who are scheduled for interviews with Mr. Dogood, then the multiplication rule can be used to find the number of possible orders in which the applicants may arrive.

The employment agency's task, supplying four applicants, can be broken down into $k = 4$ successive steps:

Step 1: Send the first applicant.
Step 2: Send the second applicant.
Step 3: Send the third applicant.
Step 4: Send the fourth applicant.

Since there are four applicants and the first applicant may be any one of the four, $n_1 = 4$. The second applicant may be any one of the remaining three; thus $n_2 = 3$. Similarly, $n_3 = 2$ and $n_4 = 1$. The total number of orders for the applicants is thus

$$N = n_1 \times n_2 \times n_3 \times n_4 = 4 \times 3 \times 2 \times 1 = 24$$

In like manner we can calculate the total number of possible orders for *five* applicants as

$$N = 5 \times 4 \times 3 \times 2 \times 1 = 120$$

From our earlier work with Mr. Dogood's problem, we know that these calculations using the multiplication rule do not solve his problem—but they do yield useful information and help us to understand the scope and complexity of the problem.

Chapter 2 presented several cryptarithmetic problems. Your experience with these problems will probably have convinced you of their complexity, but the multiplication rule can be used to discover exactly how complex they are. Consider first the cryptarithmetic puzzle:

```
  AT
+  A
 TEE
```

Solving this cryptarithm can be viewed as a sequence of $k = 3$ steps:

Step 1: Select a digit to replace A.
Step 2: Select a digit to replace E.
Step 3: Select a digit to replace T.

Ten digits—0, 1, 2, . . . , 9—are available at the start; thus $n_1 = 10$. There are then 9 digits left for step 2, so that $n_2 = 9$. Following that, there are 8 digits left for steps 3, and so $n_3 = 8$. The total number of arrangements of digits to sort through in a search for the correct solution is thus

$$N = n_1 \times n_2 \times n_3 = 10 \times 9 \times 8 = 720^*$$

Using a similar procedure, we discover that for the cryptarithm

```
  SEND
 +MORE
 MONEY
```

the maximum number of arrangements of digits to examine in a search for a solution is

$$N = 10 \times 9 \times 8 \times 7 \times 6 \times 5 \times 4 \times 3 = 1{,}814{,}400$$

Cryptarithmetic problems provide a good illustration of situations where the multiplication rule is not directly useful in solving a problem but instead provides a measure of its range and difficulty.

Exercise 1

Earlier, we considered several possible designs for license plates, examining each to see if it would yield 1 million plates. One design that we did not consider is three letters followed by two digits. Evaluate this possibility; does it yield enough plates?

Exercise 2

One of the officials in the license bureau has a bright idea: residents of each of the state's 17 counties could receive license plates bearing one or more letters that would identify the county. He thinks that this will help identify stolen or abandoned vehicles as well as automobiles of people's neighbors and friends. The 17 counties and their estimated "automobile populations" are given in Table 10-2.

a Develop a design scheme for each county so that its license plates begin with the first letter of its name and are followed by several digits. How many digits are needed in each case?

b What disadvantages might this idea have? What questions would you ask before you would recommend it?

TABLE 10-2

COUNTY	AUTOMOBILES
Adams	30,000
Bedford	30,000
Columbia	20,000
Dover	200,000
Essex	40,000
Franklin	30,000
Gloucester	15,000
Hancock	50,000
Indiana	30,000
Juniata	5,000
Keene	70,000
Lowell	80,000
Madison	50,000
Northumberland	15,000
Orange	75,000
Penn	250,000
Roanoke	10,000
	1,000,000

*If we make use of the information that leading digits cannot be zero, then we can reduce N to a lower value.

Exercise 3

Is it possible to manufacture 1 million different license plates if each must contain a maximum of four symbols (letters *or* digits)? (If letters and digits are mixed, how will you avoid difficulties in distinguishing between "oh" and "zero" and between "eye" and "one"?)

Exercise 4

In the state of Caltucky there are approximately 1.5 million motor vehicles registered.

a Which of the following design formats will yield enough different plates for all Caltucky vehicles? (1) 2 letters followed by 3 digits; (2) 3 letters followed by 2 digits; (3) 2 letters followed by 4 digits.

b Suppose that Caltucky requires a license plate design in which no letters or digits are repeated on a plate. In this case, which of the plate designs 1, 2, and 3 will yield enough different plates for all 1.5 million Caltucky vehicles?

Exercise 5

Kevin will soon graduate with a double major in computer science and accounting. He is considering four types of jobs: database analyst, systems programmer, accountant, and tax analyst. He has identified eight major cities in which he would like to live and work. If all jobs are available in all cities, how many different choices—of job type and city—are available to Kevin?

Exercise 6

Rosanne is trying to complete her schedule for next semester. She would like to schedule three courses on Monday, Wednesday, and Friday mornings at 9, 10, and 11. At 9 A.M. six courses are available that meet her requirements. At 10 A.M. four courses are available, and at 11 A.M. four courses are available. All these courses are different. How many different sequences of three courses are possible for Rosanne's schedule?

Exercise 7

A telephone company routes calls overseas in the following manner. A placed call is routed to 1 of 200 transmitting stations. Each transmitting station relays the message to 1 of 3 satellites. Each satellite in turn relays the call to 1 of 60 receiving stations, which in turn sends the message directly to the receiver. In how many different ways can a call be routed from a caller to a receiver?

Exercise 8

The registrar's office at Ticonderoga University keeps on file the following background information on students:

Sex: Male or female
Class: Freshman, sophomore, junior, senior, nondegree, or graduate
Major: 1 of 67 possibilities

If two students are in the same category only when they are of the same sex, belong to the same class, and have the same major, how many different categories of students are there at Ticonderoga?

Exercise 9

Pocono University recently had a new telephone system installed in its dormitories. Each number in the network will be of the form 983-*xxxx,* where each x may be replaced by any digit. What is the maximum number of phones that the system can handle if each phone must have a different number?

Exercise 10: Factorial Notation

When certain combinations of symbols occur over and over again, it is convenient to use abbreviations. When the multiplication rule is being used, combinations that occur often include products like $5 \times 5 \times 5 \times 5$ and $5 \times 4 \times 3 \times 2 \times 1$. For repetitions of a single factor, you probably are familiar with the exponential notation 5^4; use it whenever convenient to do so. For products like $5 \times 4 \times 3 \times 2 \times 1$, the *factorial notation* 5! is used. The symbol 5! is read "5 factorial" and designates the product $5 \times 4 \times 3 \times 2 \times 1 = 120$. Likewise $3! = 3 \times 2 \times 1 = 6$, and $n! = n \times (n - 1) \times (n - 2) \times \cdots 2 \times 1$. Become acquainted with the factorial notation by completing the following calculations; use it whenever it will save time and effort.

a 4!
b 7!
c 9!
d 10!
e $\frac{8!}{6!}$
f $\frac{20!}{18!}$
g $\frac{12!}{9!}$
h $\frac{1000!}{998!}$

Exercise 11

"I like to take true-false tests," Veronica confided to her roommate. "It's so easy to guess and get all the items right."

a How many different possible answer lists are there for a true-false test containing 3 questions? For one containing 5 questions? For one containing 10 questions? For one containing 20 questions? For one containing 100 questions?

b Do you agree with Veronica? Explain why or why not.

Exercise 12

In Morse code the letter E is represented by a dot and T is represented by a dash. All other letters of the alphabet are represented by two or more symbols.

a Is it possible to represent every one of the 26 letters of the alphabet using four or fewer symbols (each being either a dot or a dash)? Justify your answer with calculations.

b The American Morse code differs from the international Morse code in several respects, one of which is that the letter P is represented by · · · · · in American Morse and by · — — · in international Morse. Suggest possible reasons for the use of five (rather than fewer) symbols for the letter P.

Exercise 13

a Tom, Dick, and Harry each toss a penny. Use the multiplication rule to deduce that there are eight possible outcomes from their three tosses. List the eight possible arrangements of three coins that can result.

b Suppose that Mary and Jane join Tom, Dick, and Harry in tossing pennies. Find the number of different arrangements of the five coins that can result.

c Tom and Dick each toss a fair (balanced) die. How many possible outcomes are there for the pair of tosses? List them.

d Tom, Dick, and Harry each toss a balanced die. How many possible outcomes are there for the group of three tosses?

e Mary and Jane have a deck containing twelve cards—the face cards (jacks, queens, and kings) from an ordinary deck. Mary draws a card, notes it, and returns it to the deck. Jane then draws a card, likewise notes what it is, and returns it to the deck. How many different pairs can result from this two-draw process?

f Suppose that, in *e*, Mary does not replace her card before Jane draws. How many different pairs can result from the two-draw process in this case?

Exercise 14

At its grand opening the Pizza Palace gave each customer a ticket containing a three-digit number. The customer won a free drink with the purchase of a pizza if the first and last digits of the ticket number were odd (the middle digit could be odd or even). How many three-digit numbers were winners?

Exercise 15

a If eight people have purchased tickets for eight seats in a row at a hockey game, how many different arrangements of these people in the seats are possible?

b Suppose that four of the eight are women and four are men. How many different seating arrangements are there that alternate men and women?

c Discuss whether the answers to *a* and *b* are larger than seems reasonable.

d (Hard.) If the eight people are four married couples and if they want to increase sociability by using an arrangement that alternates males and females and separates each from his or her spouse, how many arrangements are possible?

Exercise 16

In genetics the letters A, G, T, and C stand for four nitrogeneous bases of DNA: namely, adenine, guanine, thymine, and cytosine. The genetic code consists of sequences of three (not necessarily different) nitrogeneous bases. How many possible code words are there? (Actually, these code words do not all describe different substances; for example the DNA code words CGA, CGG, CGT, and CGC all specify alanine.)

Exercise 17

A methodical Scrabble player with seven different letters on the rack decides to test all possible arrangements of five letters—chosen from the seven letters—before making a play. If it takes 1 second to test each arrangement, how long will it take to be ready to play?

Exercise 18

Amoebas reproduce by binary fission—that is, they divide into two parts, each of which is a new amoeba. If we start with a single amoeba and if amoebas undergo binary fission, on an average, three times in 24 hours, how many amoebas will there be at the end of 1 day? At the end of 1 week? At the end of 1 month? (Assume that all amoebas which are produced survive.)

Exercise 19: Invent

Describe a situation, different from those considered so far, to which the multiplication rule can be applied.

Exercise 20: Journal

Describe a personal situation in which calculation of the number of possible ways that some task can be completed is helpful in understanding or solving a problem.

Exercise 21: Journal

Use of the multiplication rule requires us to view an overall task as a sequence of steps. Think of a problem—perhaps one unrelated to the multiplication rule—that you have solved in which the key to a solution was breaking the problem down into a sequence of simple steps. Are there problems that you currently face for which this is an appropriate solution technique?

Note The Appendix contains complete answers for Exercises 1, 3, 4*b*, 6, 10*b* and *g*, 15*a* and *b*, and 17. It contains partial solutions or hints for Exercises 2*a* and *b*, 8, 11*a*, 13*a*, 15*c*, and 16.

BINARY CODING OF NUMBERS

The widespread use of computers has increased the importance of the binary number system. This system is described as a *base 2* system; it makes use of only two symbols: 0 and 1. By contrast, our ordinary decimal number system is called a *base 10* system; it includes ten symbols: 0, 1, 2, 3, 4, 5, 6, 7, 8, and 9.

In the decimal system, digit positions carry values that are powers of 10. For example, when we write 5432, we mean

$$(5 \times 10^3) + (4 \times 10^2) + (3 \times 10^1) + (2 \times 10^0)$$

In the binary system, when we write 1011, we mean

$$(1 \times 2^3) + (0 \times 2^2) + (1 \times 2^1) + (1 \times 2^0)$$

or

$$8 + 0 + 2 + 1 = 11$$

This idea can be applied to determine a binary code—or a "binary representation"—for any whole number. The representations begin as shown in Table 10-3.

The rule that is critical in developing and interpreting binary representations is this: In a binary representation,

The rightmost digit carries its own value (multiplied by 2^0 or 1).
The second-from-right digit should be multiplied by 2^1 or 2.
The third-from right digit should be multiplied by 2^2 or 4.
The fourth-from-right digit should be multiplied by 2^3 or 8.
And so on. . . .

TABLE 10-3

NUMBER (DECIMAL NAME)	BINARY CODE
Zero	0
One	1
Two	10
Three	11
Four	100
Five	101
Six	110
Seven	111
Eight	1000
Nine	1001
Ten	1010
Eleven	1011
Twelve	1100

From Table 10-3, we see that the binary representations of numbers as small as eight and nine require four digits. One difficulty with the binary representation of numbers is that large numbers require many digits.

The multiplication rule is useful in the study of binary representations because it enables us to determine how many numbers can be represented with a given number of digits. The construction of a binary representation can be viewed as a sequence of steps:

Step 1: Choose the rightmost digit; since this digit may be 1 or 0, the choice may be made in 2 ways—that is, $n_1 = 2$.
Step 2: Choose the second-from-right digit; $n_2 = 2$.
Step 3: Choose the third-from-right digit; $n_3 = 2$.
Step 4: Choose the fourth-from-right digit; $n_4 = 2$.
And so on. . . .

Using the multiplication rule we can calculate that the number of possible two-digit binary numbers is

$$n_1 \times n_2 = 2 \times 2 = 4$$

The number of possible three-digit binary numbers is

$$n_1 \times n_2 \times n_3 = 2 \times 2 \times 2 = 8$$

The number of possible four-digit binary numbers is

$$n_1 \times n_2 \times n_3 \times n_4 = 2 \times 2 = 2 \times 2 = 16$$

The 16 possible four-digit binary numbers are:

0000 (zero)	1000 (eight)
0001 (one)	1001 (nine)
0010 (two)	1010 (ten)
0011 (three)	1011 (eleven)
0100 (four)	1100 (twelve)
0101 (five)	1101 (thirteen)
0110 (six)	1110 (fourteen)
0111 (seven)	1111 (fifteen)

A peculiarity of this listing is the use of "unneeded" zeros. Because the value of 0001 is the same as the value of the single digit 1, we think of both as representations of the same number. As with fractions, we choose the format that is most convenient for a particular use.

Exercise 22

In the text, binary representations of the whole numbers through 15 were given. Continue that list and provide binary representations for numbers through 40.

Exercise 23

What decimal numbers correspond to the following binary representations?

a 1011
b 11001
c 11111 and 100000
d 1011100
e 111111 and 1000000
f 11111111 and 100000000

Exercise 24

Give a binary representation for the numbers with the following decimal (ordinary) representations:

- **a** 48
- **b** 63
- **c** 64
- **d** 100
- **e** 127 and 128
- **f** 255 and 256

Exercise 25

Binary representations of numbers offer the simplicity of only two symbols. However, as was noted in the text, many binary digits are required to represent even modest-sized numbers. For example, the number ordinarily represented as 73 requires seven binary digits (1001001).

Use the multiplication rule to complete Table 10-4, which summarizes the number representations that are possible with various numbers of binary digits. Observe the patterns shown by table entries.

TABLE 10-4

NUMBER OF BINARY DIGITS	NUMBER OF NUMBERS REPRESENTABLE	LARGEST NUMBER REPRESENTABLE
1	(2)	2 − 1 = (1)
2	2 × 2 = (4)	4 − 1 = (3)
3		
4	2 × 2 × 2 × 2 = (16)	16 − 1 = (15)
5		
6		
7		
8		
9		
10		
11		
12		

Exercise 26

Check whether your values in Table 10-4 support the following statement: "With k binary digits, 2^k numbers can be represented; the smallest of these is 0 and the largest is $2^k - 1$."

Exercise 27

How many binary digit positions would be required to write any number up to 1 million? Up to 10 million? Up to 1 billion?

Exercise 28

In describing computer storage capacity—"memory"—a frequently used unit is $K = 2^{10}$ words. Although the meaning of the term *word* varies from system to system, a typical word length is 16 bits—where *bit* is an abbreviation for *binary digit.*

a What is the largest whole number that can be stored in a 16-bit word?

b What is the total number of bits of storage available in a computer that is advertized as having a memory of 256K?

c A device that can store 1 trillion bits of information is called a *terabit* memory. If a terabit memory can store the number of characters that would fill 300 million pages of paper, how many 300-page books is this equivalent to? Why not replace a local book library with one (or more) computers with terabit memory?

Exercise 29: Invent

a Develop a "base 5" or "quinary" number system using the symbols 0, 1, 2, 3, and 4. In such a system the position values, reading from right to left, will be: units, 5s, 25s, 125s, etc. List the quinary representations of the whole numbers from 0 to 30.

b Complete Table 10-5.

TABLE 10-5

NUMBER OF QUINARY DIGITS	NUMBER OF NUMBERS REPRESENTABLE	LARGEST NUMBER REPRESENTABLE
1	5	4
2	25	24
3		
4		
5		
6		
7		

Exercise 30: Invent

Develop an octal (base 8) number system.

Exercise 31: Test Yourself

Develop a list of new terms and key ideas presented in this chapter. Supply an example of each item on your list.

Note The Appendix contains complete answers for Exercises 23*e*, 24*d*, and 28*b*. It contains partial solutions or hints for Exercises 22, 25, 27, and 29*a*.

CHAPTER 11

VISUALIZING THE STRUCTURE OF INFORMATION WITH A TREE DIAGRAM

Once upon a tree
I came upon a time.

—Theodore Roethke

AN INTRODUCTION TO TREE DIAGRAMS

A versatile tool for organizing information and visualizing its structure is a *tree diagram.* Such a diagram resembles its biological counterpart; an example is provided in Figure 11-1.

As background for understanding Figure 11-1, consider the following information.

> The housing office at Jersey College separates student records in its files on the basis of two criteria—sex (male or female) and residency (on campus or off campus). Labels along the four possible left-to-right paths in Figure 11-1 identify the four mutually exclusive categories of students: (1) males who live on campus; (2) males who live off campus; (3) females who live on campus; (4) females who live off campus.

To discuss tree diagrams easily we need some basic vocabulary. Each of the heavy dots (where one or more line segments meet and end) is called a *vertex.* Each vertex that is the end point of only one segment is called a *terminal vertex.* Each (bent or straight) line segment joining a pair of vertices is called an *edge.* A connected sequence of edges is called a *path.*

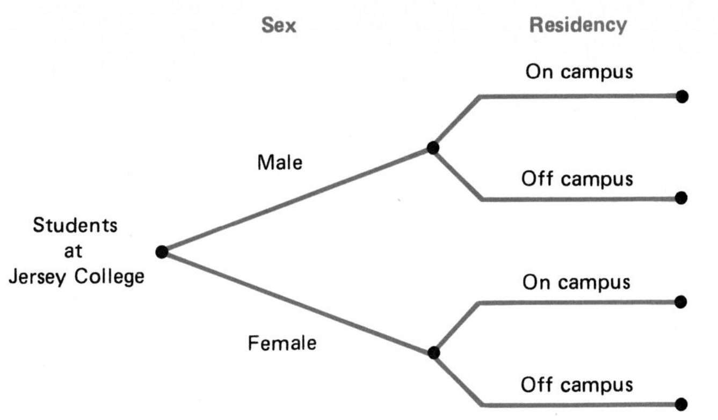

FIGURE 11-1
Tree diagram showing Jersey College Housing Office classifications

In the construction, labeling, and interpreting of tree diagrams, the following rules are observed:

T1 Each heading above the tree designates an attribute, category, or event that is completely specified by the mutually exclusive categories which label each cluster of edges beneath the heading.

T2 The cluster of edges attached to and to the right of a particular vertex must designate (by their labels) a complete list of mutually exclusive categories.

T3 Each vertex denotes a collection of objects—those objects that satisfy all labels along the left-to-right path leading to that vertex.

T4 The group of terminal vertices provides a complete collection of mutually exclusive categories for the set designated by the initial (left-hand) vertex of the tree.

Let us now consider two examples that illustrate the construction and use of tree diagrams.

Example 1 ▶

Greta teaches tennis. She thinks that three factors are of primary importance in becoming a good tennis player:

1. Mastering basic skills
2. Learning to concentrate
3. Developing confidence

If she is to be a successful teacher, she must correctly identify which types of development each student needs.

The possible categories for Greta's tennis students are shown in the tree diagram in Figure 11-2. Once Greta has determined the correct category for a given student, she can then use specific teaching techniques to meet that student's needs. ◀

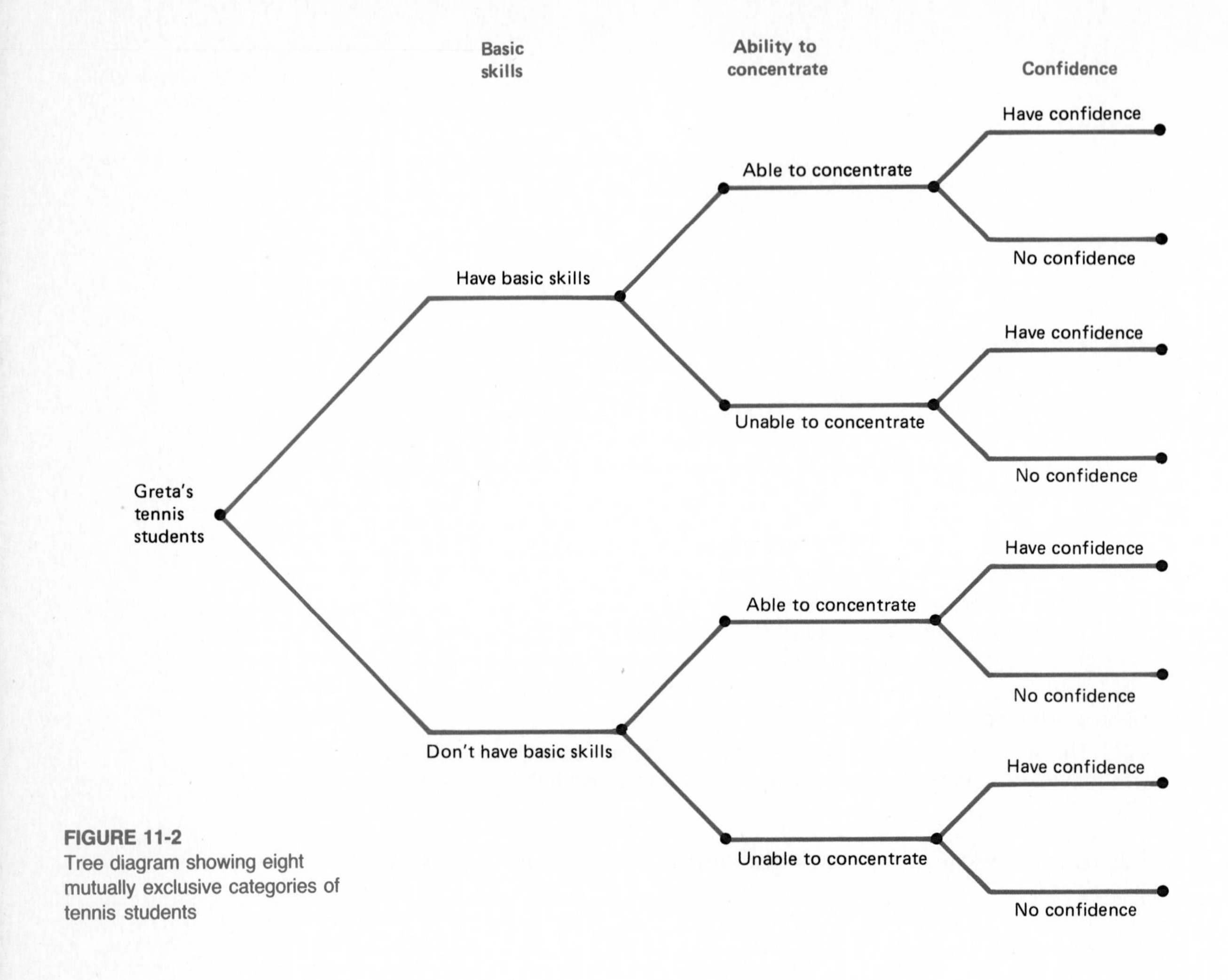

FIGURE 11-2
Tree diagram showing eight mutually exclusive categories of tennis students

---

Example 2 ▶

The Rydell Employment Agency classifies résumés that it has on file on the basis of two criteria: education and experience. The educational levels Rydell uses are:

No high school diploma
High school diploma but less than 1 year of college
At least 1 year of college but no degree
College graduate

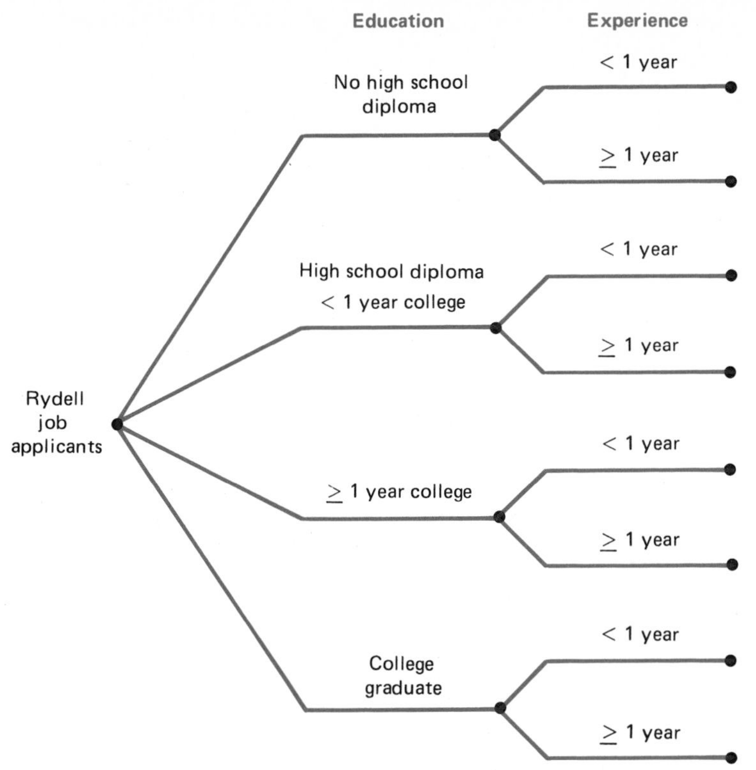

FIGURE 11-3
Tree diagram of Rydell's scheme for classifying applicants

The levels of experience it uses are:

Less than 1 year
1 year or more

With the multiplication rule we can determine that there are eight categories of applicants, since the educational level offers four possibilities and the experience level offers two possibilities and $4 \times 2 = 8$. These eight categories correspond to the eight terminal vertices in Figure 11-3.

The Rydell Agency makes use of the tree diagram to see its classification categories clearly. It then makes use of the list of categories in the following way. An employer is asked to choose and rank the categories of applicants it considers acceptable. For instance, one employer's choices of categories are shown in Table 11-1. Rydell then selects applicants within those categories and sends them to the employer. ◀

TABLE 11-1
One employer's preference ranking for Rydell's categories

CATEGORY	RANKING
No high school diploma, $<$ 1 year experience	Unacceptable
No high school diploma, $\geq$ 1 year experience	Unacceptable
High school diploma, $<$ 1 year college, $<$ 1 year experience	4th
High school diploma, $<$ 1 year college, $\geq$ 1 year experience	2d
$\geq$ 1 year college, no degree, $<$ 1 year experience	3d
$\geq$ 1 year college, no degree, $\geq$ 1 year experience	1st
College graduate, $<$ 1 year experience	Unacceptable
College graduate, $\geq$ 1 year experience	Unacceptable

Exercise 1

Examine the tree in Figure 11-4 to see that it illustrates the same four categories as the tree in Figure 11-1. Draw a tree diagram that is related to the tree in Figure 11-3 in the same way that Figure 11-4 is related to Figure 11-1.

Exercise 2

The office of academic affairs at Jersey College classifies students according to year (freshman, sophomore, junior, senior) and academic achievement (probation, satisfactory, honors). How many categories of students are there in this classification scheme? Draw a tree diagram that shows all the possibilities.

FIGURE 11-4

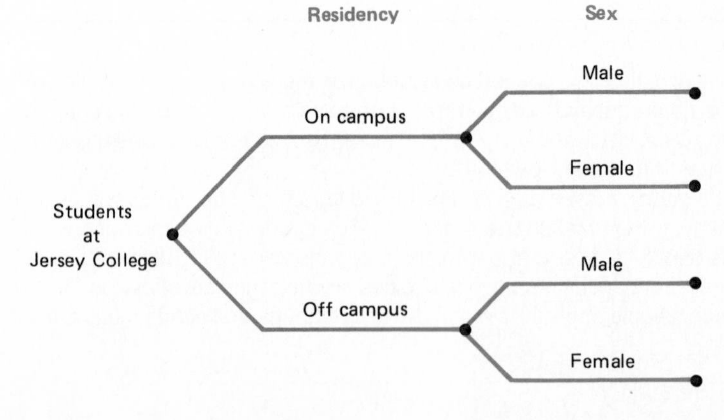

Exercise 3

Susan is trying to decide on a job. She considers the following criteria important: the job should be interesting, offer a good salary, and offer opportunity for advancement.

a Draw and label a tree diagram that shows the eight mutually exclusive categories into which a job possibility for Susan may fall.

b If you were using Susan's criteria, which of the eight job categories would you find unacceptable? Rank the remaining categories as first, second, third, etc., in order of their acceptability. (Imitate Table 11-1 in Example 2.)

Exercise 4

Problem solvers may be characterized by the presence or absence of the following characteristics: knowledge, daring, caution, and persistence. Sketch a tree diagram that shows the 16 mutually exclusive categories of problem solvers that result. In which category are you?

Exercise 5

Explain why the tree diagram in Figure 11-5 would be unsuitable for classifying tennis students using Greta's criteria (see Example 1).

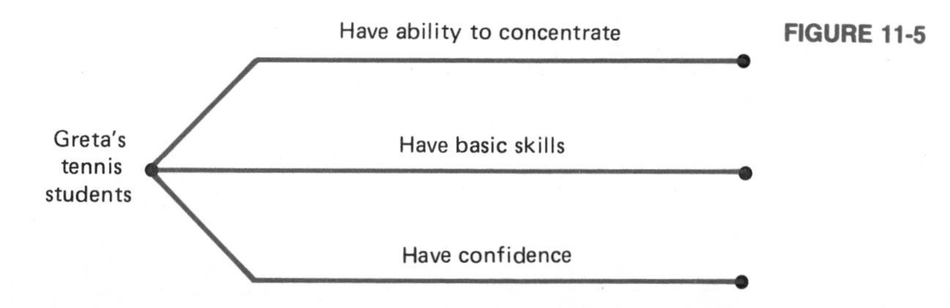

FIGURE 11-5

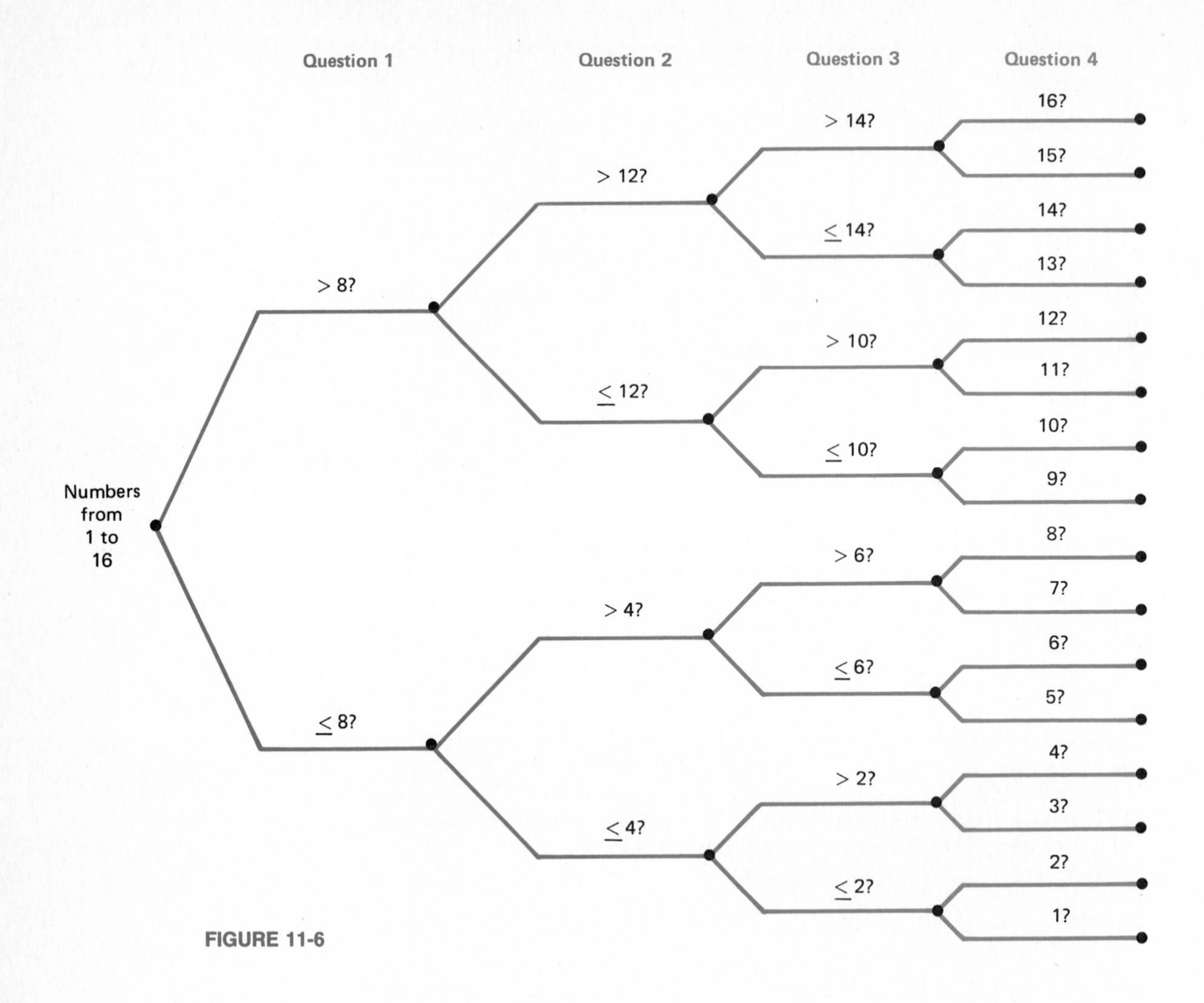

FIGURE 11-6

Exercise 6

a Suppose that you are challenged by a friend to guess a number n between 1 and 16 using as few "yes or no" questions as possible. The format described by the tree diagram in Figure 11-6 shows a way to find n using exactly four questions. Study the tree diagram to see how it works. Suppose that your friend has the number 9 in mind. What four questions will you ask and what will the answers be?

b Could you correctly identify a number between 1 and 32 by asking only five "yes or no" questions? Explain.

c State a generalization of this process.

Exercise 7: Invent; Journal

Suppose that you are selecting courses for next semester. Develop a list of criteria to use in deciding whether or not to take a course. Sketch a tree diagram to illustrate all possible mutually exclusive categories that can result from your criteria. To the right of each terminal vertex, note whether a course in the category designated by that vertex would or would not be acceptable to you. Rank the acceptable categories in order of acceptability—first, second, third, etc. (Example 2 illustrates these steps.) Do you actually use a method like this in selecting courses? Why or why not?

Exercise 8

Some texts for elementary students use tree diagrams to help the students understand the factoring of positive integers into primes. As Figure 11-7 illustrates, the procedure is to start with a vertex labeled with the number to be factored. If factors other than the number itself and 1 are possible, then a pair of edges are drawn and their terminal vertices are labeled with the two factors. The process continues from the new terminal vertices until each terminal vertex is labeled with a number having no further factors (except itself and 1). At this point, the labels for the terminal vertices are the prime factors of the original number.

a Draw factor trees for the following positive integers: 15, 19, 48, 108.

b As Figure 11-7 indicates, there can be more than one factoring diagram for a given integer. Do all factoring diagrams for the same integer lead to the same list of prime factors?

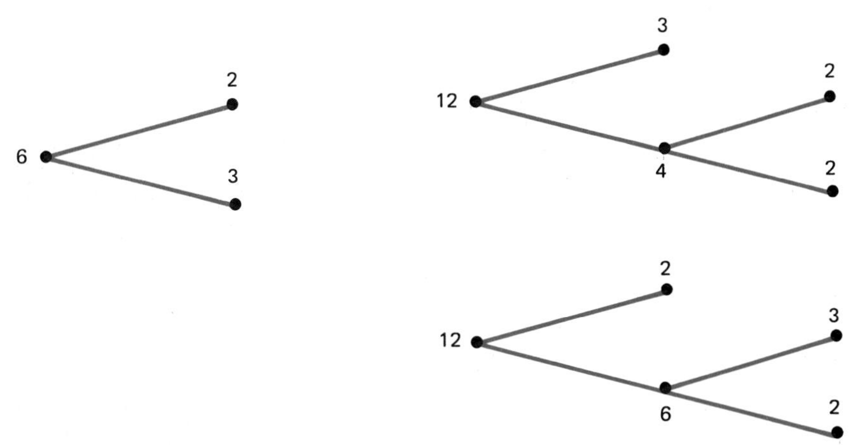

FIGURE 11-7
Factor trees for 6 and 12

Exercise 9: Invent; Journal

Describe in detail a situation where a tree diagram would be useful to you in organizing information about a project or problem.

Exercise 10: Journal

In an area of study called *decision analysis,* tree diagrams are used to illustrate the various choices that a decision maker faces and the possible consequences of these choices.

For example, suppose that Barbara is a college sophomore trying to decide what to do next summer. She is considering three choices: summer school, a job in her hometown, and a job at a beach resort. Barbara starts thinking through her situation with the tree diagram shown in Figure 11-8. But she soon discovers that the labels on the tree are not mutually exclusive choices, and she begins to sketch another tree. After several false starts, she develops the diagram in Figure 11-9, which outlines her summer possibilities and also takes into account the fact that she would prefer a seaside job to one in her hometown. (Note that in decision trees, vertices representing choices for the decision maker are usually squares, while nondecision vertices are circles.)

a Develop a tree diagram that outlines your decision process for a recent or future summer.

b Many decisions fall into a "yes-no" category; and these are often made with a great deal of uncertainty about which is the better decision, "yes" or "no." The tree diagram in Figure 11-10 describes this situation. From your own experience, supply three examples of decisions that had this structure. Did the decisions later prove to be correct ones?

c Think of a complex decision that you face, have faced, or might face—one that involves several choices and several possible consequences of each choice. Describe this decision with a tree diagram.

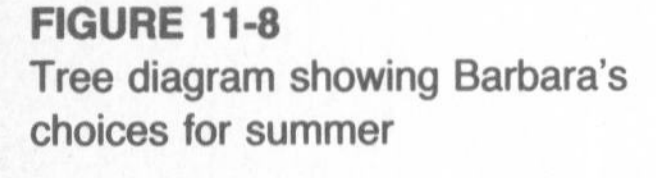

FIGURE 11-8
Tree diagram showing Barbara's choices for summer

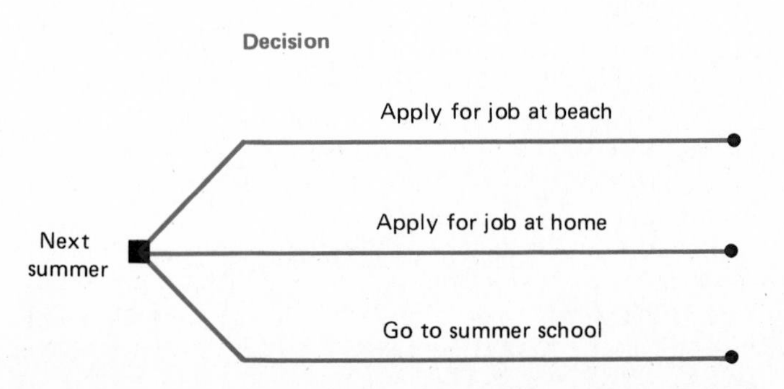

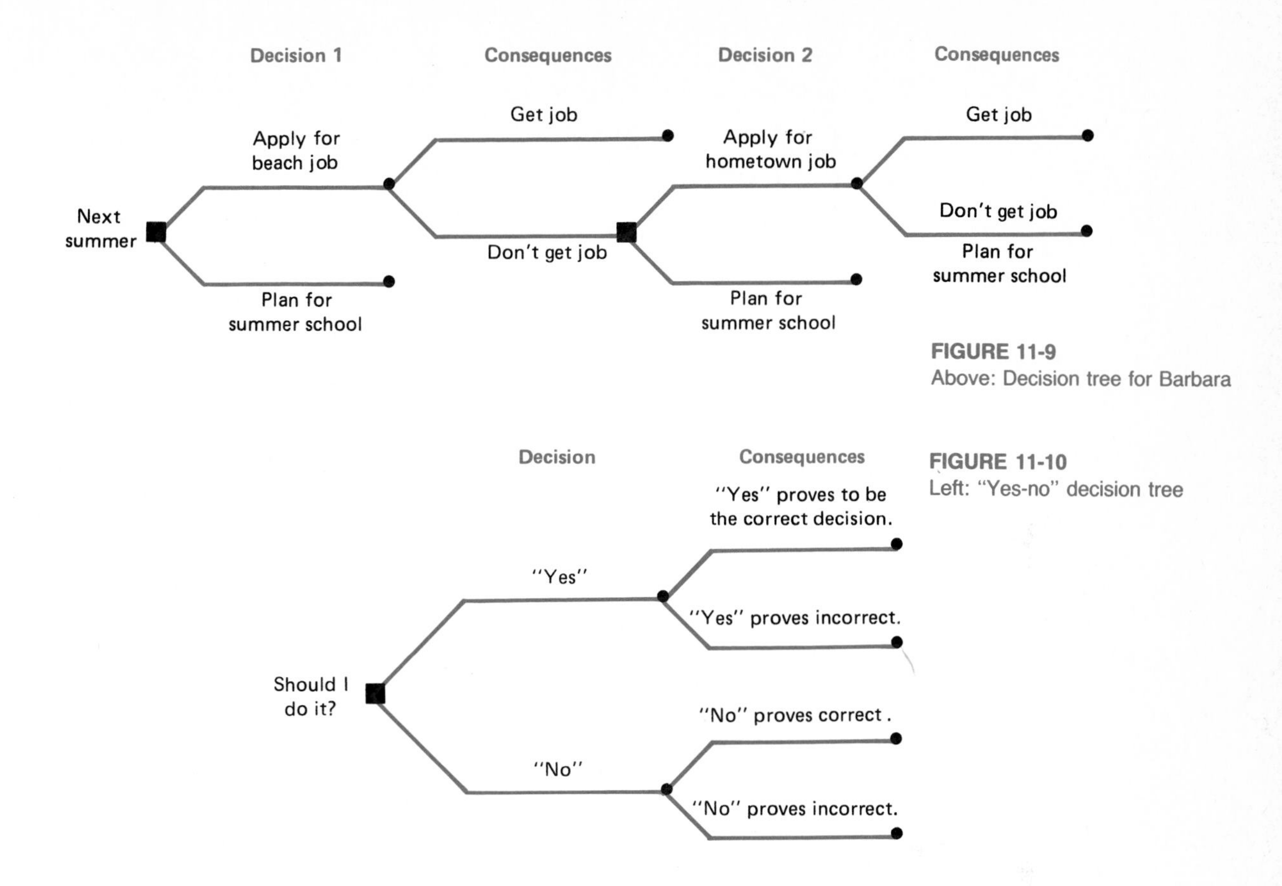

FIGURE 11-9
Above: Decision tree for Barbara

FIGURE 11-10
Left: "Yes-no" decision tree

Note The Appendix contains a complete answer for Exercise 8*b*. It contains partial solutions or hints for Exercises 2, 4, 6*b*, and 10*b*.

NUMBERED TREE DIAGRAMS

When we organize objects into categories, we may want to know the *number* of objects in each category. This information can be included in the tree diagram. To see how, let us return to Jersey College (refer back to Figure 11-1).

Suppose that the housing office has supplied the following information about Jersey College students. Total enrollment at the college is 4800 students. Of these, half are female and half are male. Of the males, 60 percent live on campus; of the females, 70 percent live on campus. That is, 60 percent of 50 percent (or 30 percent) of the students are on-campus males and 70 percent of 50 percent (or 35 percent) are on-campus females.

We can enlarge the vertices of the tree given in Figure 11-1 and, within each vertex, display the proportion or number in the category that the vertex represents. Figures 11-11 and 11-12 show two types of numerical labeling; Figure 11-11 shows percentages in each category, and Figure 11-12 shows actual numbers.

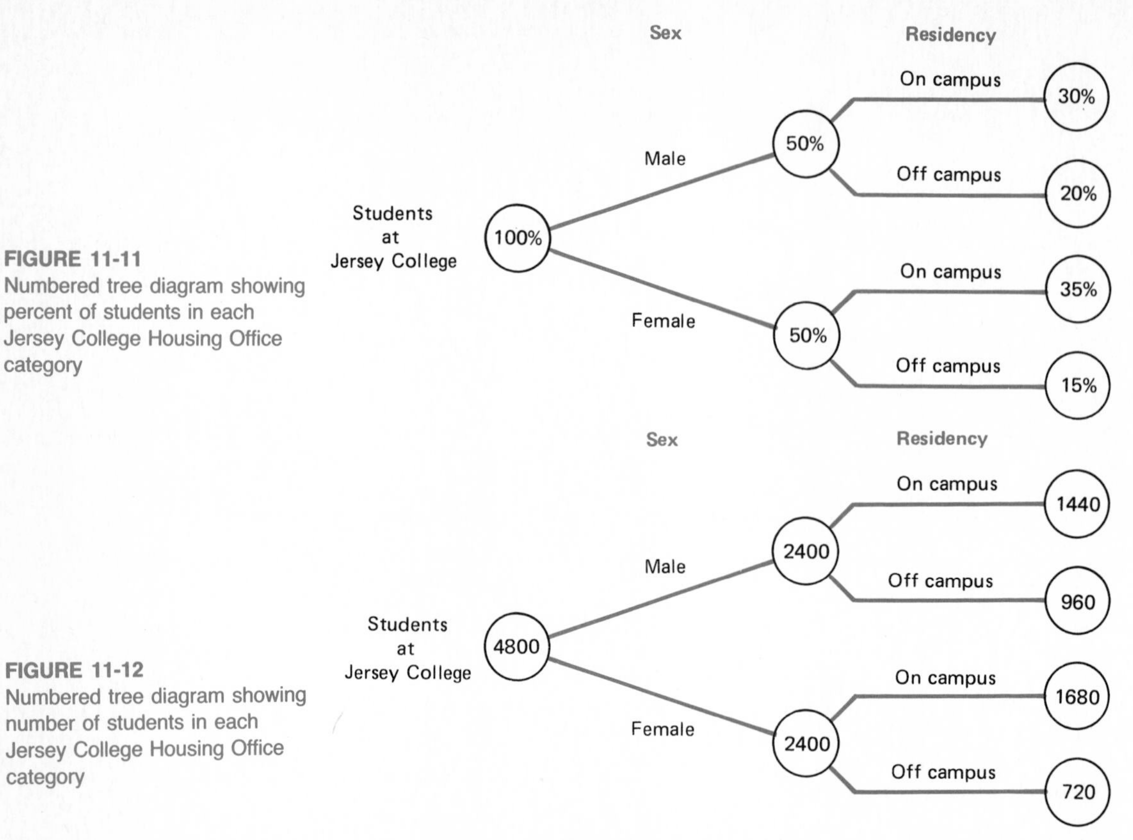

FIGURE 11-11
Numbered tree diagram showing percent of students in each Jersey College Housing Office category

FIGURE 11-12
Numbered tree diagram showing number of students in each Jersey College Housing Office category

The following rules should be followed for any numbered tree diagram:

N1 If several edges have a common left vertex, then the sum of the vertex values at the right-hand end of these edges must equal the value that labels their common left-hand vertex.

N2 Each vertex value should give the number (or fraction or percent) of objects that simultaneously satisfy all labels on the left-to-right path leading to that vertex.

Now let us consider further examples of use of tree diagrams—this time, numbered tree diagrams.

Example 3 ▶

Possible results of a random experiment may easily be visualized with a tree diagram. Tossing a pair of dice is a good illustration. Suppose that Tom and Dick each toss a balanced die. The possible outcomes are illustrated by Figure 11-13. In this tree, the edge labels are the mutually exclusive outcomes of a toss. The vertices are labeled with fractions that estimate the proportion of a large number of tosses that would be expected to yield the given outcome. (The initial vertex label 1 is analogous to the 100 percent in Figure 11-11.) ◀

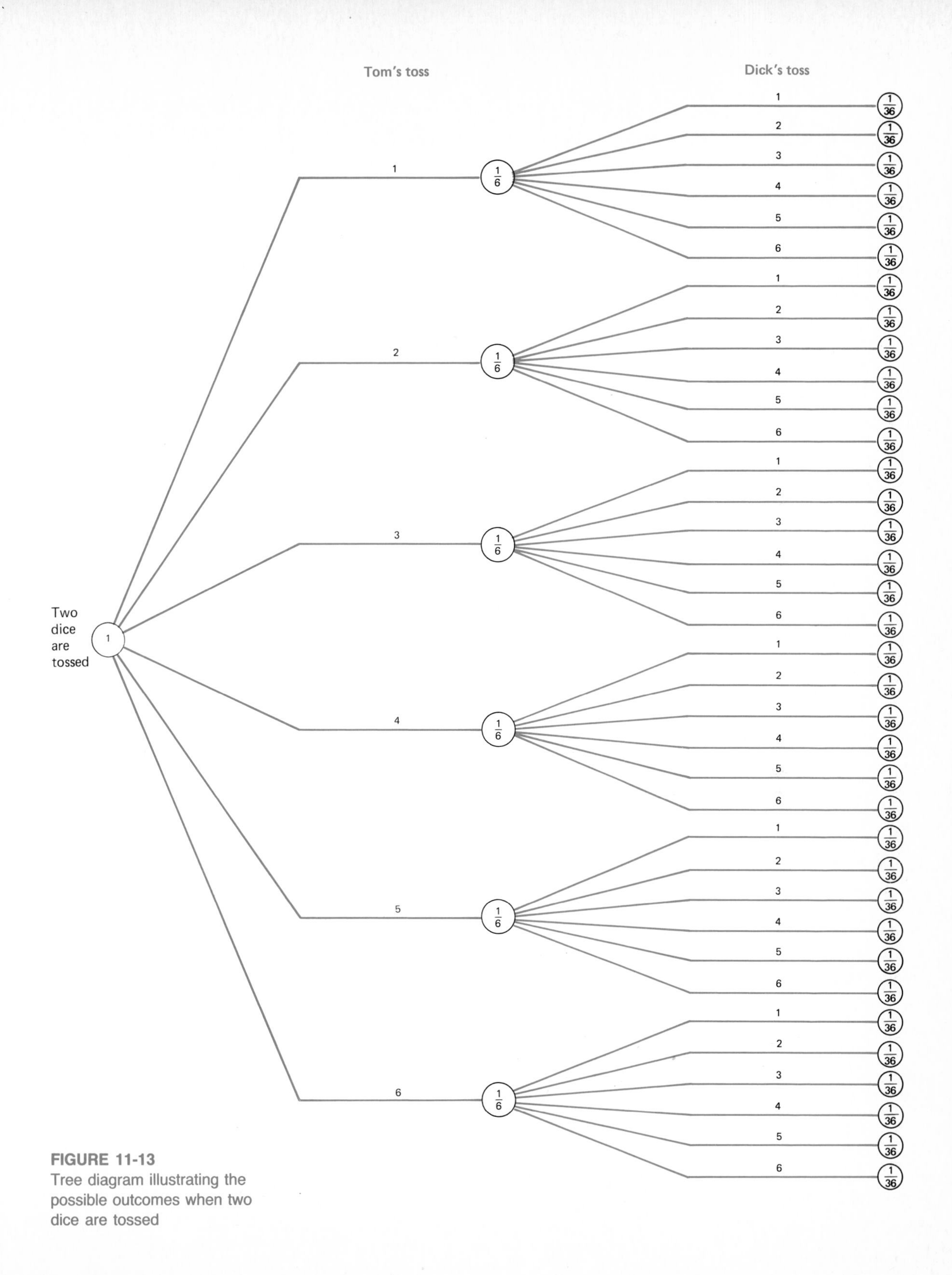

FIGURE 11-13
Tree diagram illustrating the possible outcomes when two dice are tossed

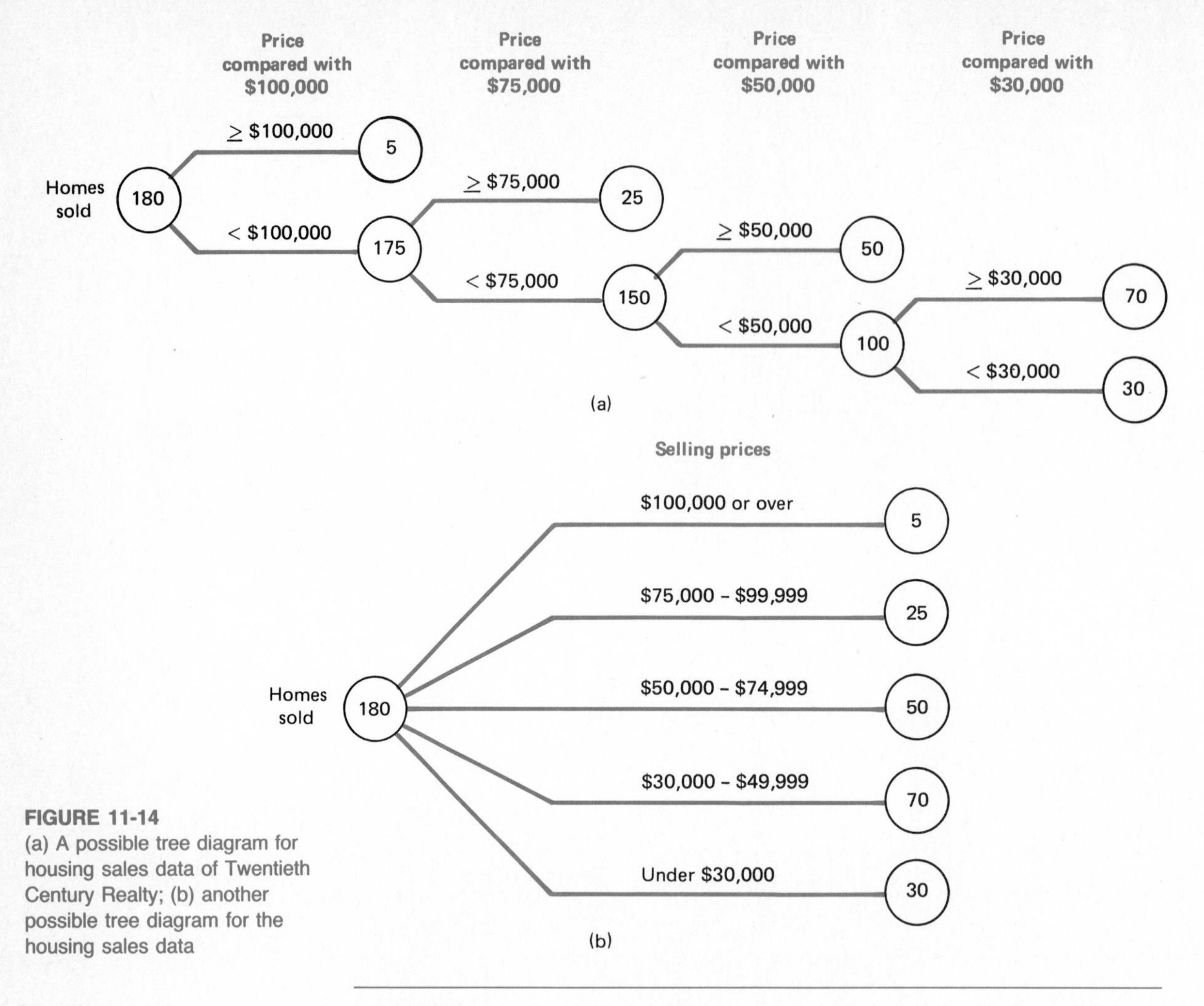

FIGURE 11-14
(a) A possible tree diagram for housing sales data of Twentieth Century Realty; (b) another possible tree diagram for the housing sales data

---

Example 4 ▶

Numbered tree diagrams are further illustrated by the following problem (which has already been considered in Chapters 6 and 9) involving reported housing sales. Twentieth Century Realty Associates has disclosed the following information about the prices of homes it sold in New Jersey communities in August:

30 homes sold at prices under $30,000.
100 homes sold at prices under $50,000.
150 homes sold at prices under $75,000.
175 homes sold at prices under $100,000.
5 homes sold at prices of $100,000 or more.
How many homes did the agency sell in August?

The tree diagrams in Figures 11-14*a* and *b* illustrate ways of organizing the problem information into mutually exclusive categories. The categories, in each tree designated by the five terminal vertices, are the same in both trees. ◀

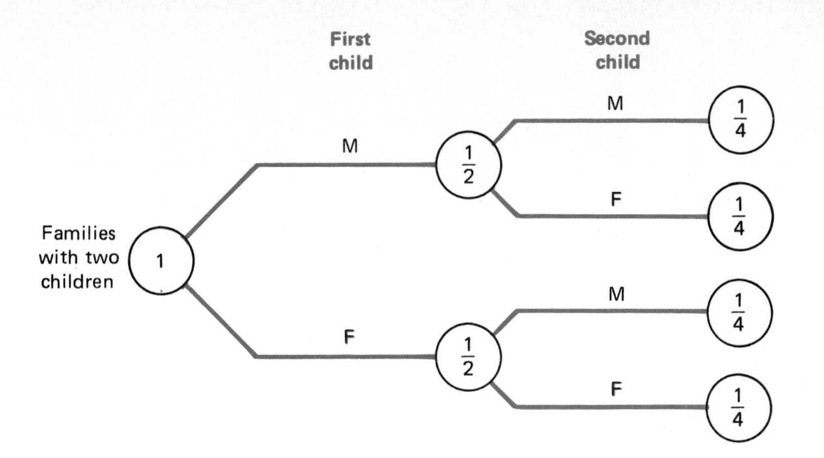

FIGURE 11-15
Left: Tree diagram showing distribution of sexes in two-child families

FIGURE 11-16
Below: Tree diagram showing outcomes for a family-planning rule

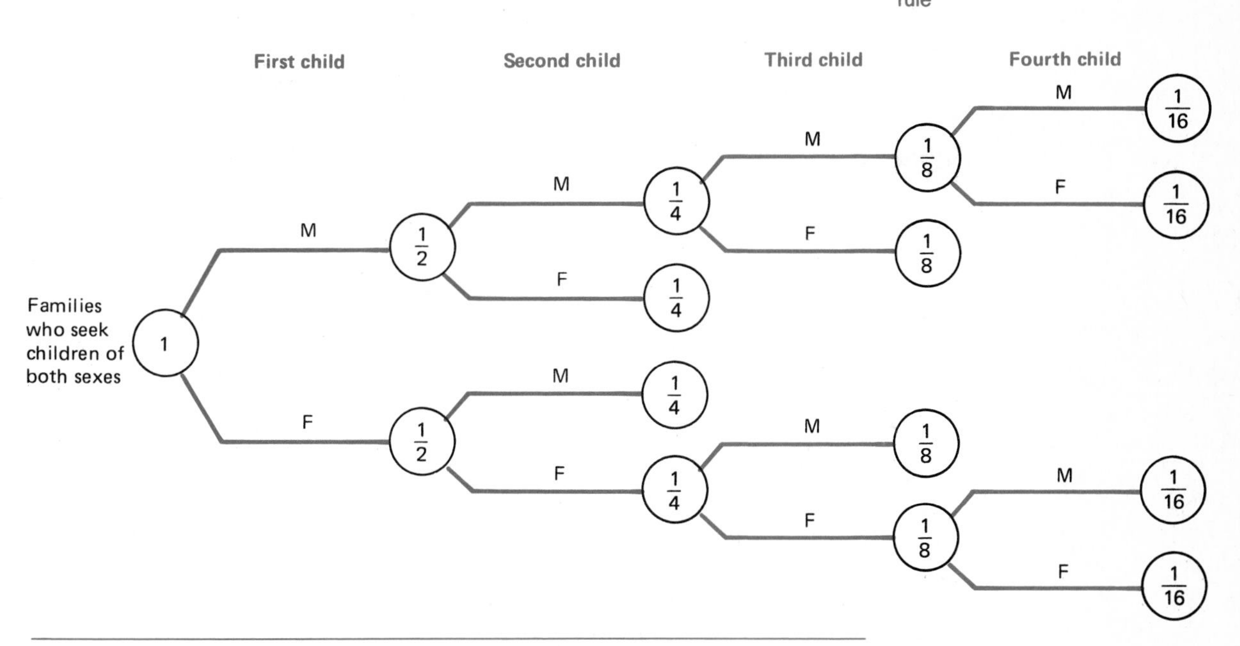

Example 5: Sex Distribution of Children in Families ▶

In Chapter 4, simulation was used to consider the effects of certain family-planning rules. Tree diagrams can also give insight into the patterns that occur in such situations.

For a couple who are planning to have two children, the possible outcomes are displayed in the tree diagram in Figure 11-15. As in Figure 11-13, the tree diagram in Figure 11-15 has been given numerical labels that correspond to the fraction of the time each outcome is expected to occur. When analyzing sex distributions of births, we will continue to make the (approximately true) assumption that each conception offers an equal chance of producing a boy or a girl.

Suppose that a couple (like Faye and Rick in Chapter 4) want to have a child of each sex and will stop when they achieve that goal or when they have four children (whichever comes first). The tree diagram in Figure 11-16 describes the possible families that can result from following this rule.

Using Figure 11-16, we observe the following:

Half of the families are expected to have two children, including one of each sex.

One-fourth of the families are expected to have three children, including one of each sex.

One-fourth of the families are expected to have four children; half of these will have all children of the same sex. ◀

Example 6 ▶

For the final example, reconsider Exercise 7 in Chapter 9:

> Clarissa provides the following information about 13 friends who live with her in the same dormitory suite: 6 of them are intelligent, 8 are entertaining, and 8 have high moral standards. Furthermore, 4 are both intelligent and entertaining, 6 are entertaining and have high moral standards, 3 are intelligent and have high moral standards, and 2 have all three virtues. List eight mutually exclusive categories for Clarissa's friends and tell how many belong in each category.

It is tempting to start a tree diagram for this problem in the manner of Figure 11-17. But Figure 11-17 ignores an important rule (T2): the edges to the right of a given vertex must designate mutually exclusive categories.

To arrive at a correct diagram, we should consider Clarissa's criteria one at a time. For intelligence, we can find a pair of mutually exclusive categories:

Friends who are intelligent
Friends who are not intelligent

Likewise, Clarissa has friends who are entertaining and friends who are not; and friends who have high moral standards and friends who do not. Consideration of these mutually exclusive categories leads to the tree diagram in Figure 11-18.

Figure 11-18 includes the numerical labels that were given in the statement of the problem. Figure 11-19 supplies the rest of the labels; but before you go on to Figure 11-19, you may want to try to figure the numbers out yourself. This is not an easy puzzle, even though all the information needed is given. Remember to follow rules N1 and N2.

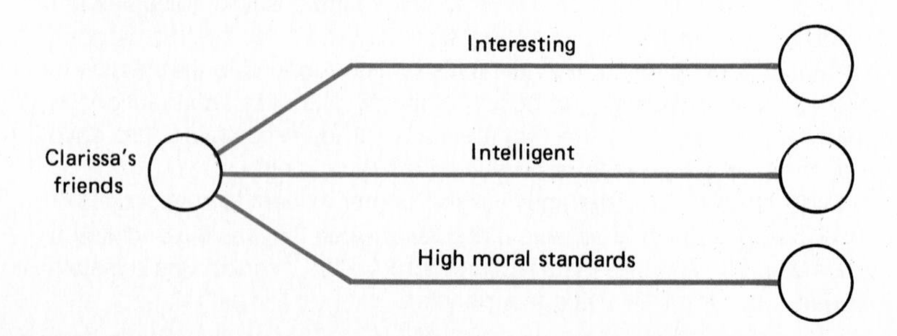

FIGURE 11-17
An incorrect start for a tree diagram showing categories of Clarissa's friends

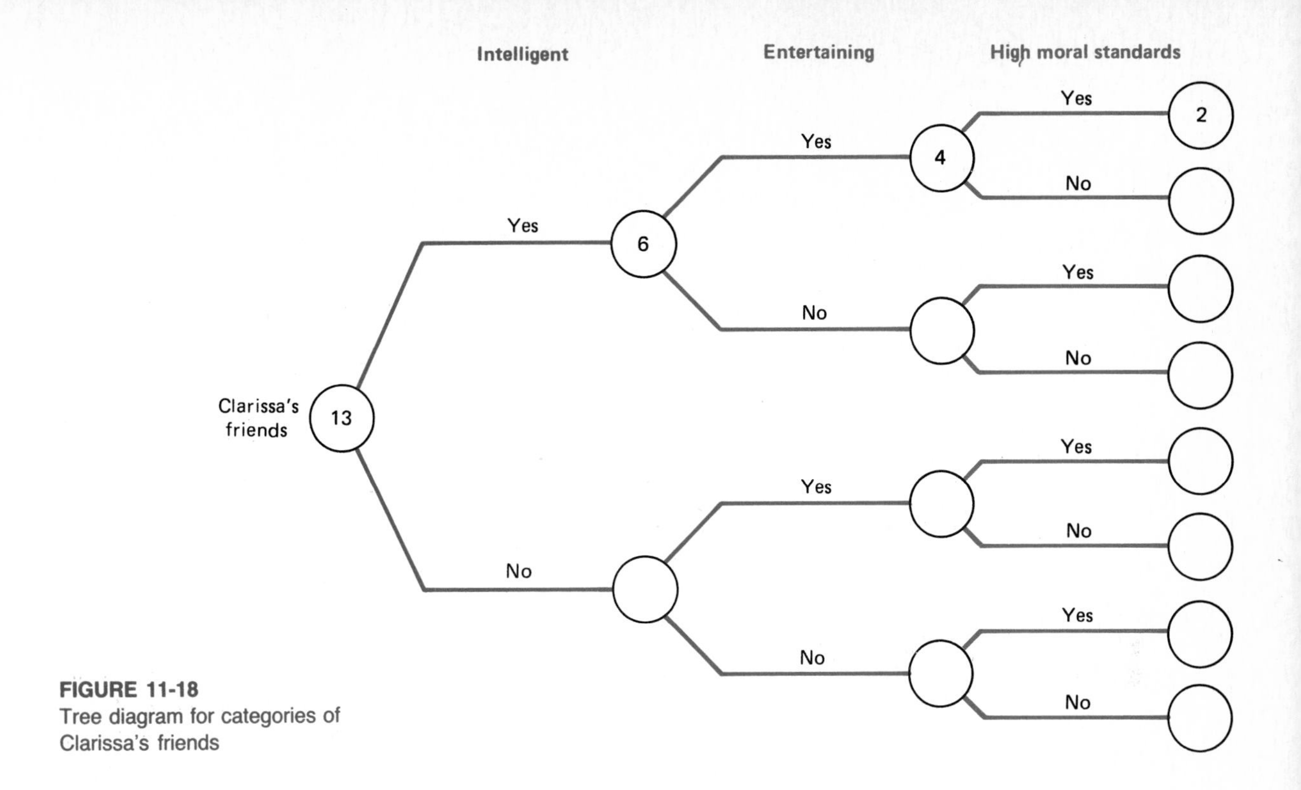

FIGURE 11-18
Tree diagram for categories of Clarissa's friends

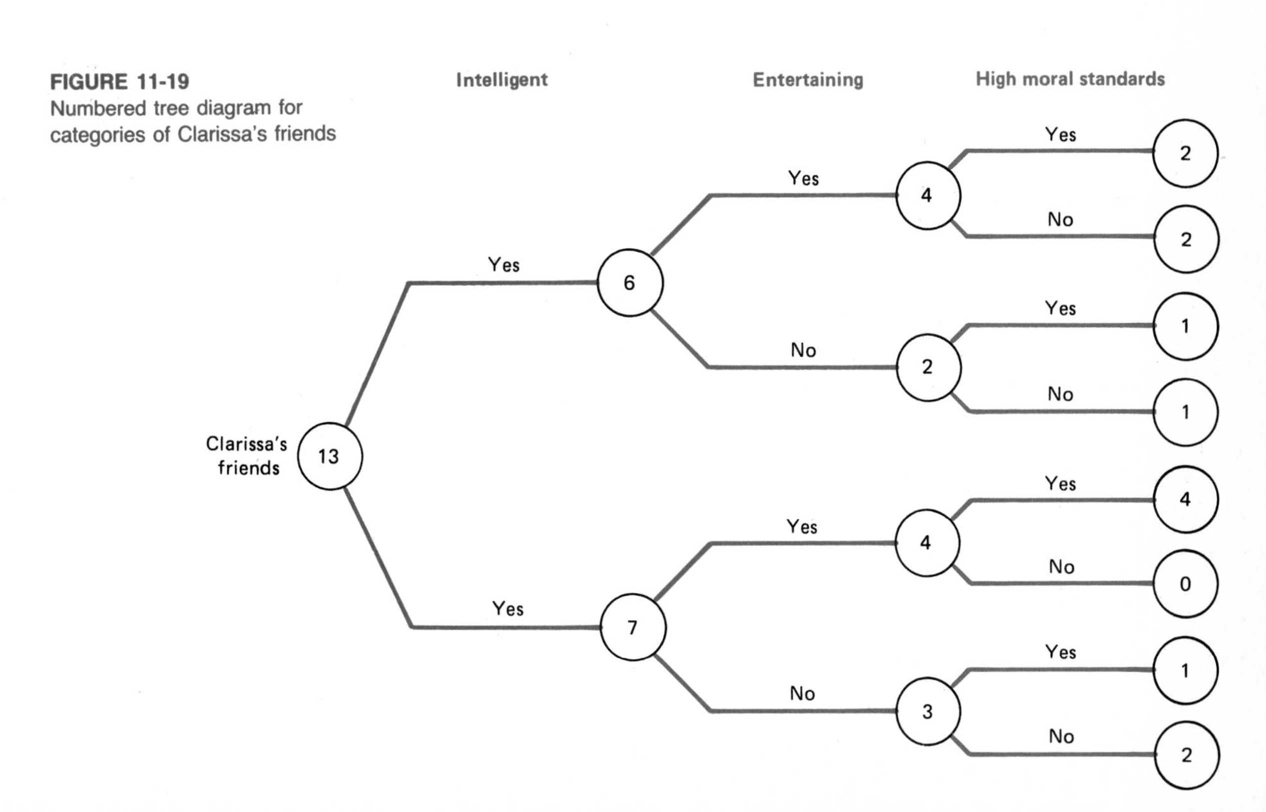

FIGURE 11-19
Numbered tree diagram for categories of Clarissa's friends

The difficulty of supplying values for Figure 11-19 reinforces a point which has been made before: information is confusing unless organized into mutually exclusive categories. If a real Clarissa wanted to categorize her friends, she should *start* with a list of mutually exclusive categories; to visualize these, she might use a tree diagram. ◀

SUMMARY

Tree diagrams provide a way of seeing structure in a collection of information. The most important guideline for any use of tree diagrams is this: The group of edges that emanate from a given vertex must be labeled with a *complete list of mutually exclusive categories.*

Exercise 11

Yesterday evening 640 people attended the junior class play at Nanticoke High School. Some were males and some were females; some were adults and some were students.

a Draw a tree diagram that shows the four possible categories of playgoers.

b Suppose that there were 390 males at the play and 290 students, 140 of which were girls. Label the vertices of your tree diagram with the number in each category. How many men (adult males) were at the play?

Exercise 12

In a large group of couples planning to have children, girls are most desired. Each couple plan to stop having children when they have a girl or whenever they have three children—whichever comes first. Draw a tree diagram that shows all possible sex distributions and family sizes that can result from this plan. Label the vertices with the fraction of the group of families that will be expected in each category. Use your numbered tree to answer the following questions: What fraction of the families will have only one child? Two children? Three children?

Exercise 13

Suppose that blond children are favored over dark-haired children. Suppose also that hair color is determined not by heredity but by chance, and that two-thirds of all children born have dark hair and one-third are blond.

Consider a group of 27,000 couples who adopt the following policy:

Have children until a blond is born or until we have three children, whichever comes first.

Represent this situation using a numbered tree diagram. Use your tree to answer the following questions.

a How many children will be born all together to the 27,000 families who follow this policy? How many blonds? How many with dark hair? What fraction of the total number of children are blonds? What fraction have dark hair?

b How many families achieve the goal of having a blond child? What fraction of the 27,000 families achieve this goal? Is it surprising that the latter answer exceeds one-third? Discuss.

Exercise 14

a Draw a tree diagram that illustrates the possible categories of children's sexes for families with four children. Label the vertices with fractions that tell the proportion of families that would be expected in each category. (As in the earlier examples, suppose that in a large group of births half are expected to be boys and half girls.)

b One might reason that since half of children born are boys and half are girls, then half of four-child families will have two boys and two girls. Does your tree diagram support this reasoning? What proportion of four-child families would you expect to have two children of each sex? Defend your answer.

Exercise 15

Molly is intrigued as she watches people play a carnival game. This is how the game works:

A customer pays $1 to play. The customer then gets to select a coin from his or her own pocket (to ensure fairness) and to toss it. If the coin turns up heads, the game is over and the customer loses the dollar. If it turns up tails, the customer is still in the game and gets to toss again. As before, heads loses and tails enables the customer to continue. Any customer who can achieve four tails in a row wins $5.

"How can the carnival man afford to pay out $5?" Molly wonders.

a Analyze this game using a tree diagram. Draw and label a tree that shows the possible outcomes of the game. Label the vertices with fractions that tell what proportion of the time you expect each outcome to occur.

b What fraction of the games would you expect a customer to win? Is this a profitable game for the carnival? How profitable?

Exercise 16

Suppose that half of all high school graduates are male and half are female. Of the male graduates, 60 percent go to college. Of the female graduates, 50 percent go to college. Of the males who go to college, 30 percent graduate. Of the females who go to college, 40 percent graduate.

a Illustrate the various categories of high school graduates using a tree diagram.

b Which is higher, the percent of high school graduates who are female and graduate from college or the percent of high school graduates who are male and graduate from college?

Exercise 17

Use the tree diagram in Figure 11-13 to answer the following questions about tosses of a pair of dice.

a For a large number of tosses, what fraction would you expect to show the same number on both dice?

b For what fraction of the tosses would you expect Tom's die to show a larger number than Dick's? Dick's to show a larger number than Tom's?

c What fraction of the tosses would you expect to yield each of the following sums: 2? 3? 4? 5? 6? 7? 8? 9? 10? 11? 12?

Exercise 18

A certain music school has only two departments—instrumental and vocal. The instrumental department admits about 80 percent of female applicants and about 40 percent of male applicants. The vocal department, which has stricter requirements, admits about 12 percent of female applicants and about 6 percent of male applicants. Of all applicants to the school, women seem to be more interested in vocal music: about 90 percent of the female applicants apply to the vocal department and 10 percent apply to the instrumental department. Among males the reverse is true: 90 percent apply to the instrumental department and 10 percent to the vocal department. On the basis of these statistics, determine whether male or female applicants have a better chance of being admitted to the music school.

Hint Use a pair of tree diagrams.

Exercise 19

A survey of 100 magazine subscribers yielded the following information:

70 subscribe to *TV Guide.*
22 subscribe to *TV Guide* and *Newsweek* but not to *Reader's Digest.*
25 subscribe to *TV Guide* and *Reader's Digest* but not to *Newsweek.*
10 subscribe to all three magazines.
45 subscribe to *Newsweek.*
10 subscribe only to *Newsweek.*
10 subscribe to none of these magazines.

Organize this information in mutually exclusive categories using a numbered tree diagram. (This problem was considered in Exercise 3 in Chapter 9.)

Exercise 20

Chris's father is going for a medical checkup that will include an x-ray for tuberculosis. Chris is thinking about the possible states of his father's health. Among men his father's age, tuberculosis is rare; only about 1 percent have it. Chris is aware that x-rays are not 100 percent reliable. Consequently, he suggests that his father inquire about the chances that the x-ray will give a false reading. The medical examiner states that the x-ray is 95 percent reliable: that is, if you do not have tuberculosis, the x-ray will give you a clean bill of health 95 times out of 100; on the other hand, if you do have it, the x-ray will report that 95 times out of 100. (This problem was considered earlier; see Exercise 14 in Chapter 3.)

a Organize this information using a numbered tree diagram.

b A few days after the checkup, Chris's father receives the results of the x-ray. It indicates that he has tuberculosis. From all the information given, what are the chances that the x-ray result is correct and that Chris's father actually has tuberculosis?

Exercise 21

In Chapter 9, Venn diagrams were introduced as a tool for organizing information into mutually exclusive categories. Here, tree diagrams have been used for the same purpose. Compare and contrast the two types of diagrams. Describe a situation in which the Venn diagram is more useful than the tree diagram. Describe a situation where the reverse is true. In general, do you prefer one to the other? What are the advantages of the diagram that you prefer?

Exercise 22: Invent; Journal

Although you may not have a formal budget, there is probably some structure for your spending. Select a fixed time period or a specific circumstance (the opening week of a semester, buying Christmas gifts, or whatever) and illustrate your spending allocations using a tree diagram.

Exercise 23: Invent; Journal

What personal applications can you see for the tree diagram? Can it help you organize your time? Select the best from among several choices? Outline an assigned paper? Study for a test? Look for ways that organizing information with a tree structure can be useful *to you.* Try them. Sometimes you will find it unnecessary to actually draw the tree; a mental image will suffice.

Exercise 24: Test Yourself

Develop a list of new terms and key ideas presented in this chapter. Supply an example of each item on your list.

Note The Appendix contains complete answers for Exercises 12, 16*b,* and 20*b.* It contains partial solutions or hints for Exercises 13*a* and *b,* 14, 15*b,* 18, and 21.

CHAPTER 12

SUMMARIZING INFORMATION USING A FEW NUMBERS

Do not put your faith in what statistics say until you have carefully considered what they do not say.

—William W. Watt

LOCATING THE "CENTER" OF A COLLECTION OF DATA ITEMS

When we want to summarize a great deal of information, we may try to describe it in terms of a "center." Three different types of "center" or "average" are commonly used: the mode, the median, and the mean. These will be illustrated with examples, followed by formal definitions.

Bentwell, Inc., is a small firm that manufactures of replacement parts for textile machinery. The firm has 11 employees whose titles and salaries are given in Table 12-1.

The mode of these salaries is $15,000. In any set of data values, the *mode* is that value which occurs most frequently. When two values are equally most frequent, the data set is called *bimodal.* When neither one nor two data values are most frequent, we say that the data set has no mode.

The median of the salaries in Table 12-1 is $18,000. When a list of data values is arranged in ascending order, the *median* is the midpoint. If there is an *odd* number of data values in the ordered list, the median is the middle value; if there is an *even* number of values, the median is midway between the middle pair.

TABLE 12-1
Salaries at Bentwell, Inc.

EMPLOYEES	SALARIES, DOLLARS
Secretary	12,000
Bookkeeper	19,000
Machinist, level I	15,000
Machinist, level I	15,000
Machinist, level I	15,000
Machinist, level II	18,000
Machinist, level II	18,000
Machine supervisor	22,000
Sales representative	20,000
Sales representative	20,000
Owner and general manager	60,000

The mean of the salaries in Table 12-1 is

$$\frac{\text{Sum of the salaries}}{\text{Number of employees}} = \frac{\$234{,}000}{11} \approx \$21{,}273$$

In general, if there are n data values

$$x_1, x_2, \ldots, x_n$$

then the *mean,* commonly denoted by $\bar{x}$, is calculated thus:

$$\bar{x} = \frac{x_1 + x_2 + \cdots + x_n}{n}$$

That is, add all the values and divide the sum by the number of values.

When the word *average* is used, usually the mean is intended. To avoid confusion, when an average is used its type should be identified.

As the salaries at Bentwell illustrate, the mode, median, and mean may all have different values. It thus is important to ask, "When I want to summarize data by use of a single number, which is the best 'average' to use?" The general answer to this question is, "Choose the 'average' that best describes the 'center' of the data." It is possible, however, to be more specific; below, more detailed guidelines are given.

Since the *mode* records only the most frequent value, it may be far from the "center" of the data values and thus may not be typical. But an important advantage of the mode is that it can be used (and is in fact the only one of our averages that can be used) for nonnumerical data. We can say, for example, that the modal sex of United States senators is male and that their modal race is white. It does not, however, make sense to speak of the median sex or mean race of senators.

The *median* makes use only of the *order* of data values, not their magnitudes. Thus it can be used with data values that are measured on an ordinal scale as well as with measurements on interval and ratio scales. The insensitivity of the median to extreme values is often an advantage. In the determination of "average" salaries for Bentwell, Inc., the median most nearly described the "center" of the list

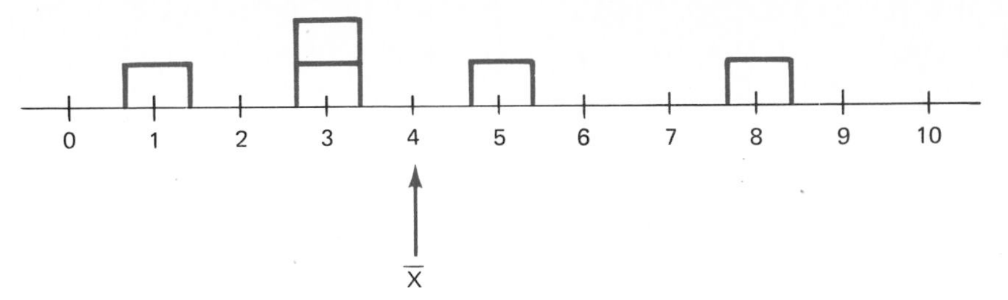

FIGURE 12-1
The mean as balance point

of salaries. The mean salary was "pulled" to a value higher than all but two salaries by inclusion of the extreme salary of $60,000. To summarize data distributions—such as income—that involve a few nontypical extreme values, the median is usually the best average to use.

The *mean* alone makes use of the actual numerical values of the data. Because it utilizes all the information given, in general it is the best and most commonly used measure of the "center." Strictly speaking, the mean makes sense only for measurements based on an interval scale or a ratio scale. In actual practice, however, mean rankings are sometimes computed for ordinal measurements. (When this is done, there is the implicit use of the difference between ranking numbers as a uniform unit of measure.) The mean also has a geometric interpretation. Consider the list of values

$$x_1 = 1, x_2 = 3, x_3 = 3, x_4 = 5, x_5 = 8$$

Imagine that each of these values represents the location of a unit of weight on a numbered seesaw (Figure 12-1). The mean gives the location of the balance point for the seesaw. (In this case, the seesaw itself is treated as if it were weightless.)

The exercises that follow provide practice in determining and interpreting the mean, median, and mode.

Exercise 1

Determine the mode, median, and mean for each of the following lists of data values. Before you calculate the mean, guess whether it will be larger than, smaller than, or the same as the median.

a 4, 2, 6, 7, 4, 5, 8

b 3, 2, 2, 6, 13, 17, 5, 5, 7, 8

c 0, −3, 4, −10, 7, 6, 89

d Look back over your results. Were your guesses correct? Are the mean and median close or far apart? Can you see reasons in your data for the differences?

Exercise 2

Determine which measure of center (mean, median, or mode) is the appropriate meaning of "average" in each of the following situations.

a A newspaper filler item tells us, "The average American is a white married female."

b A sociologist studying the standard of living in Birmingham, Alabama, produces an estimate of the average family income there.

c School subsidies are based on the total income of the citizens in a school district. If the income-earning population and the average income are known, then the total income can be calculated.

d "If at first you don't succeed, you're running about average" (M. H. Alderson).

e "Then there is the man who drowned crossing a stream with an average depth of 6 inches" (W. I. E. Gates).

Exercise 3

The mean age of five persons in a room is 21 years. A 27-year-old person joins them. Now what is the mean age of persons in the room?

Exercise 4

Some years ago a controversial public official in California quit his job and moved to Texas. An elated critic—both of the official and of Texas—said that the move raised the average IQ in both states. Suppose that the average intended in this case is the mean. Explain how this could be possible.

Exercise 5

a Jerry looked over a paper he had just completed for sociology. He noticed that he had used words varying in length from 1 to 13 letters. Jerry calculated the mean of 1 and 13: $(1 + 13)/2 = 7$. Why isn't it appropriate to say that the mean word length in Jerry's paper is 7 letters?

b For Exercise 6, just below, find the mean word length, the median word length, and the modal word length. Which of these best represents the "center" of word lengths in that paragraph? Why?

Exercise 6

While driving on an interstate highway, you adjust your speed until the number of vehicles passing you is approximately equal to the number you are passing. You then feel satisfied that you are driving at the "average" speed. Which measure of the center are you using—the mean, the median, or the mode?

Exercise 7

A real estate newsletter listed the mean and median prices of new homes sold in the Wyoming Valley in the last year. The two values listed were $58,900 and $67,000. Which of these numbers is the mean and which is the median? Explain your choice.

Exercise 8

Nancy drove 10 miles through a congested area at 30 miles per hour and then 10 more miles in the country at 50 miles per hour. At what *average* speed did Nancy make the 20-mile trip? (To find average speed, divide the total distance by the total time taken to drive it.) Don't be hasty. The answer is *not* 40 miles per hour. Why not?

Exercise 9

Myron Thomas sells used cars for Economy Motors. During the single year he has worked there he believes he has established an outstanding sales record and he is getting set to ask his boss, Mr. Economy, for a raise. To prepare his case, Myron is currently looking over the past year's sales records to see exactly how many cars he sold. He discovers that he sold between 0 and 6 cars each week. The number of weeks in which he sold each different number of cars was as follows:

0 cars sold	5 weeks	(This included 2 weeks vacation.)
1 car sold	7 weeks	
2 cars sold	3 weeks	
3 cars sold	8 weeks	
4 cars sold	12 weeks	
5 cars sold	12 weeks	
6 cars sold	5 weeks	
	52 weeks	

a What is Myron's median weekly car sales? What is his mean weekly car sales? Which of these measures of "center" more fairly represents his ability as a salesman? Which will he want to emphasize when he meets with Mr. Economy? Explain.

b Myron seems to be using the average of his past sales as an estimate of what his employer can expect in the future. Discuss the value of using a past average as a predictor in this way.

Exercise 10

In conversations about the likelihood of events, the "law of averages" is sometimes mentioned. What is this law?

Exercise 11

Is a baseball player's "batting average" one of the types of averages we have discussed? If so, which of the three is it?

Exercise 12

In Chapter 7, weighted averages were discussed. Actually, the weighted average is a slight modification of the mean. If, for example, Dan's biology teacher weights the final examination twice as much as the two 1-hour examinations and if Dan's three grades were 70, 90, and 86, then the *mean* of the four values,

70, 90, 86, 86

is the same as the *weighted average* of the original three scores. Use the mean or weighted average to solve the following problems.

a Geoffrey bought 10 gallons of gasoline at $1.20 per gallon and 15 gallons at $1.25 per gallon. Compute the mean cost per gallon of the gasoline he bought.

b Paywell Insurance Company estimates—on the basis of an extensive study of accident data—that for each 1000 auto insurance policies it sells,

900 will result in no claims for accidents.

50 will result in claims of around $500.

25 will result in claims of around $1000.
15 will result in claims of around $2000.
10 will result in claims of around $10,000.

What is the mean claim per policy sold?

c Reconsider Exercise 8. Find Nancy's average speed as a weighted average of the two speeds at which she traveled.

Exercise 13: Invent

a Make up a list of 10 numbers in which the mean is smaller than all but one of the numbers.

b Make up a list of 10 numbers in which the mean is larger than all but one of the numbers.

c Explain why the mean cannot be larger than all the numbers from which it is computed.

d Can *a* or *b* be done if *mean* is replaced by *median*?

Exercise 14: Invent

In determining the median income of a group, certain federal agencies do not consider persons with zero income. Make up an example to show how, when such a method is used, the median income can go *down* when the members of the group become better off.

Exercise 15

In the discussion of the mean, the text states that this average makes sense only for measurements based on an interval or ratio scale. Why is this true? Do the median and mode require an interval or ratio scale? Explain why or why not.

Exercise 16

a Figure 12-1 (page 185) gives a geometric interpretation of the mean. Make a sketch similar to Figure 12-1 for the following set of data values: $x_1 = -4$, $x_2 = -3$, $x_3 = -1$, $x_4 = -1$, $x_5 = -1$, $x_6 = 2$, $x_7 = 2$, $x_8 = 3$, $x_9 = 3$.

b Locate the median and the mode on the sketch you drew for *a*, Describe, in general, how to locate the median and the mode on a sketch of that sort.

Exercise 17

Look for mentions of averages in your reading of newspapers and magazines. Is each average identified as a mean, median, or mode?

Exercise 18

What are situations in which you use an "average" as a way of summarizing information? What type of average do you use most?

Exercise 19: Journal

In Exercise 9 Myron Thomas used the average of last year's weekly car sales as an estimate of future sales. Similarly, most of us can use averages based on past experience to estimate, say, the time or money needed for certain future situations.

Make a list of situations for which you can use an average of past values to estimate amount of time or money needed in the future. Note which type of average you will use in each case.

For at least one of the situations on your list, gather the necessary information and find the average.

Note The Appendix contains complete answers for Exercises 2*a* and *b*, 5*a*, and 10. It contains partial solutions or hints for Exercises 2*c*, 5*b*, 8, and 9*a*.

MEASURING VARIATION FROM THE "CENTER"

The Kramer family are planning a July vacation trip to Yellowstone National Park. They consult a travel guide for weather information and learn that the mean temperature for July days in Yellowstone is 73°F. "That sounds pleasant," they think. Fortunately, before they take off on their trip with only lightweight clothing, they speak of their plans with a neighbor who has been there. "That average is deceiving," the neighbor warns. "Some of the days are quite warm—in the high 80s. But at night it often gets down into the 50s. Be sure to take warm clothing."

This example points to a common fault in the use of a mean or median. The average alone gives only the "center" of the data; it does not tell how far the data may spread on either side of it.

Following are several simple examples to introduce two tools for describing variability of data: percentiles, with emphasis on quartiles, and standard deviation. When we know both the center and the variability of a data set, then we have a useful summary of it.

Consider the following lists of data:

Data set A

5, 12, 18, 20, 21, 22, 23, 24, 25, 26, 29, 34, 40

Data set B

5, 7, 8, 11, 13, 17, 23, 28, 30, 32, 35, 38, 40

The two data sets, A and B, have the same median (23) and the same range (35). (The *range* is the difference between the largest and smallest data values.) However, the two sets of data values are distributed differently. In A, half of the remaining data values are within 3 of the median; in B, none of the other values is within 3 of the median.

Percentiles

To distinguish between different sets of data with the same median, the concept of *percentile* is useful. The *pth percentile* of a set of data values is a value such that (approximately) *p* percent of the data numbers lie at or below it and the rest lie above. The median is, of course, the 50th percentile. Two other percentiles are important enough to have special names: the 25th percentile is called the *first quartile;* the 75th percentile is called the *third quartile.* (An infrequently used name for the median is the *second quartile.*)

The full range of percentiles 0 to 100 is generally used only for large sets of data values. Scores on standardized tests (such as the SAT college entrance examinations) are frequently reported in terms of percentiles.

Let us restrict our attention to quartiles. In data set A, which contains 13 values, the first quartile boundary is determined by calculating 25 percent of 13, which is $3\frac{1}{3}$. To include 25 percent of the data at or below the first quartile, we select the fourth data number, 20, as the first quartile value. Similarly, 75 percent of 13 is $9\frac{3}{4}$, and so the tenth data value, 26, is chosen as the third quartile.

Summarizing the center and spread of data set A, we have:

First quartile:* 20
Median: 23
Third quartile: 26

Likewise, we can summarize data set B as follows:

First quartile: 11
Median: 23
Third quartile: 32

These summaries show clearly how different the variability of the two data sets is.

When, as above, the *median* is used as the indicator of the center of a data set, information about variability can be conveyed by giving certain percentiles. Half of the data lie between the first and third quartiles and one-fourth of the data lies at each end, beyond the quartile values.

Standard Deviation

In contrast, when the *mean* is used as the indicator of the center of data, the variability of the data is measured in terms of a calculated quantity called the *standard deviation.*

For a list $x_1, x_2, \ldots, x_n$ of data values with mean $\bar{x}$, the standard deviation σ is calculated using the following formula:

$$\sigma = \sqrt{\frac{(x_1 - \bar{x})^2 + (x_2 - \bar{x})^2 + \cdots + (x_n - \bar{x})^2}{n}}$$

The next group of exercises will give you a chance to calculate a few values of σ. The main emphasis, however, will be on its interpretation.

The Russian mathematician Tchebycheff (1821–1894) established the following theorem concerning the standard deviation: For any set of data,

At least 75 percent of the data values are no more than 2 standard deviations away from the mean.

At least 89 percent of the data values are no more than 3 standard deviations away from the mean.

*Methods of quartile selection vary. Quartiles need not be actual data values, but they are often the most convenient choice.

At least 94 percent of the data values are no more than 4 standard deviations away from the mean.

In general, at least $(1 - 1/k^2) \times 100$ percent of the data values are no more than k standard deviations away from the mean.

To see how Tchebycheff's theorem works, suppose that for some large set of data values

$$\bar{x} = 50$$
$$\sigma = 5$$

We then know that

At least 75 percent of the data values lie between 40 and 60.

At least 89 percent of the data values lie between 35 and 65.

At least 94 percent of the data values lie between 30 and 70.

For a second set of data values, suppose it is given that

$$\bar{x} = 50$$
$$\sigma = 10$$

In this second case

At least 75 percent of the data values lie between 30 and 70.

At least 80 percent of the data values lie between 20 and 80.

At least 94 percent of the data values lie between 10 and 90.

It is criticial to remember that a larger standard deviation describes data which are spread more widely on either side of the mean.

Tchebycheff's theorem gives *conservative* estimates about how near data are to the mean. A formal study of statistics reveals that many distributions of data are *normal* distributions, in which data tend to be near the mean. Height data for a large group of people chosen at random from the population would probably provide such a distribution: most people have heights that are close to the mean height, and the number who do not grows smaller as distance from the mean increases. For normal distributions most data values (around 95 percent) lie within 2 standard deviations from the mean and nearly all (over 99 percent) lie within three standard deviations from the mean.

Although the examples here involve short lists of data values, these measures of center (median and mean) and variability (percentiles and standard deviation) apply to data sets of any size. They are especially useful in cases in which there are too many data values for easy examination of the entire list.

Exercise 20

What effect does changing the 9 to a 10 in the set of data 5, 3, 4, 3, 9 have on each of the following; will it increase, decrease, or remain the same?

a Mean
b Median
c Mode
d Standard deviation

Exercise 21

a Calculate the standard deviation for the following data sets. Each of these data lists has the same mean, $\bar{x} = 25$.

Data set 1: 21, 23, 25, 27, 29
Data set 2: 15, 20, 25, 30, 35
Data set 3: −5, 10, 25, 40, 55

b In each of these data sets, all the data lie within 2 standard deviations from the mean. How can this be related to the 75 percent value predicted by Tchebycheff's theorem?

Exercise 22

Give the first and third quartiles for each of the following data sets. Each data list has the same median, 25.

Data set 1: 9, 9, 11, 13, 15, 18, 19, 22, 22, 24, 26, 30, 31, 33, 35, 38, 40, 45, 47, 47
Data set 2: 5, 10, 15, 20, 25, 30, 35, 40, 45
Data set 3: 1, 2, 2, 3, 3, 3, 4, 4, 4, 4,
4, 5, 6, 6, 6, 7, 8, 8, 9, 9,
9, 9, 9, 10, 10, 11, 11, 12, 12, 12,
12, 13, 14, 15, 15, 15, 15, 16, 17, 17,
18, 18, 19, 20, 21, 22, 23, 23, 24, 25,
25, 25, 25, 25, 25, 26, 26, 26, 26, 27,
27, 27, 27, 27, 27, 27, 27, 28, 28, 28,
28, 28, 28, 28, 28, 28, 28, 29, 29, 29,
29, 29, 29, 29, 29, 29, 29, 29, 29, 29,
29, 29, 29, 29, 29, 29, 29, 29, 29, 29

Exercise 23

The quartiles on a recent statistics test taken by 240 students were 67, 77, and 84.

a How many scores were between 67 and 84? Above 84? Below 77?

b Bill had a score of 70. How many scores were above his? How many were below his?

c If the range of scores was 38 and the highest score was 92, what was the lowest score?

Exercise 24

Consider a list of 100 data values that contain 10 zeros, 10 ones, 10 twos, . . . 10 nines. This list has mean $\bar{x} = 4.5$ and standard deviation $\sigma \approx 2.87$. Use this information to illustrate that although Tchebycheff's theorem is true, it gives "conservative" estimates concerning the nearness of data to the mean.

Exercise 25

The following list of data values has mean $\bar{x} = 8$ and standard deviation $\sigma \approx 4.5$: 2, 3, 6, 7, 10, 13, 15.

a What are the mean and standard deviation of the following list of data values (obtained by adding 2 to each member of the given list): 4, 5, 8, 9, 12, 15, 17?

b Change two of the seven numbers to obtain a new list which also has $\bar{x} = 8$, but which has a larger standard deviation than the given list.

c Change two of the seven original numbers to obtain a new list which also has $\bar{x} = 8$, but which has a smaller standard deviation than the given list.

d Modify the original list to obtain a new list of seven numbers with $\bar{x} = 12$ and the same standard deviation as before, $\sigma \approx 4.5$.

Exercise 26: Invent

For each of the following, select eight numbers from the whole numbers 0, 1, 2, . . . , 9. (Repeats are allowed.)

a Choose the numbers so that the standard deviation is as large as possible.

b Choose the numbers so that the standard deviation is as small as possible.

Exercise 27

A survey of book publishers reveals that the mean height of current publications is 9 inches with a standard deviation of 0.6 inch. Using this information as a guide, how far apart would you recommend that library shelves be set in order to accommodate most new book purchases? If your recommendation is followed, what percent of new books do you estimate will require "oversize" shelving?

Exercise 28

The community of Pleasant Grove has been discussing whether a school remedial reading program should be funded for next year. It has been focusing on statistics about reading in the fifth grade. Testing has revealed that the mean reading level for students leaving fifth grade is 5 years and 6 months* (exactly equal to the amount of schooling the students had acquired at the time of the test). The standard deviation is 3 months. (In this context, 9 months is a "school year.") Do the facts support a need for the remedial reading program?

Exercise 29: Journal

Return to Exercise 19. For one of the situations you listed there, estimate both an average (mean or median) and a measure of the variability of the data. Why is this more useful than knowledge of the average alone?

---

*A child has a reading level of 5 years and 6 months if material rated (by experts) at that level is the most difficult reading material that the child can read fluently.

Exercise 30: Test Yourself

Develop a list of new terms and key ideas presented in this chapter. Supply an example of each item on your list.

Note The Appendix contains complete answers for Exercises 20 and 21*a*. It contains partial solutions or hints for Exercises 22, 25*d*, and 27.

PART FOUR

VISUALIZING RELATIONSHIPS AND ROUTES

PART CONTENTS

CHAPTER 13

PICTURING RELATIONSHIPS

Friends, Enemies, and Communication
Finding an Efficient Order for Activities
Developing a Project Schedule

CHAPTER 14

FINDING ROUTES

The Rabbit under the Bush
Graphs and Routes
An Application of Euler Circuit Techniques

CHAPTER 13

PICTURING RELATIONSHIPS

Artists can color the sky red because they know it's blue. Those of us who aren't artists must color things the way they really are or people might think we're stupid.

—Jules Feiffer

FRIENDS, ENEMIES, AND COMMUNICATION

Derek Lewis had a summer job as recreation supervisor at the Stevens School playground. One morning he observed a lengthy hassle as 10 boys tried to divide themselves into two basketball teams. Despite their intense and varied efforts, nothing seemed to work. As Derek drew closer to the group, one of the boys, Joe, ran over to him and tried to explain the difficulty. Joe supplied the following information:

Rick and Fred are my friends, but Todd and I don't get along.
Al and Rick are friends, but Bob and Rick don't like each other.
Eric and Bob don't like each other, but Mark is Bob's friend.
Fred and Scott don't like each other.
Chuck and Eric are friends.

To try to make sense of this tumble of information, Derek took a piece of playground chalk and began to draw on the cement. For each boy he drew a heavy dot and labeled it with the initial of the boy's first name. He then drew solid lines between pairs of boys who were friends and dotted lines between pairs who didn't like each other. Figure 13-1 shows a diagram like the one Derek made.

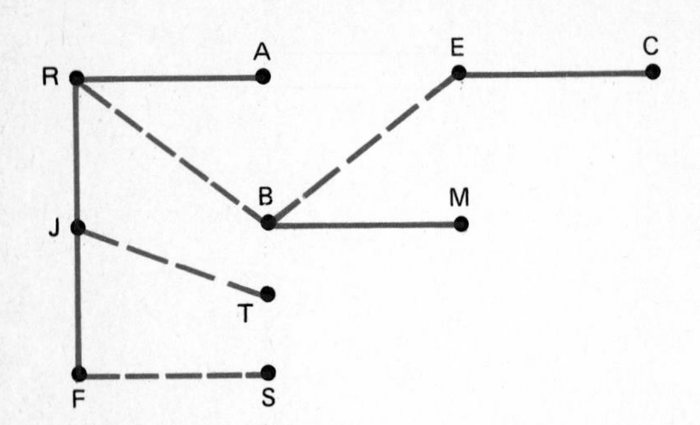

FIGURE 13-1
Playground friends and enemies

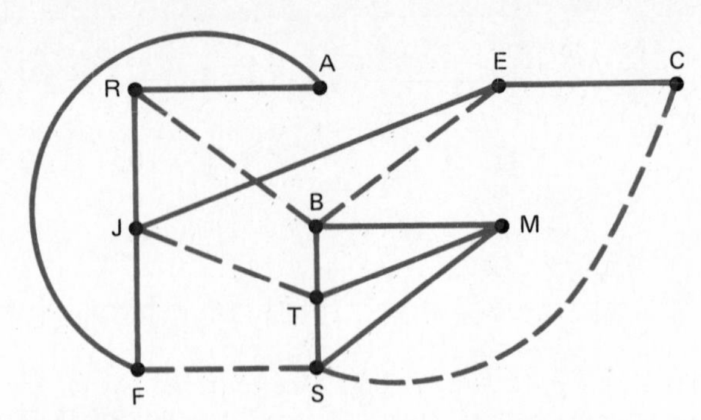

FIGURE 13-2
Playground friends and enemies, with additional information

However, Derek wasn't able to make much sense of his picture. Joe kept supplying more information:

Eric and I are friends.

Fred and Al are friends.

Bob and Mark and Scott are Todd's friends.

Scott and Chuck don't like each other.

Derek added this information to his diagram (Figure 13-2).

Muttering, "Who says a picture is worth 1000 words?" Derek concluded that his diagram still wasn't helping him understand the boys' situation. He solved their problem by arbitrarily assigning them to two teams of 5 and told them to "get out there and play ball" instead of wasting their time in disputes.

While the boys were playing, Derek stood nearby and puzzled over their problem. Suddenly he had an idea; he knelt down and erased the dotted lines in his sidewalk sketch. The resulting diagram is shown in Figure 13-3. This new picture helped Derek see the boys' friendship situation more clearly. Among the boys there were two groups of friends—one with 6 members and the other with 4 members. No wonder the boys couldn't divide themselves into two teams of 5!

FIGURE 13-3
Playground friends

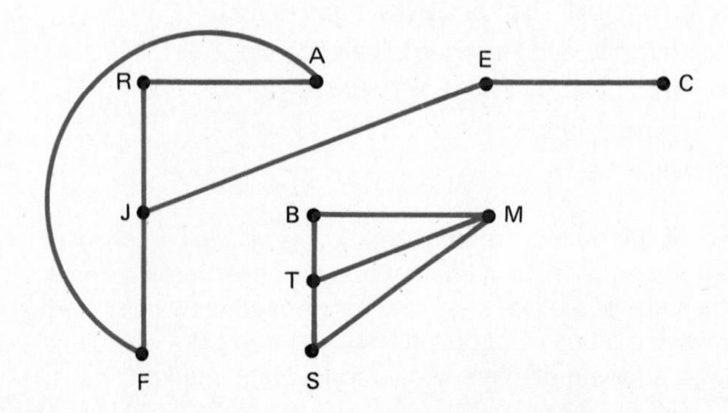

Still watching the boys play, Derek continued to think about the friendship relationships among them. He conjectured that the following generalities might hold. For each boy:

1 The friends of his friends are also his friends.
2 The friends of his enemies are also his enemies.
3 The enemies of his friends are also his enemies.
4 The enemies of his enemies are his friends.

Derek made one more sketch. This time he put the dots for the group of 6 friends near each other and the dots for the other 4 friends near each other. He supposed that generalities 1 to 4 actually held. Figure 13-4 is the result. Satisfied at last, Derek moved on to his other playground duties. The final diagram seemed to be a good description of the situation. On the basis of his analysis, Derek concluded that the boys might never have divided into two teams of 5 by themselves. His intervention had been genuinely needed to solve their problem.

The diagram that Derek drew in his effort to solve the boys' problem is called a *graph.* More specifically, suppose that we have a (finite) set of points; call them *vertices.* An arc or line segment joining two different vertices is called an *edge.* A *graph* is a finite set of vertices (one or more) together with a finite number of edges (possibly none). (The tree diagrams that were considered in Chapter 11 are examples of graphs.)

Graphs may be used to describe relationships that associate pairs of objects. The objects are denoted by vertices; and when a certain relationship holds for a given pair of objects, their vertices are joined by an edge. Derek used vertices to represent boys; an edge between two vertices indicated that the boys liked each other. (In Derek's first diagram—Figures 13-1 and 13-2—he unwisely tried to picture two relationships, "liking" and "disliking," on the same graph. Either relationship may be pictured, but each of the relationships requires its own separate graph.)

Relationships that are not symmetric can be pictured by placing arrows along the edges to give the direction of the relationship. The resulting figure is called a *digraph* (short for *directed graph*).

For example, if Ralph likes Betty but she does not like him, we can represent this by Figure 13-5. If, furthermore, Betty and Sarah like each other and Betty likes Edwin while Edwin likes Sarah and Sarah likes Ralph, this information may be pictured as in Figure 13-6.

The additional examples on the following pages will illustrate the versatility of graphs as a problem-solving tool.

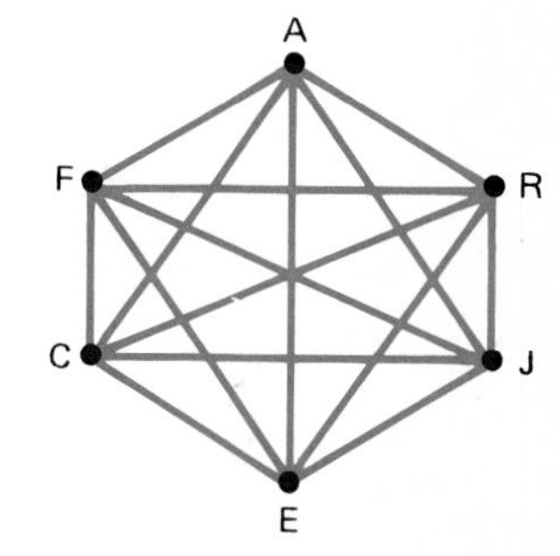

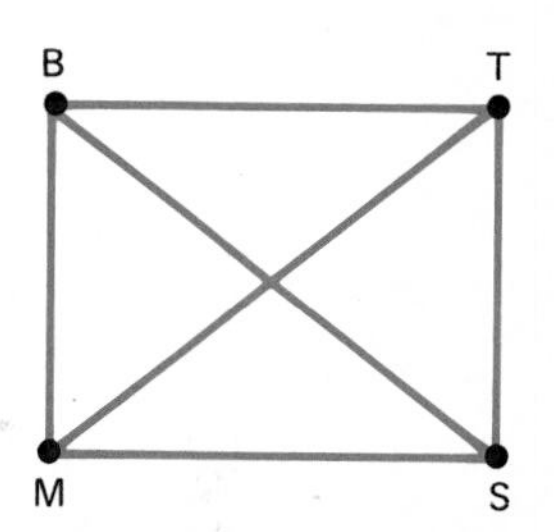

FIGURE 13-4
Completed diagram of playground friendships

FIGURE 13-5

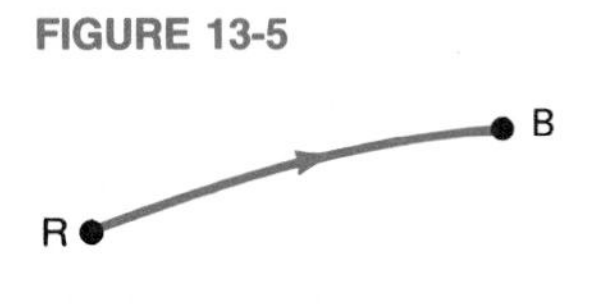

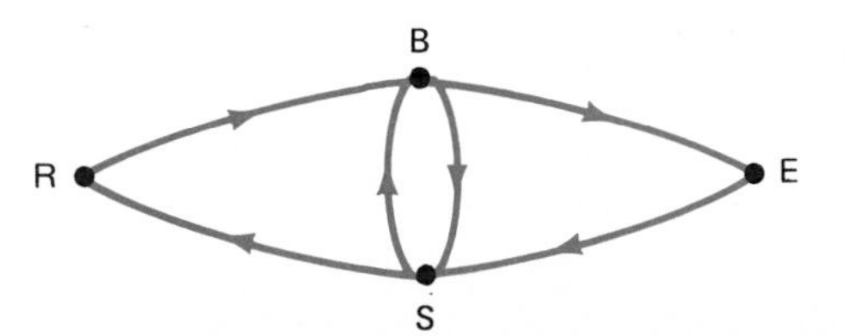

FIGURE 13-6

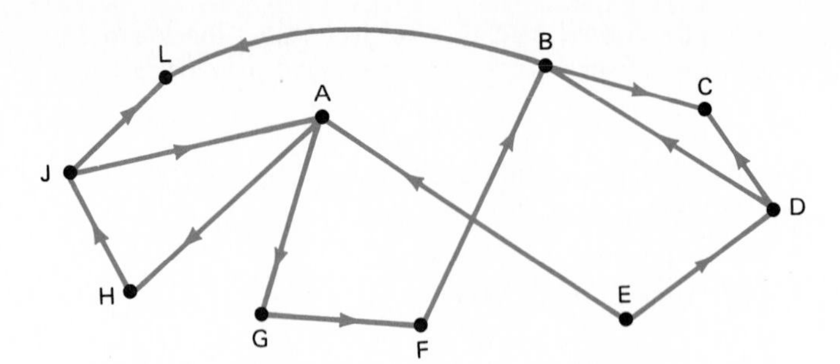

FIGURE 13-7
Digraph showing communication patterns for Alpha Omega men

Example 1: A Communication Problem ▶

Alfred, one of 10 students living on the second floor of the Alpha Omega fraternity house, hears that there is to be a "surprise" quiz tomorrow morning in economics, a course in which all 10 men are enrolled. Here are the facts:

Alfred shares all he knows with Greg and Horace.
Ben shares all he knows with Clarence and Lawrence.
Clarence keeps what he knows to himself.
Dominic shares all he knows with Ben and Clarence.
Edwin shares all he knows with Alfred and Dominic.
Fowler shares all he knows with Ben.
Greg shares all he knows with Fowler.
Horace shares all he knows with James.
James shares all he knows with Alfred and Lawrence.
Lawrence keeps what he knows to himself.

After late-night convesations have taken place between all pairs who customarily share their news, will everyone have learned about the economics quiz?

If we let vertices denote men and draw a directed edge (arrow) from one man to a second if the first shares all he knows with the second, we obtain the digraph of Figure 13-7.

By following the directions of the arrows, we can find directed paths from Alfred to Ben (a path 3 edges long), to Clarence (a path 4 edges long), to Fowler, to Greg, to Horace, to James, and to Lawrence. But there are no directed paths leading from Alfred to Dominic or to Edwin; thus, these two men will not get news of the quiz. ◀

Example 2: Designing a Zoo ▶

In most modern zoos animals are no longer caged individually. Instead, large areas are enclosed and a number of different species are allowed to roam freely together. Species that are natural enemies are kept in different enclosures. Mr.

Orwell, director of the Grimm Zoo, wishes to provide enclosures for 10 new species. The following restrictions prevail:

Animal	Cannot be placed with
Antello	Eleraffe, monkilla, panthanzee, wild bore
Bare	Crocogator, rhinocerich
Crocogator	Bare, eleraffe
Eleraffe	Antello, crocogator, wild bore
Hippophant	Rhinocerich, zebralope
Monkilla	Antello, panthanzee, rhinocerich
Panthanzee	Antello, monkilla
Rhinocerich	Bare, hippophant, monkilla, zebralope
Wild bore	Antello, eleraffe
Zebralope	Hippophant, rhinocerich

What is the minimum number of new enclosures that Mr. Orwell needs to provide so that he can keep natural enemies separate?

The given information can be represented by a graph in which the vertices denote animals. Since the problem emphasizes enmity, our graph will also. We will join two vertices with an edge if they represent animals that may *not* be enclosed together. One graph that can result is shown in Figure 13-8.

A different choice of positions for the vertices results in a different-looking graph; Figure 13-9 shows a simpler-looking picture. However, despite the fact that this second graph looks different from the first one, the two graphs are structurally the same. That is, vertices with the same labels are joined in both graphs.

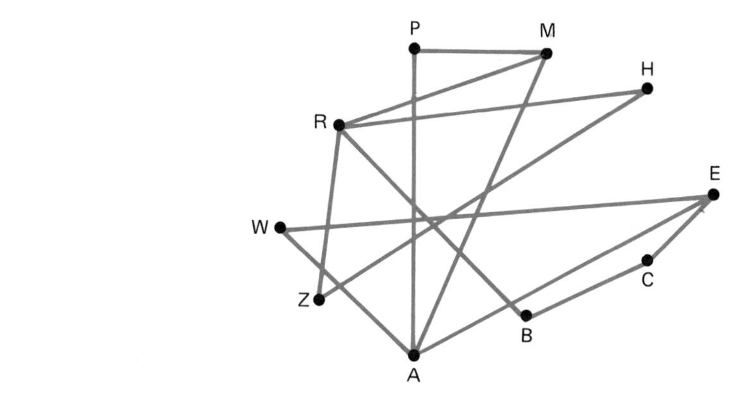

FIGURE 13-8
Graph showing animal enmity for species at the Grimm Zoo

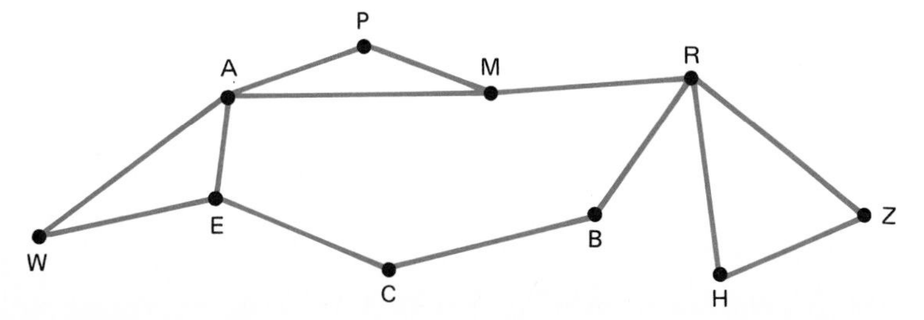

FIGURE 13-9
Redrawn graph of animal enmity for species at the Grimm Zoo

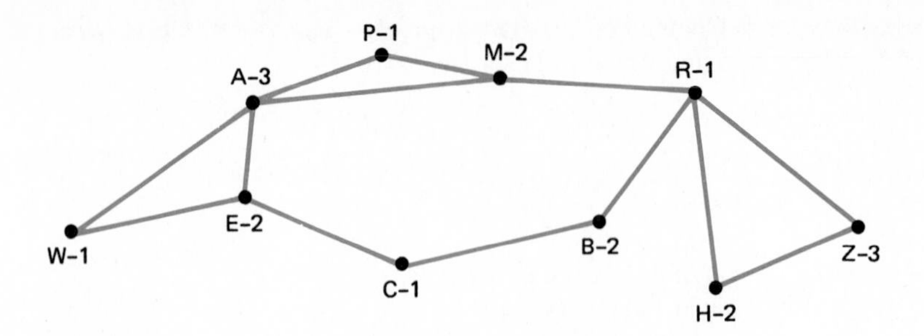

FIGURE 13-10
Graph showing enclosure assignments for species at the Grimm Zoo

In drawing a graph, it is preferable to position the vertices so that as few as possible edges must cross. But this is not always easy; often it requires some trial-and-error experimentation.

Let us now use the second version of our graph to solve the problem for Mr. Orwell. Two enclosures will not be sufficient; we deduce this from the triangles in our graph. Triangle R-H-Z indicates, for example, that the rhinocerich, hippophant, and zebralope must all be separated. Thus at least three enclosures will be required. Suppose we put the rhinocerich in enclosure 1, the hippophant in enclosure 2, and the zebralope in enclosure 3. These assignments can be indicated on the graph by labeling vertices R, H, and Z with the given numbers (see Figure 13-10).

Continuing, we see that it is possible to label the other vertices as well—using 1, 2, and 3—in such a way that no two vertices joined by an edge have the same label. Figure 13-10 shows one possible labeling scheme.

The labeling in Figure 13-10 provides the following solution to Mr. Orwell's problem. Three enclosures are sufficient:

In the first enclosure put the crocogator, panthanzee, rhinocerich, and wild bore.

In the second enclosure put the bare, eleraffe, hippophant, and monkilla.

In the third enclosure put the antello and the zebralope.

Check Is this a solution? Are all the restrictions satisfied? ◀

Exercise 1

An intramural volleyball program at Fitness State College involved four teams. Last Thursday evening the teams—for simplicity designated here as T_1,T_2, T_3 and T_4—engaged in a round-robin tournament. In the first round, T_1 beat T_3 and T_2 beat T_4. In the second round, T_2 beat T_1 and T_3 beat T_4. In the third round, T_1 beat T_4 and T_2 beat T_3. Represent the tournament results using a graph or digraph. (Which is more appropriate?) How should the four teams be ranked, on the basis of the tournament results?

Exercise 2

Morton's grandfather George has 3 children—Richard, Thomas, and Ursula. Richard has 1 son, Zeke. Thomas has 3 children—Karl, Susan, and Roberta. Ursula has 2 children—Cindy and Morton.

a Make a diagram of the "family tree" relationship among the 10 people named above. Is the appropriate diagram a graph or a digraph? What relationship do edges represent?

b Now consider only George's 6 grandchildren. Form a graph in which two vertices are joined when they represent cousins.

Exercise 3

"Janice is my niece," says David to his sister Suzanne. "She is not my niece," says Suzanne. Both are telling the truth. If Janice is a blood relative of David, what relationship has she to Suzanne?

Exercise 4: Invent

Think of at least three situations familiar to you that may be represented by graphs or digraphs. (Possible examples include: friendship patterns, authority structure in an organization, committee and subcommittee structure in an organization, predator-prey relationships among various species, communication patterns, and cause-and-effect relationships.)

For one of these situations, draw and label a graph. Explain what the vertices denote and what relationship is designated by the edges that join pairs of vertices.

Exercise 5

A road map is somewhat similar to a graph. In what ways are they alike? In what ways do they differ?

Exercise 6

Return to the Grimm Zoo example. Find a solution that has at least three animals in each enclosure.

Exercise 7

Return to Example 1. If you wanted to get some information to *every* student living on the second floor of the Alpha Omega house, what *one* person could you tell and expect that the word would spread to all the others?

Exercise 8

Carla, Donna, Heidi, Karen, and Margie were seated, in that order, around a circular table in the Snack Bar drinking Cokes. They all got up—to get some nibbles—and when they sat down again, they changed their seating arrangement so that no one was sitting next to a person beside whom she had sat before. What was the new seating arrangement?

Exercise 9

To catch up on the latest news, a Hollywood gossip columnist has invited 12 celebrities—all women—to a lunch. However, she faces a problem with seating arrangements, since some of the women are not on speaking terms with others. In particular, the following hostilities hold:

Celebrity	Is not on speaking terms with
Ursula	Anastasia, Yvette, Camille, Lenore, Sasha, Francine, Diana, and Paulette
Germaine	Lenore and Camille
Martina	Anastasia and Diana
Imogene	Sasha and Paulette
Yvette	Camille, Francine, and Ursula
Diana	Francine, Ursula, Anastasia, and Martina
Lenore	Sasha, Camille, Ursula, and Germaine
Paulette	Anastasia, Sasha, Imogene, and Ursula

a Using the information given above, draw a graph with a vertex for each celebrity and with edges joining pairs that are not on speaking terms.

b Because the hostess wants to encourage conversation, she should seat women who aren't on speaking terms at different tables. Find a satisfactory seating arrangement with only three tables. If possible, have four women at each table. (Use your graph; imitate the solution process used in the Grimm Zoo example.)

c Suppose that the hostess wants to seat her 12 guests around one large circular table. Find an arrangement where no woman is seated next to someone with whom she is not on speaking terms.

Exercise 10

A foreign language department at the university plans to offer seven introductory language courses next semester—French (F), Spanish (S), Italian (I), German (G), Chinese (C), Russian (R), and Japanese (J). Twelve language majors have indicated their intent to schedule more than one of these courses. The department wants to develop a schedule that will use as few time periods as possible but will not have two courses at the same time if some student wants to take both of them. After each student's name, below, are listed the first letters of the languages that he or she wants to schedule.

Alvin: C, R, J
Barbara: C, G, S
Carol: G, F
David: C, R
Ed: R, F
Frank: C, G
Gayle: J, R
Heather: G, R
Irving: C, J
Joe: C, S, J
Karen: I, S
Leon: I, J

a Represent the problem information using a graph. What should edges and vertices represent?

b Use your graph to determine the minimum number of time periods needed for scheduling these seven courses if all student conflicts are to be avoided.

c Find at least two different schedules for the courses that avoid student conflicts. (You may wish to refer to the time periods as period 1, period 2, etc.) Discuss possible criteria that might make one solution "better" than another.

Exercise 11*

The United Nations has approximately 150 member nations, two of which are the United States and the Soviet Union. These two may be classified as "enemies." Suppose that the United States counts 65 of the members nations as its friends. Suppose also that the relations in the United Nations satisfy the four general rules that Derek conjectured to be applicable in the playground situation.

a If each member of the United Nations is represented by a vertex, and if edges join the vertices of friendly nations, describe the graph that will result.

b Which of Derek's conjectured rules do *not* apply to this situation?

*Adapted from an example in C. A. Lave and J. G. March, *An Introduction to Models in the Social Sciences,* Harper and Row, New York, 1975, p. 67.

Exercise 12

Imagine a graph with a vertex for each person living in the United States. Two vertices are joined by an edge if they denote people who are acquainted. (For specificity, let's say that two people are *acquainted* if, when they meet on the street, they can name each other.) Consider the following conjecture: The *acquaintanceship graph* for the United States (formed as described above) is connected: that is, from any vertex there is a path to any other vertex.

a Develop an argument that supports this conjecture.

b If two United States residents are selected at random, how long a path (how many edges?) would you expect to need to get from one person's vertex to the other's?

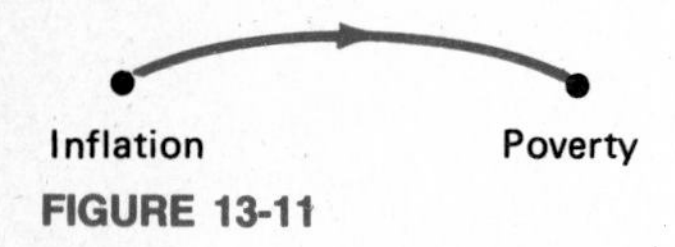

FIGURE 13-11

Exercise 13

One popular use of digraphs is to depict cause-and-effect relationships. For example, the simple digraph in Figure 13-11 describes the relationship "inflation causes poverty." Consider the following items—higher prices, inflation, pollution controls, higher taxes—and draw on your knowledge of their causes and effects to draw a reasonable cause-and-effect digraph that shows how they are related.

Exercise 14

Three economists were asked what factor they thought would have the greatest influence on the economy in the coming year. Each drew a digraph (see Figure 13-12), using four major economic factors as vertices and drawing an arrow from one vertex to another if the first was thought to have a greater influence than the second. On the basis of these graphs, which factor do the economists as a group think will most affect the economy? Explain.

Exercise 15: Invent; Journal

Ned wants to find a good balance among studies, relaxation, and sleep. He observes that studying long hours leaves insufficient time for sleep, which in turn makes him so inefficient in his studies that long hours are required for study. A cause-and-effect digraph for this situation would be as shown in Figure 13-13. This digraph contains what we might call a *vicious circle.* (Do you see a reason for this name?)

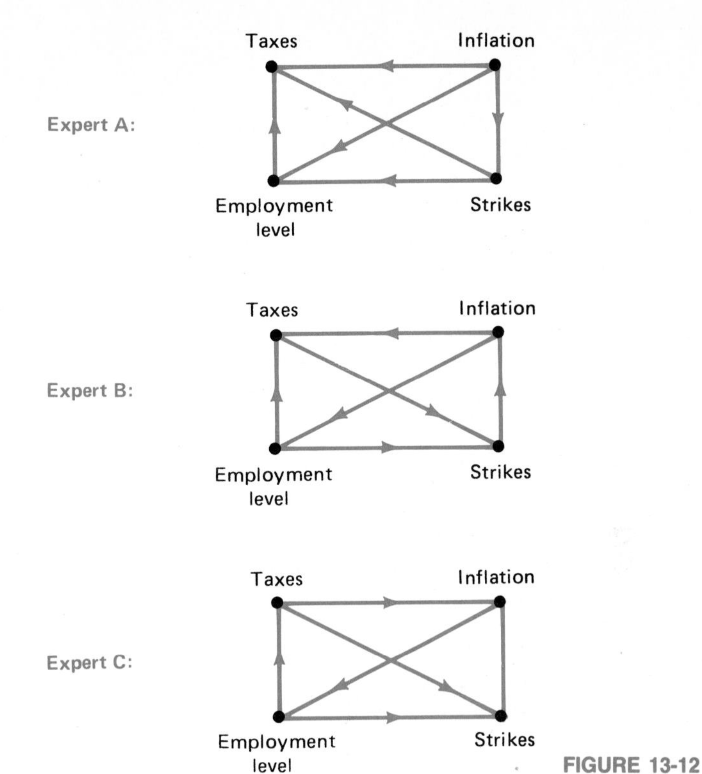

FIGURE 13-12

FIGURE 13-13

Study long hours

Insufficient sleep

Invent a cause-and-effect digraph to illustrate how certain of *your* activities affect others. Does your graph contain any vicious circles like the one in Figure 13-13? Are there simple ways that you can allocate your time better?

The Appendix contains complete answers for Exercises 5 and 14. It contains partial solutions or hints for Exercises 1, 2, 6, and 10*a*.

FINDING AN EFFICIENT ORDER FOR ACTIVITIES

This section will deal with a particular type of scheduling problem: scheduling activities so that the time between them is kept low. The solution process will include a graph. First, in Example 3, finding an efficient schedule for producing different products in a manufacturing process will be considered. Then, in the exercises, you will be asked to imitate the procedure used in Example 3 to develop an itinerary for a sales trip.

TABLE 13-1
ASM conversion times

	A	B	C	D	E	F	G	H
A	—	30	20	20	30	20	20	40
B	30	—	30	35	30	35	20	20
C	20	30	—	35	20	25	30	30
D	20	35	35	—	25	30	30	20
E	30	30	20	25	—	40	30	20
F	20	35	25	30	40	—	35	20
G	20	20	30	30	30	35	—	40
H	40	20	30	20	20	20	40	—

Example 3 ▶

A Massachusetts engineering firm uses a complex apparatus called an *automatic screw machine* (ASM) to produce a wide variety of products, small fittings for many different types of machinery. In a given week eight different products (let us refer to them simply as A, B, C, D, E, F, G, H) are manufactured in sufficient numbers to meet orders placed the previous week. Each different product requires a substantial change of settings on the ASM. The times (in minutes) required for conversion depend on which pair of products one is converting between; these are given in the matrix shown in Table 13-1.

Because the conversion times vary, the order in which the products are manufactured makes a difference. Currently the products are made in the following order each week:

$$A \to B \to C \to D \to E \to F \to G \to H$$

and the machine is reset at the end of the week (on Friday) to produce product A—the product most in demand—first on the following Monday.

The present sequence of products ($A \to B \to C \to D \to E \to F \to G \to H \to A$) requires a total conversion time of

$$30 + 30 + 35 + 25 + 40 + 35 + 40 + 40 = 275 \text{ minutes}$$

(The numbers summed are from Table 13-1.) At present, then, over 4½ hours per week are spent in conversion.

While the ASM is being adjusted to produce a different product, quite a few workers are idle. This lost time is costly, and John Beech, owner of the engineering firm, is seeking a way to reduce it. Mr. Beech has called in Mr. Quick, an efficiency expert, and has asked him to find a weekly production schedule with as little conversion time as possible. Because product A is most in demand, Mr. Quick is asked to find a schedule that starts with product A.

Mr. Quick begins by examining the conversion-time matrix. He notes that the smallest conversion time is 20 minutes. On this basis, he sets as his goal a schedule that will include all conversion times of 20 minutes. To see the relevant information more clearly, he sketches a graph.

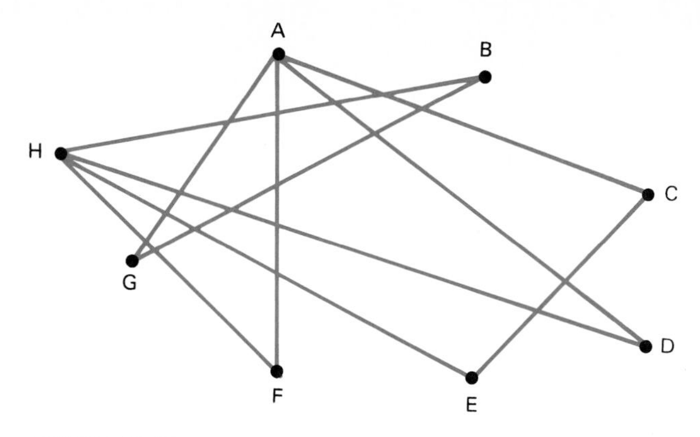

FIGURE 13-14
Graph of 20-minute conversion times

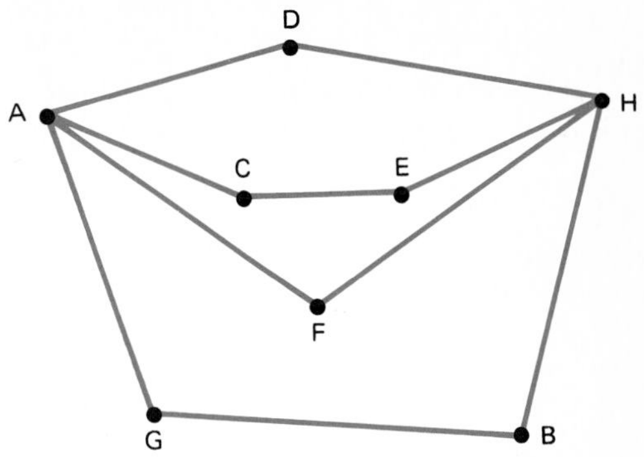

FIGURE 13-15
Graph of 20-minute conversion times, redrawn to avoid edge crossings

He uses vertices to denote products A, B, . . . , H.

He joins two vertices with an edge if the conversion time between the two products is 20 minutes.

The result is Figure 13-14.

When he sees Figure 13-14, Mr. Quick recognizes that if he had positioned his vertices differently, fewer edges would have crossed and his picture would be easier to study. After some experimentation with placement of vertices, he arrives at Figure 13-15.

He now studies Figure 13-15 with the following goal in mind:

Find a path* through the graph that starts with A, goes through each vertex once, and ends with A.

This graphical goal is the pictorial analogue of the scheduling goal:

Find a weekly production schedule that starts with product A, includes each product, and ends the week in readiness to begin production of A the next week.

But he is stymied. No matter how much he thinks and how long he examines his graph, instead of revealing a solution, it seems to show an impossible situation. He cannot find a circuit in Figure 13-15 that starts and ends at A and goes through each vertex once. (*Experiment* with Figure 13-15. Do you agree that such a circuit cannot be found?)

---

*A *path* is a connected sequence of edges. Actually, since the proposed path begins and ends at the same vertex, it may be called a *circuit*.

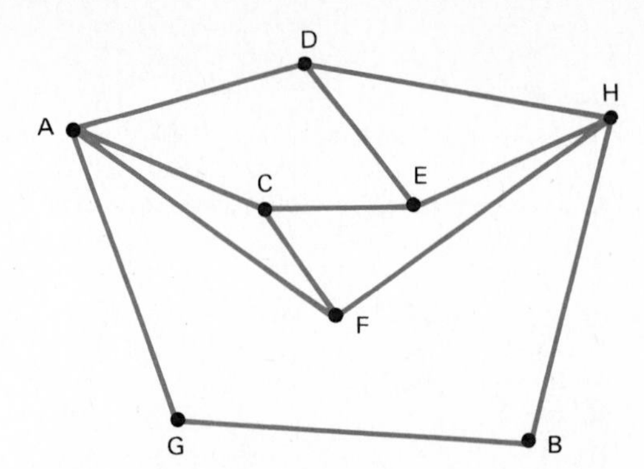

FIGURE 13-16
Graph of 20- and 25-minute conversion times

Finally, he returns to the original conversion-time matrix and considers changing his goal. "What if I allow 25-minute conversion times as well?" he wonders. Allowing both 20- and 25-minute conversion times adds two new edges, CF and DE, to the graph in Figure 13-15. The resulting graph is shown in Figure 13-16.

Success at last! In Figure 13-16, Mr. Quick is able to find the following circuit that meets his (revised) goal:

$$A \rightarrow D \rightarrow E \rightarrow C \rightarrow F \rightarrow H \rightarrow B \rightarrow G \rightarrow A$$

This circuit entails the following conversion times:

$$20 + 25 + 20 + 25 + 20 + 20 + 20 + 20 = 170 \text{ minutes,}$$
$$= 2 \text{ hours, } 50 \text{ minutes}$$

On the basis of this circuit from his graph, he proposes the following weekly product schedule to Mr. Beech:

Begin on Mondays with product A.
Next manufacture product D.
Next manufacture product E.
Next manufacture product C.
Next manufacture product F.
Next manufacture product H.
Next manufacture product B.
Next manufacture product G.
Finally, return machine settings to product A.
Total conversion time = 170 minutes.

"This is the best that can be done," he says apologetically. But Mr. Beech is grateful: "By cutting total conversion time by 105 minutes you have saved me much more than 105 minutes," he remarks. "I have at least a dozen workers who are idle while the machine is reset. A dozen times 105 minutes is 21 hours. With the wages I pay, that's an important saving." ◀

Exercise 16: Journal

In Example 3, Mr. Quick finds a production schedule that saves only 105 minutes per week. However, it turns out that the saving is multiplied by 12 because time is saved for 12 people. Furthermore, this saving will be repeated each week.

Think of situations in which you do not try to organize tasks efficiently, because only a small amount of time would be saved. Among these situations, are there some that are repeated so often that saving even a little time for each repetition would result in substantial savings in the long run? Choose one of these situations and organize its activities. Estimate your long-run time savings. (Examples you might consider include doing dishes, doing laundry, shopping for groceries, organizing class notes, studying for a test, and running errands.)

Exercise 17

a "This is the best that can be done," Mr. Quick tells Mr. Beech when he proposes a product schedule requiring a total conversion time of 170 minutes. Is this true? That is, is 170 minutes the minimum possible conversion time? Justify your answer with an explanation.

b If possible, find another production schedule with a total conversion time of 170 minutes.

Exercise 18

Suppose that Mr. Beech purchases a new ASM which requires the same conversion time as those given in Table 13-1—with two exceptions: conversion between products E and F requires only 20 minutes; conversion between products F and G requires only 20 minutes.

a Modify Figure 13-15 to obtain a graph that illustrates the 20-minute conversion times for this new machine.

b Find a production schedule for Mr. Beech that requires only 20-minute conversion times.

	BATAVIA	CALEDONIA	DUNKIRK	FRANKLINVILLE	GOWANDA	JAMESTOWN	LOCKPORT	NIAGARA FALLS	SALAMANCA	WARSAW
BATAVIA	—	20	75	50	60	90	35	40	65	25
CALEDONIA	20	—	90	60	70	95	50	65	80	30
DUNKIRK	75	90	—	45	20	30	65	60	50	80
FRANKLINVILLE	50	60	45	—	35	40	65	70	20	40
GOWANDA	60	70	20	35	—	30	60	60	30	60
JAMESTOWN	90	95	30	40	30	—	100	90	25	80
LOCKPORT	35	50	65	65	60	100	—	20	100	60
NIAGARA FALLS	40	65	60	70	60	90	20	—	100	70
SALAMANCA	65	80	50	20	30	25	100	100	—	70
WARSAW	25	30	80	40	60	80	60	70	70	—

TABLE 13-2

Exercise 19

Marilyn Kramer is a sales representative for Aurora Lamps. She lives in Lockport, New York, and her sales territory is western New York State. She is planning a sales trip next week to visit furniture stores in 10 cities and towns. She would like to visit two cities per day, leaving Lockport on Monday morning and returning Friday afternoon for appointments there. The last time she visited this group of cities, she unthinkingly scheduled the cities (except for Lockport, of course) in alphabetical order, without considering the distances between them. The total distance she traveled on that trip was 575 miles. "That was awful," Marilyn remembers. This time she wants to consider distances carefully before she schedules her appointments. She has set as a goal a schedule for which no city-to-city trip exceeds 40 miles. Can she do it?

a On the basis of the mileages given in Table 13-2, draw a graph with vertices representing cities and edges representing trips not exceeding 40 miles. If your initial graph has edge crossings (as in Figure 13-14), change the positions of the vertices and redraw the graph (as was done to obtain Figure 13-15) to eliminate edge crossings.

b Using your graph (and imitating the solution process followed in Example 3), find a schedule for Marilyn. In your schedule, list the cities that she should visit each morning and afternoon and give the travel distances to each.

c Evaluation of your solution Unless you spend quite a lot of time experimenting with this problem, you cannot determine whether your schedule is the *best* possible solution. However, you may still be able to identify your schedule as a *good* solution. Answer the following questions and use them as the basis for an argument that you have found a *good* solution to Marilyn's problem.

What is the total distance your schedule will require Marilyn to travel? Is your distance significantly less than the 575 miles required by her earlier schedule?

How well does your schedule meet Marilyn's wish for no trips that exceed 40 miles?

What is the mean (average) trip length that your schedule requires?

Note The Appendix contains partial solutions or hints for Exercises 18*a* and *b* and 19*a*.

DEVELOPING A PROJECT SCHEDULE

The examples involving Mr. Quick and Ms. Kramer illustrate the use of graphs to determine an efficient sequence of activities. Examples 4 and 5 below continue with scheduling problems, but now (1) some activities can be carried out simultaneously, and (2) some activities cannot be started until others have been completed.

Example 4 ▶

A bus pulls in to the station at New Concord. Before it leaves again, certain activities must take place:

Activity	Time required, minutes
A: Some passengers get off.	5
B: Baggage is unloaded.	5
C: Bus undergoes routine service inspection.	20
D: Baggage is loaded.	5
E: New passengers get on.	5

How long will it be before the bus is ready to leave New Concord?

To answer this question, it seems reasonable to suppose that some of the activities can take place simultaneously but that all departing passengers must get off before new passengers get on and that baggage destined for New Concord must be unloaded before new baggage is loaded. We can describe the relationship among the activities using a digraph (Figure 13-17).

In Figure 13-17, each labeled vertex denotes the point when the activity with that label is completed. A vertex Ⓢ has been added to denote the time when the activities start. A directed edge joins one vertex to another if the second activity cannot begin until the first is completed. Each directed edge is labeled with the time required for the activity at its terminus to take place.

All left-to-right paths through the graph describe sequences of activities that must be completed before the bus can leave. The sum of the times along each path gives the total time for that sequence of activities.

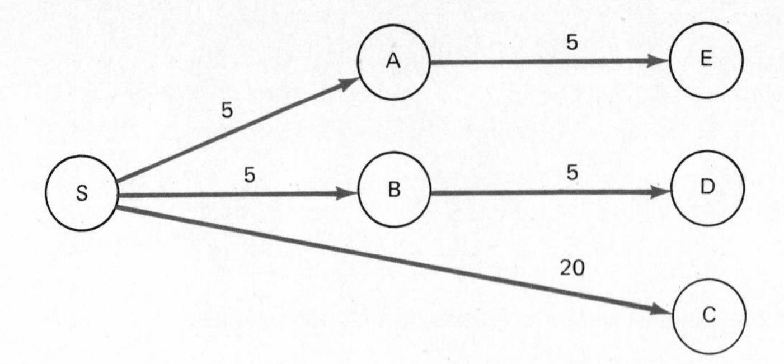

FIGURE 13-17
Activity analysis digraph for bus stop activities

Path S→A→E describes a sequence of activities requiring 10 minutes.
Path S→B→D describes a sequence of activities requiring 10 minutes.
Path S→C describes a sequence of activities requiring 20 minutes.

The length (20 minutes) of the longest path (S→C) gives the answer to our question: It will be 20 minutes before the bus is ready to leave. ◀

A digraph used, as in Example 4, to picture the sequencing of activities is called an *activity-analysis digraph.* The length of the longest left-to-right path in an activity-analysis digraph is the *minimum completion time* for the group of activities. Any longest path is called a *critical path,* and any activities on it are called *critical activities.*

In Example 4 the only critical activity is C. Any delay in the completion of C will delay the bus. Activities A, B, D, and E are not critical; changes in the time that these require need not affect the departure time of the bus. For example, if unloading baggage takes 10 minutes instead of only 5, there is still plenty of time left for loading without causing a delay. Similarly, a faster unloading of baggage—in, say, 3 minutes—has no effect on departure time.

The next example is more complicated, but it illustrates more fully the value of the activity-analysis digraph.

Example 5 ▶

Felicity is the chair of a committee that is organizing this year's community spring arts festival. It is now February 25; the festival is scheduled for the first Saturday in May (May 5). In past years the festival has included a daylong exhibit by local artists and craftspeople at which these people have demonstrated their skills and sold their wares. It has also included a series of performances by music and dance groups.

Felicity and her committee have just had their first meeting and have made the following list of activities:

A Poll community residents to obtain names of artists, craftspeople, and performers in whom they are interested.

ACTIVITY	ESTIMATED TIME REQUIRED, WEEKS	IMMEDIATE PREDECESSORS
A	2	None
B	2	None
C	3	A
D	3	None
E	1	C
F	1	C
G	2	B, D, E, F
H	3	B, D, E, F

TABLE 13-3
Activity information for spring arts festival

B Canvass local businesses for financial support through program advertising.

C Contact potential participants and obtain confirmation from those that will attend.

D Contact local fund-raising organizations to provide food stands.

E Assign locations in the town park to exhibitors and food sellers.

F Schedule entertainers.

G Have the program printed.

H Advertise the festival.

With the committee's help, Felicity has estimated the time each activity will require and has determined which activities must be completed before others can begin. This information is given in Table 13-3.

Felicity must now find different members of the committee to be in charge of the different activities. But first she wants to develop a schedule so that activities will be completed on time. It is clear that some activities must go on simultaneously—Table 13-3 describes 17 weeks of activity, but the committee has only 10 weeks from now (February 25) to the date (May 5) of the festival.

As in Example 4, we can make a digraph to describe the relationships among activities; Figure 13-18 is the result. Each left-to-right path in Figure 13-18, along with its length, is given in Table 13-4.

The critical paths (indicated by dashed lines) in Figure 13-18 are:

$$S \to A \to C \to E \to H$$
$$S \to A \to C \to F \to H$$

The critical activities are A, C, E, F, and H. Because the critical paths have a length of 9 weeks, this is the minimum preparation time for the festival.

Felicity turns to her calendar. From May 5 she counts back 9 weeks, and she makes the time line shown in Figure 13-19. Above her time line, Felicity fills in segments to describe the duration of the various activities. Figure 13-20 shows this information and, in fact, provides a schedule for the activities. Although activities B, D, and G are not critical, Felicity thinks it wise to schedule them to begin at the earliest possible times.

FIGURE 13-18
Right: Activity analysis digraph for spring arts festival

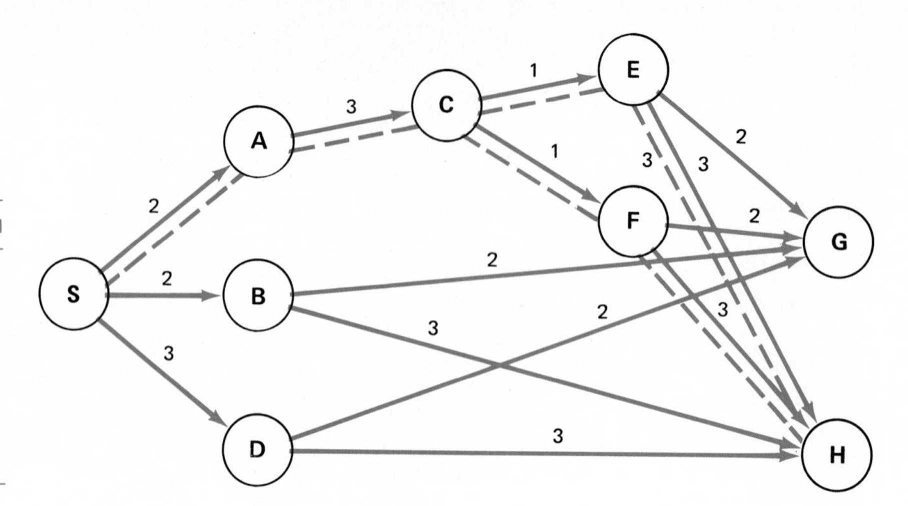

TABLE 13-4
Paths in the digraph

PATH	LENGTH
S→A→C→ E→G	8
S→A→C→F→G	8
S→A→C→E→H	9
S→A→C→F→H	9
S→B→G	4
S→B→H	5
S→D→G	5
S→D→H	6

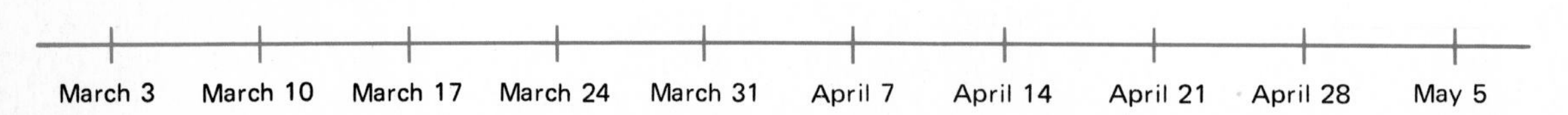

FIGURE 13-19
Arts festival time line

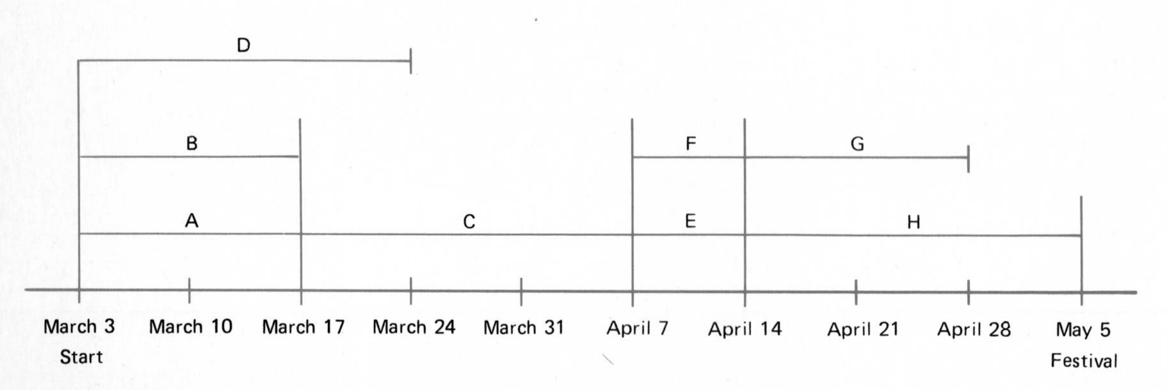

FIGURE 13-20
Arts festival activity schedule

Felicity shows her schedule (Figure 13-20) to the committee, and various members volunteer to take charge of activities A–H. Felicity provides a final reminder that the festival will not be a success unless critical activities are completed on schedule, or sooner. ◀

The task of managing major projects is, of course, not always as simple as this. In many cases the times required for activities can be estimated only roughly, and there is much uncertainty; weather, illness, and mistakes are among the unpredictable events that can cause delays. One sophisticated version of project scheduling techniques is called *program evaluation and review technique* (PERT). It was developed in 1958 in connection with the United States Navy's Polaris program and was credited with shortening by many months the time required to develop the Polaris missile.

The central idea in any use of activity-analysis digraphs is to identify critical activities, since it is these activities that determine when a project will be completed. If a scheduler wants to influence the length of time that a project requires, attention and extra effort should be focused on the critical paths.

Exercise 20

Suppose that the situation in Example 4 is modified in the following ways: (1) baggage may not be unloaded until passengers get off the bus; (2) new passengers may not get on the bus until baggage has been loaded.

a Draw an activity analysis digraph for this modified situation.
b Why is it appropriate to say in this situation that every activity is critical?

Exercise 21

In Example 5, what effect do the following changes have on the arts festival schedule?

a Activity B takes 3 weeks instead of 2 weeks.
b Activity B takes 1 week instead of 2 weeks.
c Activity C takes 4 weeks instead of 3 weeks.
d Activity C takes 2 weeks instead of 3 weeks.

Exercise 22

Consider the data shown in Table 13-5 for the six activities in a project.

a Draw an activity analysis digraph for the project.
b What is the minimum project-completion time?
c What activities are critical?
d Suppose that the project is to be completed by the end of the day on

TABLE 13-5

ACTIVITY	ESTIMATED TIME REQUIRED, DAYS	IMMEDIATE PREDECESSORS
A	2	None
B	3	A
C	4	A
D	6	B, C
E	2	None
F	8	E

August 24. Develop a schedule that indicates when each activity should begin. (Assume that work will begin first thing in the morning on the dates you specify; suppose also that work continues on weekend days.)

e What if activity F can be completed in only 6 days? Does the critical path change? Can the project be completed more quickly? How long may activity F take without affecting the critical path?

f Suppose that you are afraid your estimate for activity A might be incorrect. If A *might* take 3 days, how would this affect your planning?

Exercise 23*

In an effort to increase its circulation, *Lady,* a women's magazine, is considering some major changes in content and format. To gain information about what changes to make, the business manager has asked the circulation department to conduct a survey of subscribers and potential subscribers. Information about the activities involved is given in Table 13-6.

a Construct an activity-analysis digraph for the survey project.

b What is the minimum project-completion time?

c Which activities are critical?

d If the project starts on Monday morning, March 24, and if working days include only Monday through Friday, when will the project be completed?

*Exercises 23 and 26 are adapted from Robert V. Hogg, Ronald H. Randles, Anthony J. Schaeffer, and James C. Hickman, *Finite Mathematics with Applications to the Business and Social Sciences,* Cummings, Menlo Park, Calif., 1974, pp. 140–142.

TABLE 13-6

ACTIVITY	DESCRIPTION	ESTIMATED TIME REQUIRED, WORKING DAYS	IMMEDIATE PREDECESSORS
A	Decide on objectives for the survey	2	None
B	Decide whom to survey	3	A
C	Hire personnel to conduct survey	3	B
D	Develop and test survey questions	5	B
E	Print survey materials	2	D
F	Train survey personnel	3	C, D
G	Conduct survey	5	E, F
H	Analyze survey results and make recommendations	3	G

Exercise 24

a Return to Exercise 8*a* in Chapter 6. Decide what activities should precede others and make an activity-analysis digraph. Develop a time schedule for the dinner project.

b Return to Exercise 16*a* in Chapter 2. Make an activity-analysis digraph and develop a schedule for the basement project.

Exercise 25

Reputable Construction Company has won a bid to construct a small industrial plant on the outskirts of a city. Since the area has not been developed previously, it will be necessary for the company to clear the area, bring in utility lines, drill a well, and install a water tower before construction can begin. The activities that need to be completed before starting construction are listed in Table 13-7, with the precedence relations and times.

a Make an activity-analysis digraph for the project.

b Find the minimal time for completion of the entire project. What is the critical path? If the project is begun on Monday, July 28, when can it be completed? (Assume that five days, Monday through Friday, constitute a work week.)

c If July 28 is the starting date, when must each *critical* activity be started in order to complete the project in the minimal time? For each noncritical activity, give the earliest date on which it may be started.

d If Reputable Construction Company finds that either the well drilling or the sewer installation can be speeded up by as many as 3 days at no increase in cost, which of the two activities should be expedited? Explain.

TABLE 13-7

ACTIVITY	DESCRIPTION	TIME REQUIRED, WEEKS	IMMEDIATE PREDECESSORS
A	Clear site	2	None
B	Survey	2	A
C	Rough grade	1	B
D	Drill well	6	C
E	Install water pump	2	D
F	Install water pipes	4	E
G	Excavate sewer	3	C
H	Install sewer	6	G
I	Excavate for manholes	3	C
J	Set utility poles	2	C
K	Install manholes	4	I
L	Install electrical ducts	4	H, K
M	Bring in power feeder	2	L

TABLE 13-8

ACTIVITY	DESCRIPTION	ESTIMATED TIME REQUIRED, WEEKS	IMMEDIATE PREDECESSORS
A	Select a geographical region for the test	2	None
B	Plan advertising campaign	2	A
C	Contact retailers and request their cooperation	4	A
D	Distribute product and special displays to retailers	2	B, C
E	Advertise test product	3	B, C
F	Observe sales	8	D
G	Plan survey of consumers' satisfaction	2	None
H	Conduct consumer survey	2	E, F, G
I	Analyze results and make recommendations	2	H

Exercise 26

Table 13-8 gives information about activities required to test-market a new product. Construct an activity analysis digraph for the project; analyze your graph and prepare a summary of what can be learned from the graph about scheduling these activities.

TABLE 13-9

ACTIVITY	DESCRIPTION	ESTIMATED TIME REQUIRED, SCHOOL DAYS	IMMEDIATE PREDECESSORS
A	Review scripts	10	None
B	Select script to produce	2	A
C	Obtain production rights and script copies	5	B
D	Choose director	2	C
E	Cast major roles	5	D
F	Choose and schedule production dates	3	D
G	Design sets	5	C
H	Construct sets	20	G
I	Design wardrobes	5	C
J	Create wardrobes	20	I
K	Cast minor roles	5	E
L	Obtain volunteers for backstage tasks	5	K
M	Design programs	2	C
N	Print programs	10	M
O	Plan advertising	3	C
P	Rehearse major roles	20	E
Q	Rehearse full cast	5	K
R	Dress rehearsals	5	L, Q

Exercise 27

Table 13-9 gives information about activities included in the production of a play by a college theater department. Develop a schedule that will complete all these activities in a minimum number of school days. Identify the critical activities; for each noncritical activity examine whether it is "nearly" critical and must occur almost on schedule.

Exercise 28: Invent, Journal

One good test of whether you really understand a process is to try to invent a problem that illustrates it. Take a project that interests you—organizing a youth soccer league, arranging for the renovation of your basement into a family room, planning an election campaign for a friend who's running for office, etc. Select a project that several people will be working on, so that some tasks can be accomplished simultaneously.

a Develop a list of project activities.

b Tell what activities are immediate predecessors of what others.

c Estimate the time that each activity will take.

d Draw an activity-analysis digraph and find the critical activities.

e Develop a project schedule. Review your schedule; does it indeed describe a way to accomplish the project you have in mind?

Exercise 29: Journal

Some problem-solving processes are important because they help us tackle major problems; others are important because they help us in countless minor situations. The process involved in activity-analysis digraphs may be useful to you in the second way. For several days, keep this problem-solving method in mind each time you think about scheduling activities. Think about working backward from a target completion time to develop a schedule; focus on the completion of critical activities.

After several days of thinking about scheduling in this way, write up the results in your journal. What parts of this scheduling process are worth developing into habits for the future?

Exercise 30: Test Yourself

Develop a list of new terms and key ideas presented in this chapter. Supply an example of each item on your list.

Note The Appendix contains complete answers for Exercises 20*b*, 21*a*, 22*d*, and 23*c*. It contains partial solutions or hints for Exercises 25*b*, 26, and 27.

CHAPTER 14

FINDING ROUTES

He who doesn't lose his wits over certain things has no wits to lose.

—Gotthold Ephraim Lessing

THE RABBIT UNDER THE BUSH

A dog has been following the trail left by a rabbit in the snow. The dog comes to the edge of a woods and stops. Before it is a large field containing a large number of clumps of weeds and bushes. The rabbit is nowhere to be seen. However, there is a maze of single tracks between pairs of bushes, as shown in Figure 14-1. Start in the woods and trace out a path that goes along all the tracks. Under which bush will the dog find the rabbit?

GRAPHS AND ROUTES

When you try to trace the rabbit's path in Figure 14-1, you discover its hiding place as bush C. An interesting aspect of this problem is that the order in which you trace out the tracks, starting from the woods, does not matter; you will always end at bush C. This fact is related to another fact: C is the only bush with an odd number of sets of tracks. To see the critical importance of *odd* and *even* in tracing routes, let us return to the study of graphs.

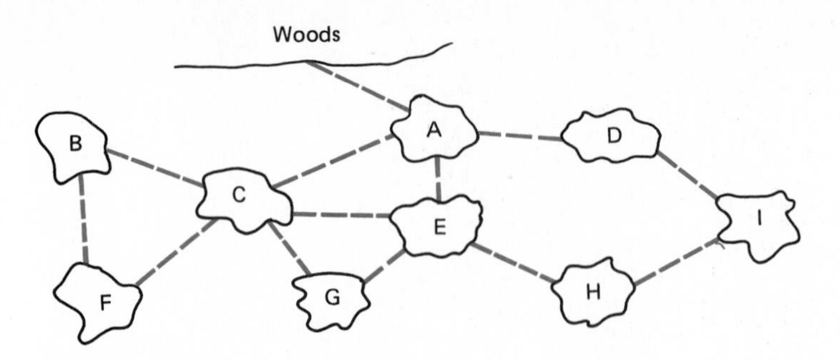

FIGURE 14-1
Bushes and rabbit tracks

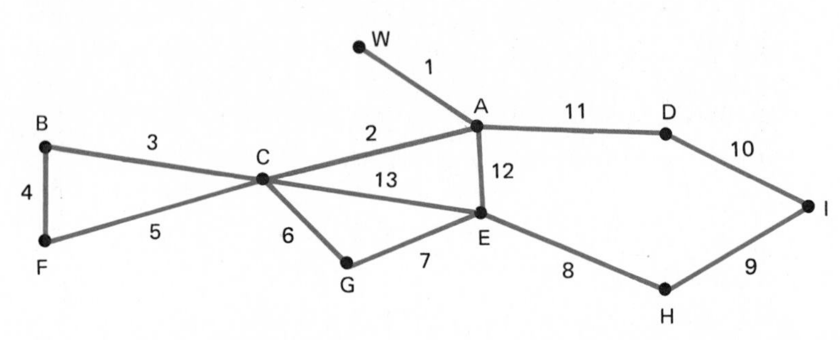

FIGURE 14-2
Graph of bushes and rabbit tracks

The rabbit's pattern of tracks in the snow can be described using a graph in which vertices denote the woods and the bushes and edges denote sets of tracks that join pairs of hiding places; see Figure 14-2. One possible path that the rabbit might have taken is described by the sequence of numbered edges 1, 2, 3, . . . , 13 in Figure 14-2. The following sequence of vertices provides another way of designating the same path:

W, A, C, B, F, C, G, E, H, I, D, A, E, C

A vertex of a graph will be called an *odd* vertex if the number of edges that meet there is odd. Likewise, a vertex at which an even number of vertices meet is called *even.* (In Figure 14-2, vertices W and C—where the rabbit starts and ends—are odd; all other vertices are even.)

To examine more closely the relationship between odd and even and routes in a graph, consider the three similar graphs shown in Figure 14-3. If we can trace through a graph (without lifting the pencil) a route that covers each edge once and returns to the starting place, the route is called an *Euler circuit.** When we can trace all the edges exactly once but end at a vertex different from the one where we started, our route is called an *Euler trail.*

Examination of the graphs in Figures 14-2 and 14-3 suggests the following rules about the possibility of finding Euler circuits and trails.†

---

*This term honors the eighteenth-century Swiss mathematician Leonhard Euler (1707–1783), who was the first to study graphs formally and to solve routing problems.

† Finding Euler circuits and trails also depends on having a connected graph. Obviously, we cannot find an Euler circuit for a graph like △ ⌑ even though all vertices are even.

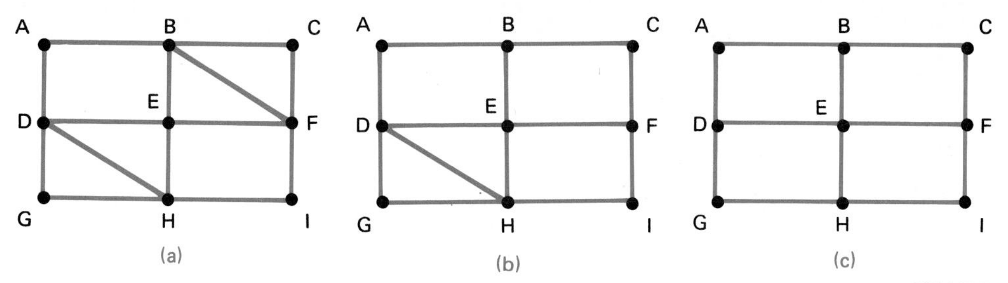

FIGURE 14-3
(a) In this graph, all vertices are *even*. Pick any vertex as a starting point and you can trace, without lifting your pencil, all the edges exactly once and return to the start. (b) In this graph, vertices B and F are *odd*. If you start with one of them, you can trace all the edges exactly once and end at the other. If any other starting-ending points are used, the tracing will either omit or duplicate edges. (c) This graph has *four odd* vertices—B, D, F, and H. No matter where you start, you cannot trace all the edges unless you repeat some of them.

1 If a graph has *all vertices even,* then we may choose any vertex as a starting point and find an Euler circuit.

2 If a graph has *exactly two odd vertices* (and the rest even), then we may choose either odd vertex as a starting point and find an Euler trail. If we do not choose to start at an odd vertex, it is impossible to trace all the edges of the graph without duplicating some of them.

3 If a graph has *more than two odd vertices,* then it is impossible to find either an Euler circuit or an Euler trail.

A consequence of these rules is:

4 If a graph has *any odd vertices,* then a route that traces all the edges of the graph and returns to the start *must* cover some edges more than once.

A critical question in this discussion is: Why are even vertices so important in finding a nonduplicating route? The key to the answer lies in the fact that each time we go through a vertex while tracing a route, we require a *pair* of edges—one edge that serves as a path into the vertex and another that serves as a path out. When a vertex is even, each time we enter a vertex along one edge, there is always another edge available on which to leave.

An odd vertex, on the other hand, requires one more entrance than exit or one more exit than entrance—if we are to cover all the edges. Thus odd vertices can occur only at the beginning or at the end of a tracing, unless we cover some edges more than once. (Reexamine Figures 14-1 and 14-2 to see that this explains the solution to the rabbit problem.)

The exercises below will help you understand these ideas by asking you to apply them to particular problems.

Exercise 1

a Check that each graph in Figure 14-4 (page 230) has all vertices even.

b For each graph find an Euler circuit that starts at vertex A. Indicate your solution in two ways: (1) by numbering the edges on the graph 1, 2, 3, . . . in the order in which you have traced them; (2) by listing the vertices along your circuit in the order in which you pass through them.

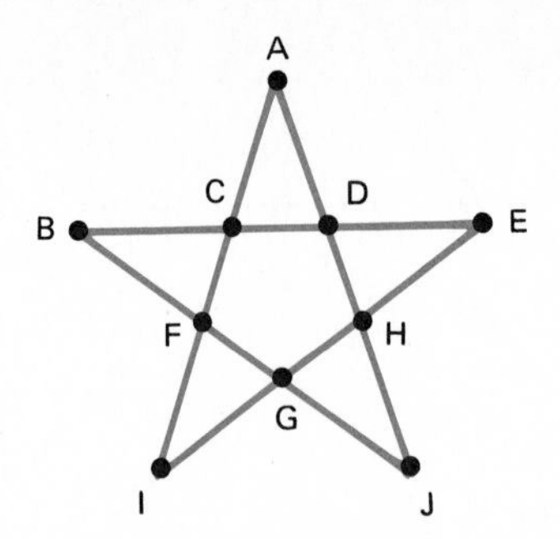

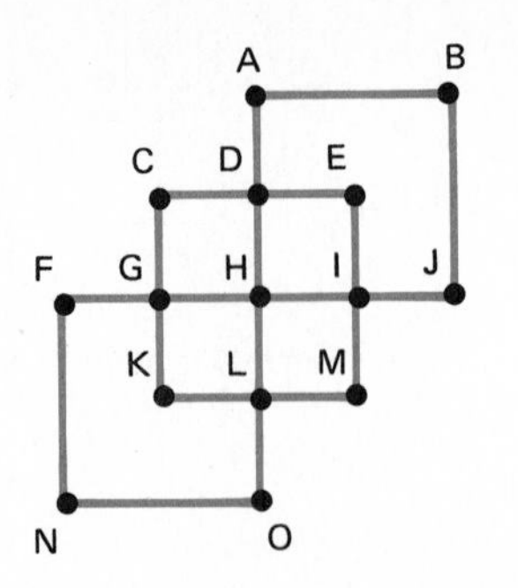

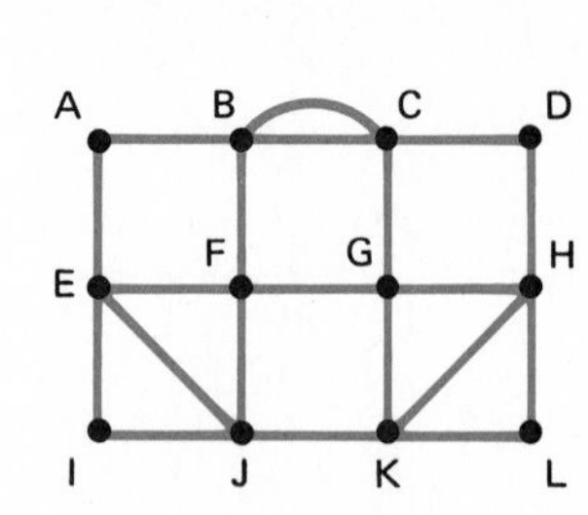

FIGURE 14-4

FIGURE 14-5

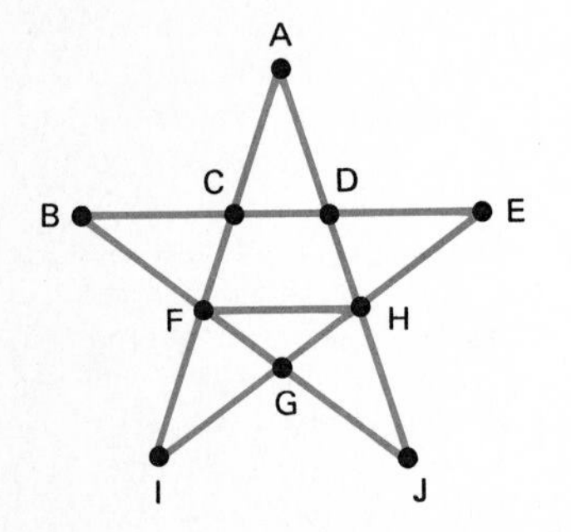

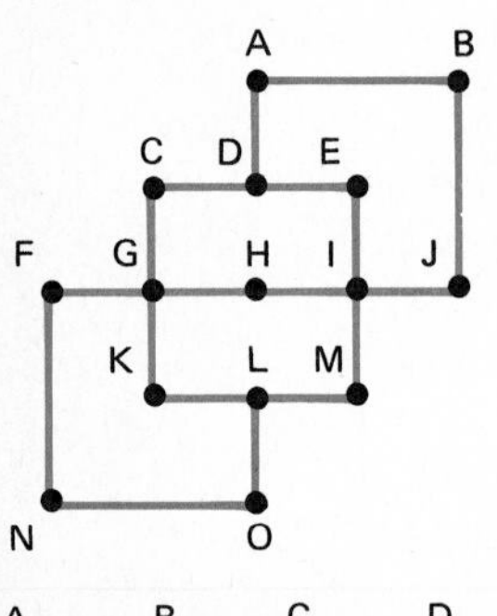

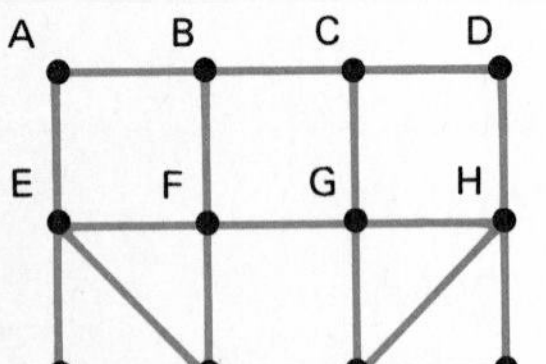

Exercise 2

a Check that each of the graphs in Figure 14-5 has two odd vertices.

b Find an Euler trail for each graph.

c For the first of the graphs in Figure 14-5, experiment to see that an Euler trail cannot be found if you start at vertex A. At what vertex do you get stuck?

Exercise 3

The graph in Figure 14-6 does not have an Euler circuit. However, if two edges are added—in the right places—the resulting graph will have an Euler circuit.

a Between what two pairs of vertices should edges be added?

b Add the two edges and find an Euler circuit for the graph, starting at vertex C. Indicate the circuit by numbering edges 1, 2, . . . , 28.

Exercise 4

An ice cream vendor wants to start at V and travel over each of the streets in the neighborhood indicated in Figure 14-7 but does not want to travel over any part of the route more than once.

a Explain why this cannot be done.

b Trace a route for the vendor that goes along every street at least once and returns to V. Number the street sections in the order traced. Make sure that your route does not make the vendor turn the truck around in the middle of an intersection. How many street sections must be covered more than once? Work to obtain a route that makes this number as small as possible.

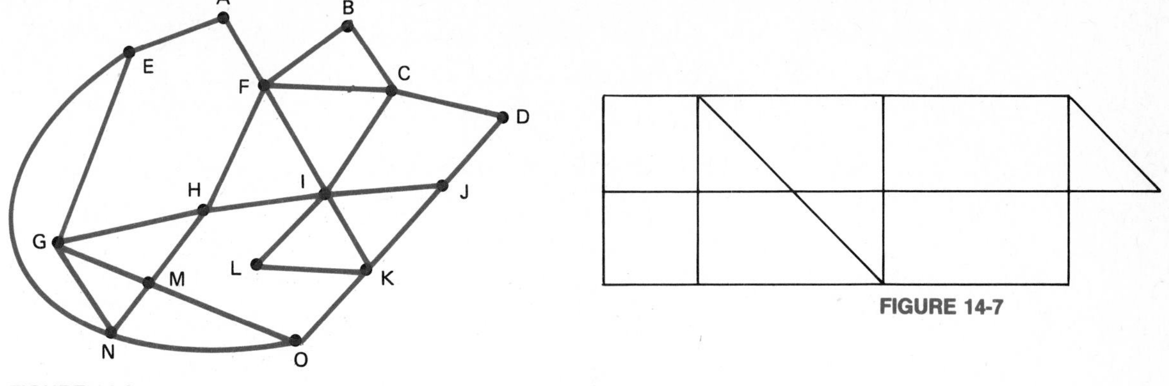

FIGURE 14-6

FIGURE 14-7

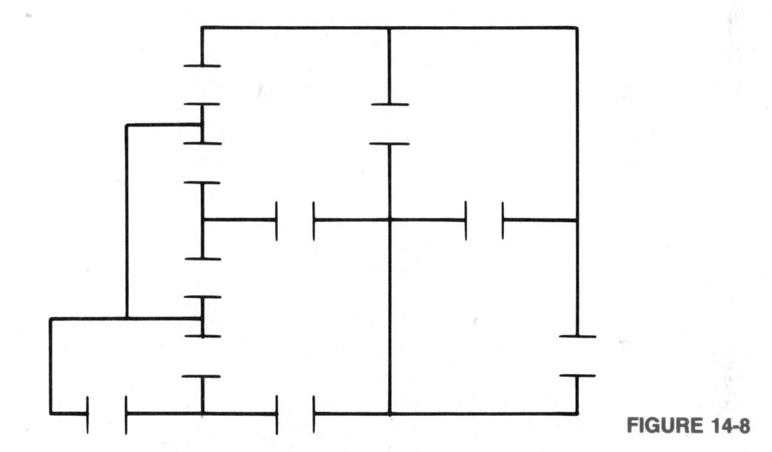

FIGURE 14-8

Exercise 5

Figure 14-8 shows a floor plan for a house. Can a person take a tour of the house and walk through each doorway once and only once? If so, show how it can be done.

Exercise 6

Teddy claims to have made the sketch of an "airplane" shown in Figure 14-9 without lifting his pencil and without tracing over any section more than once. Can he possibly be telling the truth?

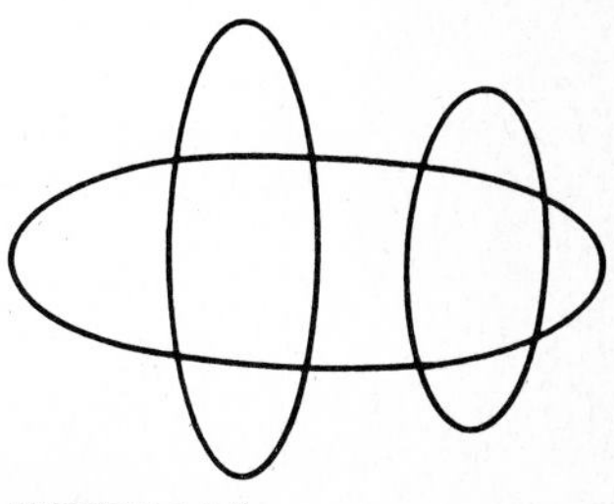

FIGURE 14-9

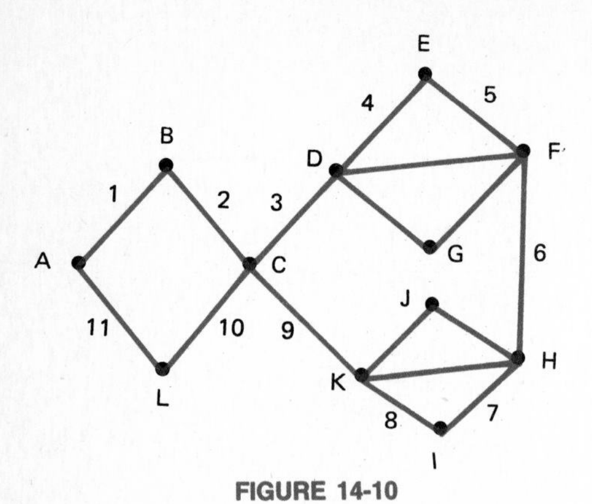

FIGURE 14-10

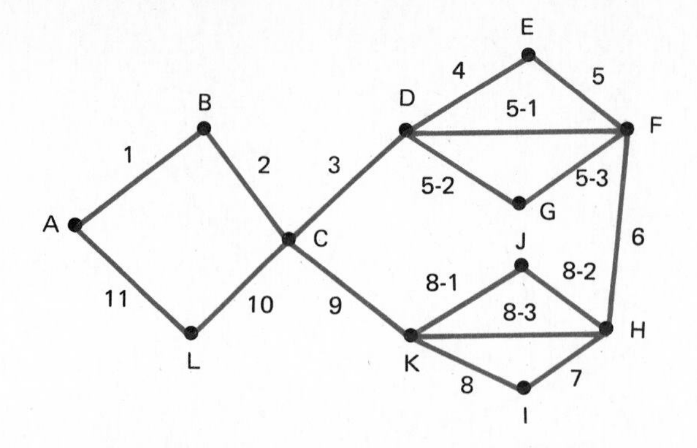

FIGURE 14-11
Edges numbered 5-1, 5-2, and 5-3 designate the first, second, and third edges following edge 5. A similar meaning is given to 8-1, 8-2, and 8-3.

Exercise 7

Suppose (as illustrated in Figure 14-10) that in trying to find an Euler circuit for a graph, my first attempt fails—that is, I return (as shown) to my starting point A without having traced all the edges. Rather than erase and start over, I may use "splicing" to complete my circuit, as follows:

> Select any vertex (such as F) at which some but not all edges have been traced. Use the vertex as a starting point for a new circuit. Trace that circuit and "splice" it into the original.

Figure 14-11 shows the results of "splicing" at vertices F and K.

With this splicing technique, incomplete circuits obtained as a first attempt may be expanded into complete circuits rather than discarded. Use splicing to expand the incomplete circuits in Figure 14-12 into complete Euler circuits.

FIGURE 14-12

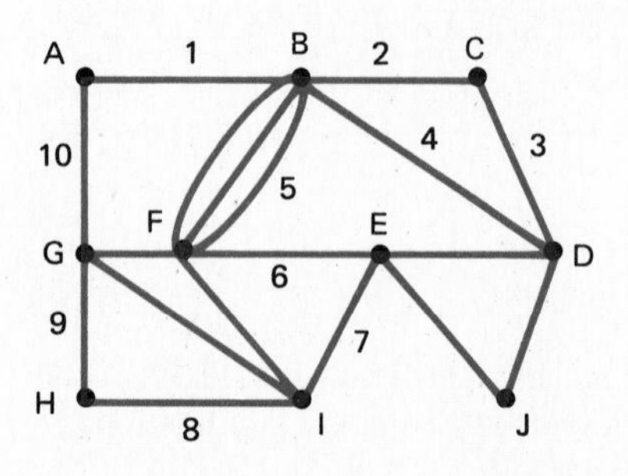

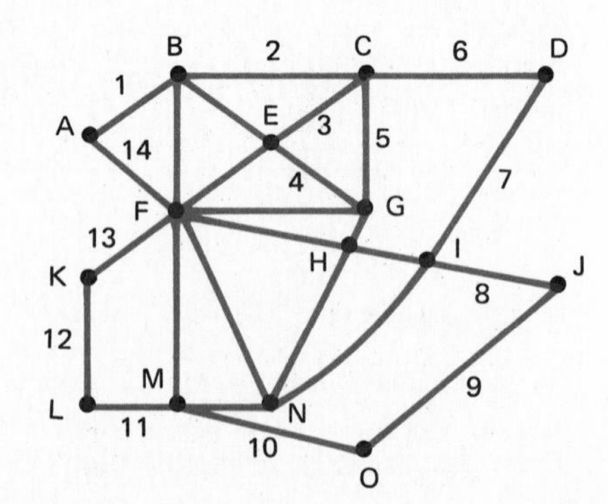

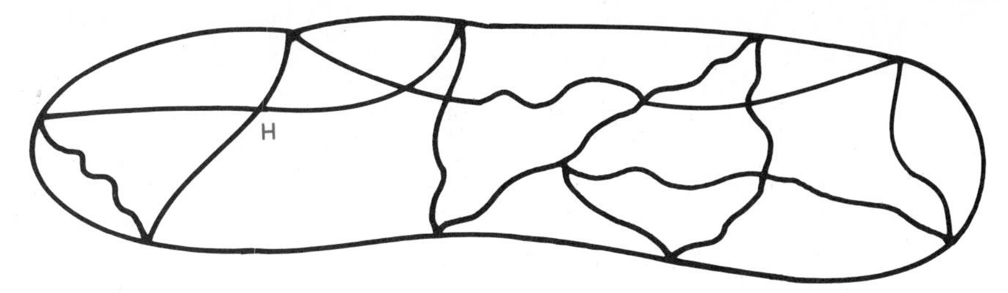

FIGURE 14-13

Exercise 8

A highway inspector for Catawba Township wants to inspect the township roadways for potholes. Using the map shown in Figure 14-13, the inspector is trying to find a route that goes over each roadway once.

a Explain how you know that the problem can be solved.

b Find a route that begins and ends at the highway office, H.

Exercise 9

The following problem provided part of the motivation for the Swiss mathematician Leonhard Euler to formulate the theory of graphs. In Euler's time the Baltic seaport that is now called Kaliningrad was known as Koenigsberg; the Pregel river flowed through the city, and there were seven bridges that joined river islands and the shore, as shown in Figure 14-14. A Sunday afternoon amusement for the citizens of Koenigsberg was to try to complete a walking tour of their city that crossed each bridge exactly once.

a Euler solved the puzzle that the citizens of Koenigsberg amused themselves with by using a graph. Represent the map in Figure 14-14 with a graph in which regions of land are denoted by vertices and bridges by edges.

b What do you learn from your graph about the possibility of completing a walking tour of Koenigsberg that crosses each bridge once?

FIGURE 14-14

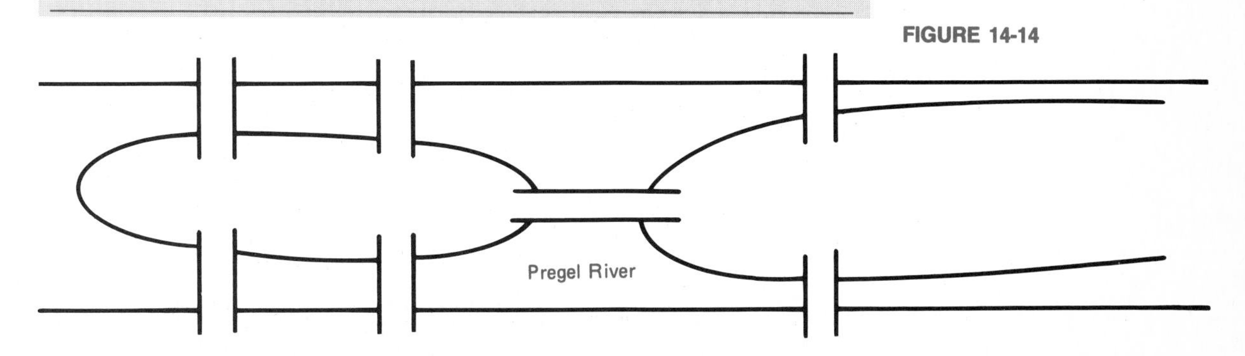

Exercise 10

a The *degree* of a vertex is the number of edges that meet there. For each of the graphs of Figure 14-3, add the degrees of all the vertices to see that the sum, in each case, is even. Explain why this is no coincidence—that is, why the sum of the degrees in *any* graph will be even.

b Miss Geography is showing her fifth-grade class a map of the 13 original colonies of the United States. Together they label each colony with the number of other colonies that share a border with it. Explain why the sum of these labels must be even. (Seek an explanation that will hold for any map—the entire United States, the nations of Europe, the Canadian provinces, or whatever—to which this procedure is applied.)

c Seven philatelists met to trade stamps. Each trade that took place was between only two people. After their meeting, each of the seven was asked how many people he or she had traded with. The answers given were 5, 6, 4, 2, 3, 2, and 3. Explain why at least one of the traders must have been mistaken.

d After a political gathering, each guest was asked how many persons he or she had shaken hands with. Why must the total of all the guests' handshakes be an even number?

Exercise 11: Just for Fun

How may 14 lumps of sugar be distributed among 3 cups of coffee so that there is an odd number of lumps in each cup?

Exercise 12: Invent

Make a list of problematic situations where finding an Euler circuit is applicable.

Exercise 13: Invent

By performing the following experiment, convince yourself that it is possible to find an Euler circuit for any graph (no matter how large) that has all vertices even. Draw a complicated-looking graph with at least 30 vertices and at least 100 edges. Make sure that all vertices are even. Then find an Euler circuit, using the splicing technique (Exercise 7) if necessary.

Exercise 14: Journal

One way that a problem-solving method is useful is *directly:* you can use it to solve a problem of interest to you. A second way is *indirectly:* you can transfer thinking patterns used in one situation to another, somewhat similar situation. (Charles Darwin, for example, adapted types of arguments used by the economist Thomas Malthus to explain facets of his theory of evolution.)

Describe in your journal either a direct or an indirect application of the ideas in this section of the chapter ("Graphs and Routes").

Note The Appendix contains complete answers for Exercises 10*a* and 11. It contains partial solutions or hints for Exercises 1*b*, 2*c*, 3*a*, 4*b*, 5, and 10*b*, *c*, and *d*.

AN APPLICATION OF EULER CIRCUIT TECHNIQUES

> My greatest strength as a consultant is to be ignorant and ask a few questions.
>
> Peter Drucker

Kleen-Up Collection, Inc., is the largest trash collector for Millmont, a rural community in southeastern Pennsylvania. Walter Miller, founder of Kleen-Up, has devoted a lot of energy to recruiting customers; now he has many customers—and he is faced with the problem of finding an efficient route for collections. He has heard of our expertise in planning nonduplicating routes and has come to us for advice. The map in Figure 14-15 shows Mr. Miller's collection region. Roads marked with / / have no customers along them. What help can we be to Mr. Miller?

An examination of Figure 14-15 reveals some intersections at which an odd number of roads meet. If we think of intersections as vertices and roadways (those not marked with / /) as edges, the result is a graph with 10 odd vertices. We thus are led to the unpleasant discovery that the problem has no solution: it is impossible to find a route that covers each section of roadway exactly once.

FIGURE 14-15
Map of Mr. Miller's trash-collection service region

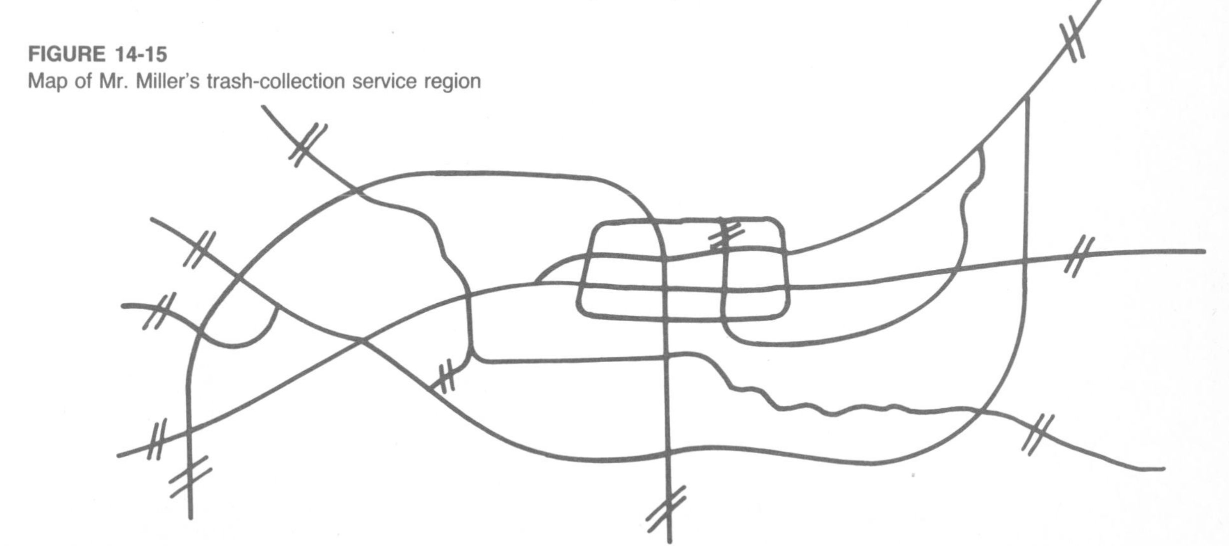

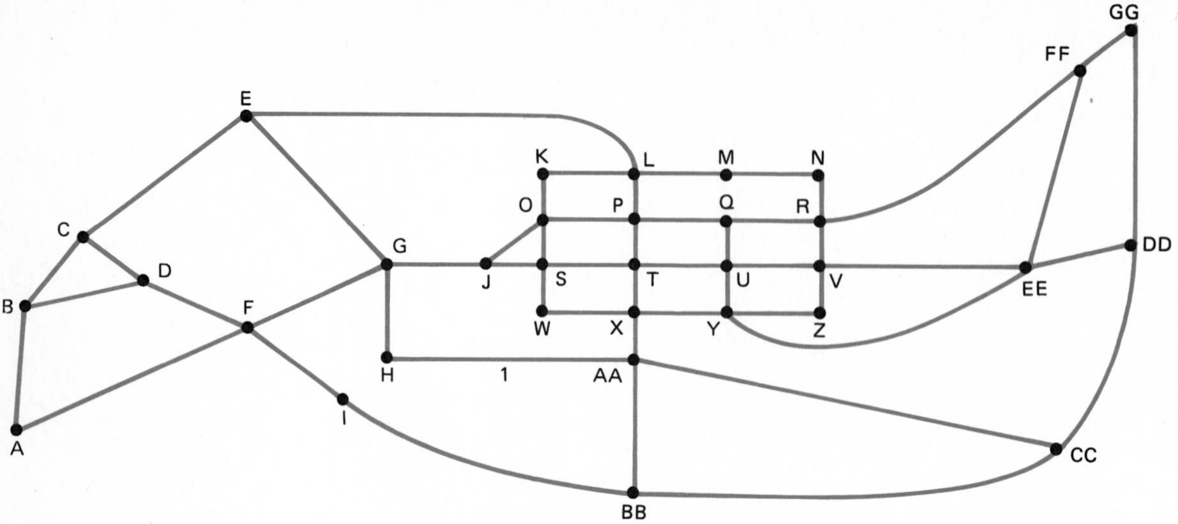

FIGURE 14-16
Graph for Mr. Miller's trash-collection service region

Since Mr. Miller must of course collect his customers' trash, we can't be content to quit work on his problem simply because the ideal solution is not possible. A reasonable next step seems to be to find a route that covers *as few roadways as possible* twice. To study this problem, we redraw Mr. Miller's map as a graph (Figure 14-16). Vertices designate intersections, and edges denote sections of roadway along which Mr. Miller has customers; we label the vertices so that we can refer to them easily. Mr. Miller's home and office are located near the intersection denoted by vertex H; his collection route must start and end there.

Exercise 15 below asks you to experiment with the graph of Figure 14-16 to find a route that starts and ends at H and covers every edge. Make a guess: How many edges must be traveled twice? Do Exercise 15 before you read further.

Exercise 15

Find a circuit for Figure 14-16 that begins and ends at H and goes over each section of roadway at least once. Indicate your circuit on the graph by numbering edges 1, 2, 3, ... How many edges are covered twice in your circuit?

One approach to Mr. Miller's problem is to modify our graph. If we add edges so that all vertices are even, we will be able to find an Euler circuit. We do this (adding dashed lines to Figure 14-16) and obtain Figure 14-17. You may be wondering whether Figure 14-17 has any sensible relationship to our original problem. It does; when we find an Euler circuit for Figure 14-17, we can use it in the following way:

> Each time a dashed edge is traveled in our circuit, it will indicate the need to travel the designated section of roadway a second time. Furthermore, since Figure 14-17 contains eight dashed edges, an Euler circuit for that graph will designate a route for Mr. Miller that travels eight sections of roadway twice.

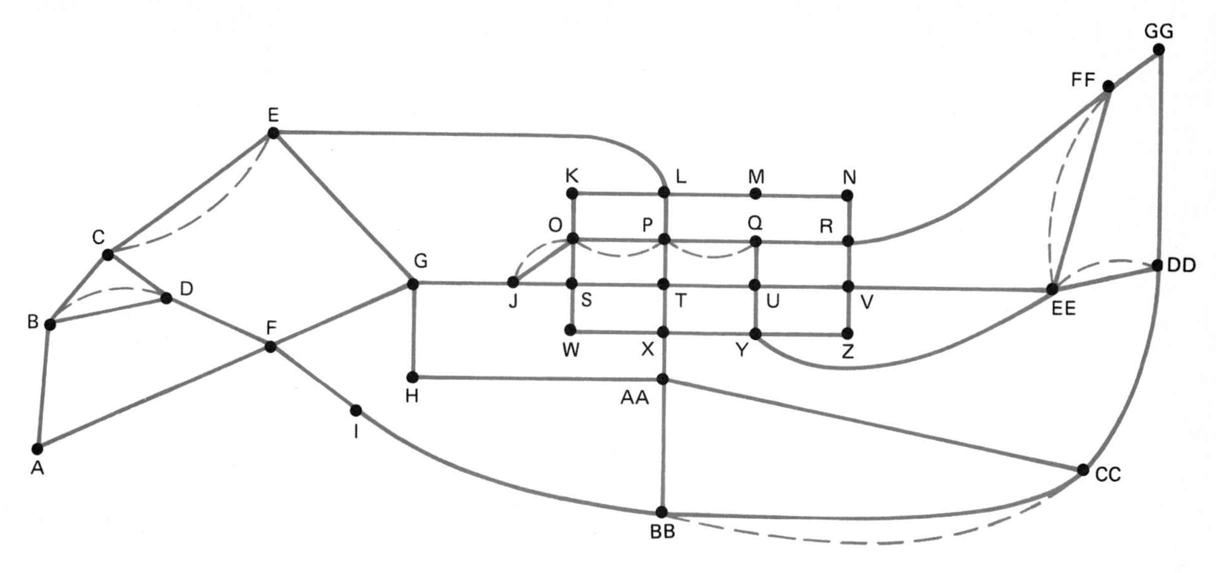

FIGURE 14-17
Graph for Mr. Miller's region with edges added to make all vertices even. (Reproduce this graph for use in Exercises 17, 18, and 19.)

Exercise 16 below asks you to experiment with Figure 14-16 to determine that eight is the smallest number of dashed edges that need to be added to make all vertices even. Do this exercise before you read on.

Exercise 16

Add edges to Figure 14-16 to make all vertices even.

a What is the smallest number of edges that can be added to Figure 14-16 and make all vertices even?

b What is the smallest number of edges that can be added to the graph and make all vertices even *if* each added edge must correspond to a single section of road?

c Do your answers to *a* and *b* support *eight* as the minimum number of sections of roadway that Mr. Miller must travel twice? Discuss.

Figure 14-18 gives one possible Euler circuit for Figure 14-17. Before you look at that solution, find one yourself by completing Exercise 17. Although the large size of the graph makes finding a circuit appear complicated, you will probably find the task quite a bit easier than it looks.

Exercise 17

Find an Euler circuit that begins and ends at vertex H in Figure 14-17.

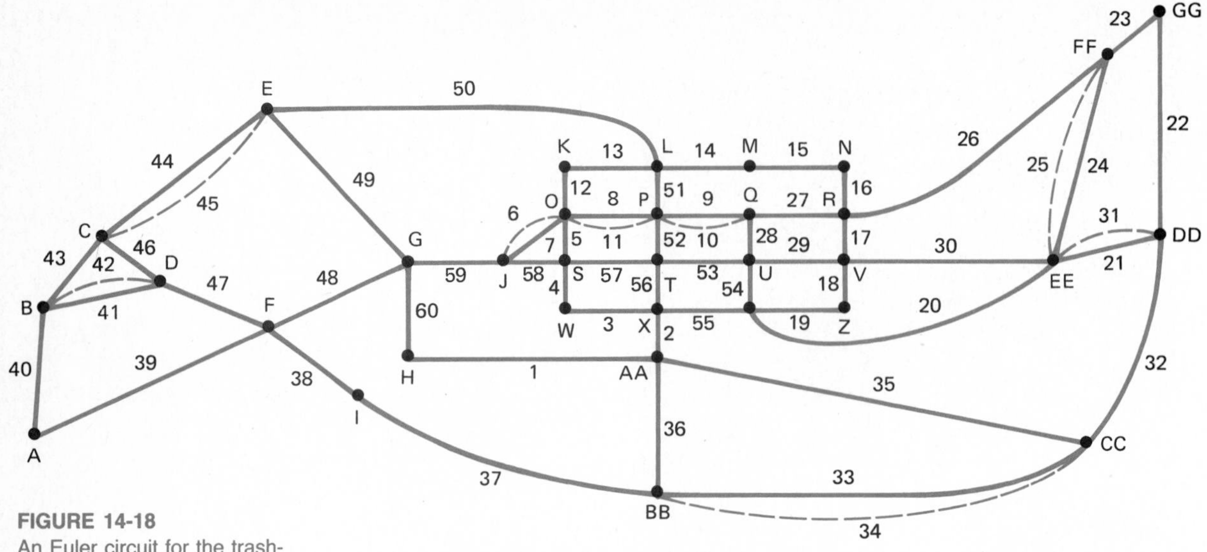

FIGURE 14-18
An Euler circuit for the trash-collection graph

Our final step is to show our circuit—Figure 14-18—to Mr. Miller and to help him transfer the circuit information to his map. He expresses pleasure at our success in finding a route that travels only eight sections of roadway twice; this is much better than he had been able to do on his own. However, as we discuss our solution with him further, we find him not entirely satisfied. Here are his objections:

FIGURE 14-19

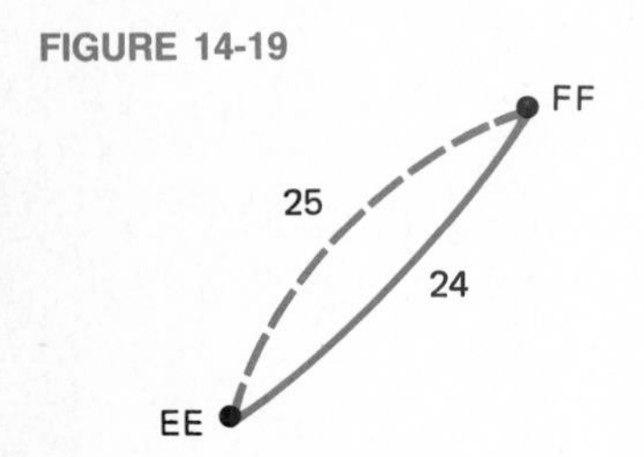

1 His large garbage truck cannot turn around easily unless a large space is provided. Parts of our route, such as the segment shown in Figure 14-19, require him to turn around in an intersection and are therefore unsatisfactory.

2 Since his route begins at 5 A.M. and since his truck is noisy, as a courtesy to his "urban" customers, Mr. Miller prefers to travel as much as possible of the section that includes vertices J, K, . . . , Z toward the end of his route, after people are awake.

Imagine the disappointment we feel as we listen to Mr. Miller's objections. "Why is it that finding solutions always seems to be such an endless process?" we wonder. "Why do new complications keep entering the picture?" Real problems are seldom solved by the simple methods we know how to use. Each step that we take reveals yet another that needs to be taken. Exercises 18 and 19 below ask you to help us take steps to overcome Mr. Miller's objections.

Exercise 18

Find an Euler circuit for Figure 14-17 that overcomes Mr. Miller's first objection.

Exercise 19

Mr. Miller tells us that he is able to collect trash along about 9 or 10 sections of roadway per hour. To avoid a lot of noise in the densely populated portion of his service area before 8 A.M., we seek to find an Euler circuit (and thus a route for Mr. Miller) in which edges joining vertices J, K, . . . , Z are numbered greater than 30. Find a circuit for Figure 14-17 that avoids turnarounds and also meets this goal.

If you cannot completely avoid "urban" Millmont with the first 30 edges of your route, compromise to make your goal more realistic. Try to find a route that has as few as possible edges numbered 1 to 30 in the "urban" region. Consider your solution "good enough" if no more than 5 of your edges numbered 1 to 30 are in "urban" Millmont and if these edges are on the outskirts of town.

Exercise 20

If a graph describes a practical problem—such as a route along which trash must be collected or mail or milk must be delivered—the existence of odd vertices and the impossibility of finding an Euler circuit do not free us from dealing with the problem. Odd vertices merely assure us that a complete circuit will need to go over some edges twice. To discover a route that travels the fewest edges twice, follow these steps: (1) Add edges to the graph to make all vertices even. Use as few new edges as possible. All new edges must parallel existing edges. For example, in modifying

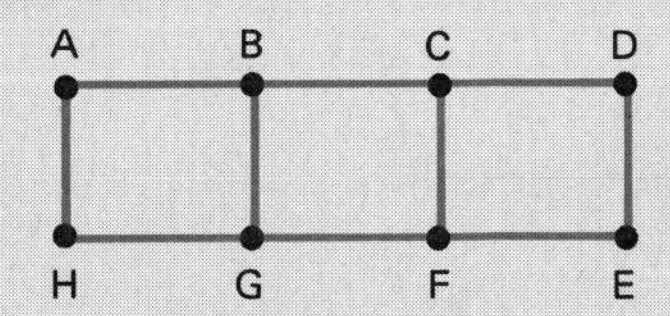

it is not permissible to add an edge from B to F or one from C to G. Instead, edges BG and CF should be added, obtaining

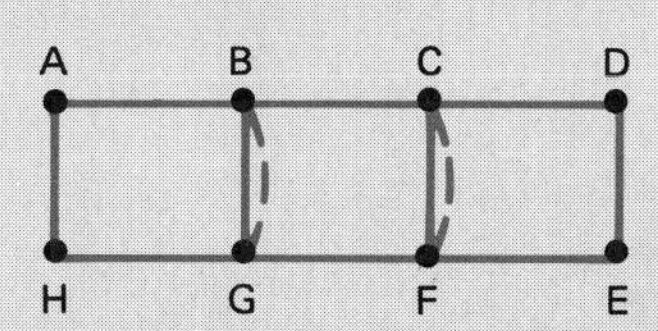

(2) Find an Euler circuit for the new graph obtained from step 1. (3) Interpret each extra edge, when it is traveled in your circuit, found in step 2, as a portion of the original route to be traveled twice.

For each of graphs *a* through *e* in Figure 14-20, find the minimum number of edges that need to be traveled twice in a circuit which starts and ends at A and covers all edges of the graph. Find a circuit that meets these conditions.

(a)

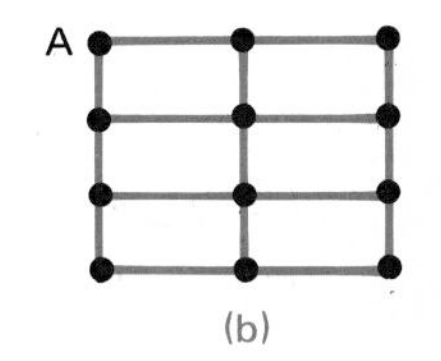

(b)

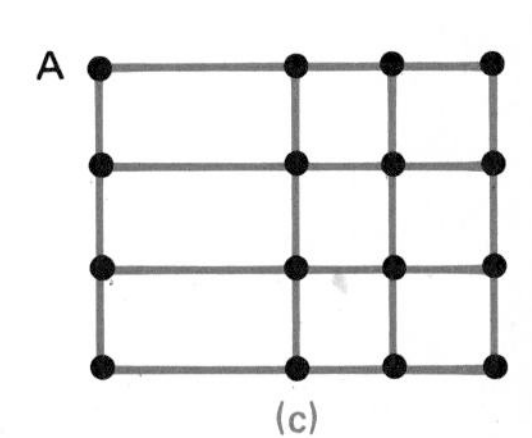

(c)

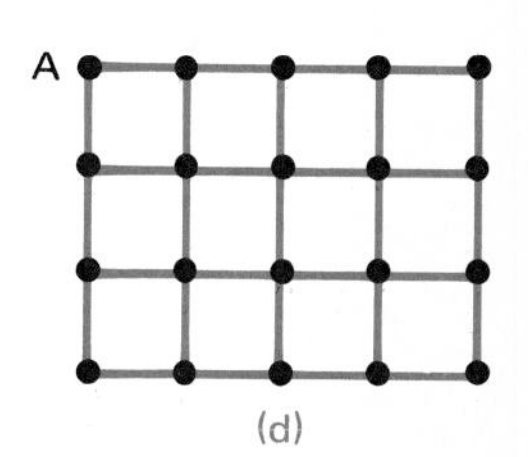

(d)

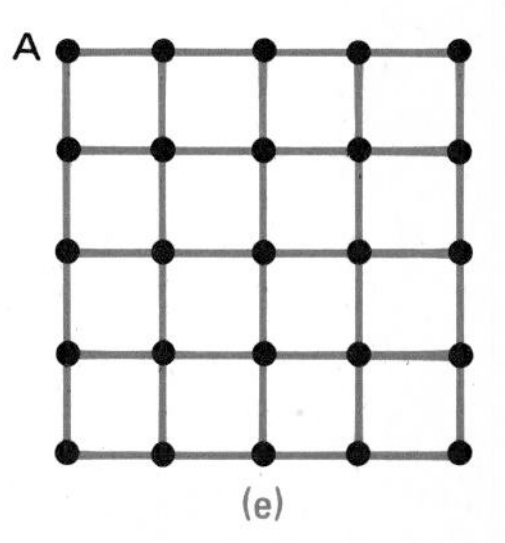

(e)

FIGURE 14-20

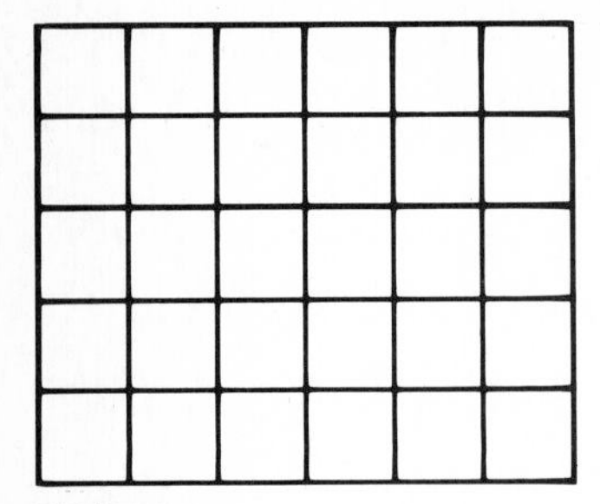

FIGURE 14-21

Exercise 21

a A milk truck makes Monday-Wednesday-Friday deliveries in the 5-block by 6-block section of Summertown shown in Figure 14-21. Find a route for the milk deliveries that has no turnarounds in the middle of an intersection or street and that requires no more than 11 blocks to be traveled twice. Is 11 the minimum number of blocks that must be traveled twice?

b Suppose the milk route in *a* is enlarged to 5 by 7 blocks. Find a route that travels each block, requires no turnarounds, and duplicates as few blocks as possible. What is the minimum number of blocks to be covered twice? Compare this answer with *a;* is the result surprising?

c State a general milk route problem for which *a* and *b* above are simply special cases. If possible, also provide a solution to the general problem.

Exercise 22

A mail carrier delivers mail to a certain neighborhood in Casper City. The carrier would like to travel each section of the route exactly once. This problem is slightly different from the others: the carrier must travel along both sides of many streets. For example, suppose the carrier needed to deliver mail to residents of blocks A, B, C, and D as shown in Figure 14-22. If we let vertices denote intersections and edges denote sidewalks, we obtain Figure 14-23.

A graph is not absolutely necessary for solving this problem; Figure 14-24 shows how a circuit can be displayed on either the block diagram or the graph. It is useful, however, at least to observe the similarity of this problem to graph problems and to recognize that, because every intersection (vertex) has an even number of sidewalks (edges) leading to it, the problem has a solution.

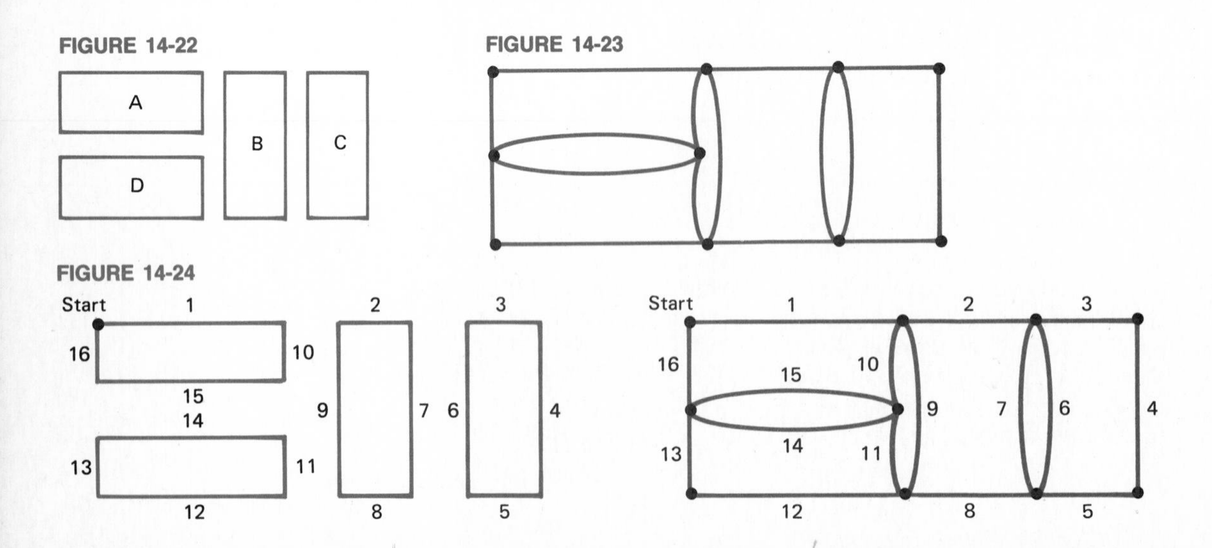

FIGURE 14-22

FIGURE 14-23

FIGURE 14-24

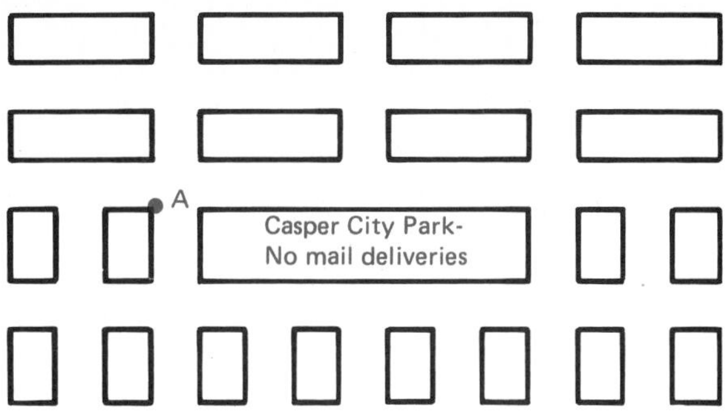

FIGURE 14-25

Now suppose that our Casper City mail carrier is assigned to the neighborhood shown in Figure 14-25. The post office, where the route must begin and end, is located at A. Mail must be delivered on all 4 sides of each block shown. Additional considerations are: (1) Most of Casper City's businesses are located in the 4-block area south of the park. These people want early mail delivery. (2) Often a day's mail is too heavy to carry. The carrier would like a route that returns to intersection A at midday to pick up a second bag of mail.

a Find a route for the carrier that includes the four-block business area in the first half and that returns to intersection A about midway through the deliveries.

b A factor that adds time and distance to a mail route is crossing streets. Examine the route you found in *a* and count the number of times it requires crossing a street. If your route has more than 40 crossings, see if you can make some simple changes that will result in fewer crossings. What is the best you can do?

Exercise 23: Invent

Suppose that you are planning to go door to door campaigning for a political candidate or selling candy to raise money for a club project, or whatever. Draw a map for the region you will be canvassing. Supply realistic details to create a problem similar to those discussed. Solve the problem.

Exercise 24: Journal

Review the numerous steps that were followed (in the text and in Exercises 18 and 19) in the effort to find a satisfactory route for Mr. Miller. Considered as a whole, this problem is quite complicated. Extract from the solution process some guidelines that can help you solve other complicated problems.

Exercise 25: Test Yourself

Develop a list of new terms and key ideas presented in this chapter. Supply an example of each item on your list.

Note The Appendix contains complete answers for Exercises 16*a* and *c*, 20*a*, and 21*a*. It contains partial solutions or hints for Exercises 19, 20*b* and *c*, 21*b*, and 22*b*.

PART FIVE

PROBABILITY ESTIMATION AND INDIVIDUAL DECISION MAKING

PART CONTENTS

CHAPTER 15

MEASURING CERTAINTY

The Meaning of Probability
A Mathematical Framework for Probability
Determining Probability Assignments

CHAPTER 16

EXPECTED VALUE

Gambling Games
Expected Value

CHAPTER 17

DECISION RULES: DECIDING HOW TO DECIDE

Introduction
Deciding What to Study
Which Rule Should I Use?

CHAPTER 18

PROBABILITIES AND ESTIMATION

An Almanac Survey
Evaluating the Survey Results
Confidence Intervals in Statistics

CHAPTER 19

REDUCING THE EFFECTS OF INTERRUPTIONS

Introduction
Simulation of Interruptions
More Interruption Problems—And a Formula
Organizing Tasks to Reduce the Effect of Interruptions

CHAPTER 15

MEASURING CERTAINTY

As for a future life, every man must judge for himself between conflicting vague probabilities.

—Charles Darwin

THE MEANING OF PROBABILITY

Some aspects of the future can be controlled by our decisions. We can decide whether to go to an amusement park on Saturday. We can decide which route to take to the park and whether or not to take a picnic lunch. We cannot decide whether Saturday will be rainy or sunny. We cannot decide whether we will get caught in a traffic jam or whether there will be long lines for the giant roller coaster.

However, even when future events are beyond our control, we may be able to estimate how likely they are to occur. Such numerical estimates are called *probabilities.* For example, we may say:

1 The probability of rain on Saturday is 30 percent.

2 The probability of getting caught in a traffic jam near the shopping mall is 50 percent.

3 The probability that on Saturday we will have to wait in line for the roller coaster at least ½ hour is 90 percent.

Probability has two aspects: First, it is associated with the strength of our belief that something is true or will occur; second, it is associated with the tendency of certain unpredictable events to occur a definite proportion of the time.

When we make a probability assessment about the truth of a statement, we are indicating how close we are to certainty about it. If a group of students, including your friend Dan, are suspected of covering a sidewalk with graffiti, then you might make a statement such as, "I'm 95 percent certain that Dan was not involved. He likes a good prank, but he is very careful never to damage the property of others."

When we make a probability estimate about a future event, the probability value can be interpreted in at least the first and often also the second of these ways:

Interpretation 1 (always) The probability value is an assessment of the degree of certainty that the future event will occur.

Interpretation 2 (sometimes) The probability value is an assessment of the relative frequency with which similar events occur under similar circumstances.

For example, statements 1, 2, and 3 at the beginning of this chapter may be interpreted as follows:

1 On Saturday, the weather is more likely to be dry than to be rainy, but there is still a significant chance of rain.

2 When traveling near the shopping mall on a Saturday, about half of the time one gets caught in a traffic jam.

3 On most Saturdays the lines for the roller coaster require a wait of at least ½ hour.

Probability values are necessarily subjective rather than objective, since they relate to events about which there is incomplete evidence. But despite the subjective nature of probability, there are important objective techniques we can use to develop probability assessments that are reasonable and consistent with the available evidence. Guidelines for use of these techniques are part of the mathematical framework for probability. In the next section of this chapter that framework will be developed.

Exercise 1

In statements *a* to *e* below, p_E designates the probability of an event E. For each statement select the probability value that best matches the given description of E; choose from the following list:

$p_E = 0$	$p_E = .3$	$p_E = .6$	$p_E = .99$
$p_E = .03$	$p_E = .5$	$p_E = .8$	$p_E = 1$

a E is impossible; it can never occur.

b It is impossible to predict whether or not E will occur.

c E is almost certain to occur; it will be a rare occasion if it does not.

d E is somewhat more likely to occur than not to occur.

e E is very unlikely, but it will occur once in a while in a long sequence of repetitions.

Exercise 2

A narrow bridge across the Susquehanna River is frequently the scene of traffic delays. The chance of a delay varies with the time of day. A regular traveler of the bridge has supplied the following statements about the likelihood of delay. How do you interpret each?

a At 1 P.M. the probability of a delay is .1.
b At 4 P.M. the probability of a delay is .7.
c At 6 P.M. the probability of a delay is .5.
d At 9 P.M. the probability of a delay is .01.

Exercise 3

When we are undecided, we may use probability values to help express which way we are "leaning." For example, if I am uncertain whether I will go to Pittsburgh this weekend but am more likely to go than not to go, I might use a probability value of .7 to express this.

Suppose that it is your habit to go out for a late-evening snack if you get your class assignments completed in time—say, by 10 P.M. Tom stops by at 7 P.M. to ask whether you will have your work done in time to go out with him for a snack at 10. Possible replies you might give him are "no," "maybe," "I don't know," "probably," and "yes." Interpret each of these replies as a probability value that could be inserted in the following statement, "I will go with probability ______ ."

Exercise 4

a When a weather forecaster gives the probability of rain as 10 percent, you usually don't bother with an umbrella or raincoat but behave as if there were no chance of rain. On the other hand, when the probability of rain is as high as 80 percent or 90 percent, you are likely to take action. When you hear a forecast of rain, what values are critical to you, causing your behavior to change? In a chart like the one that follows (see page 248), list for each probability value any ways in which your behavior differs from the way you behave when the forecast is for sunshine. (For example, at what probability do you carry an umbrella? Wear old shoes? Close your car windows? Walk instead of bike? Postpone a tennis date?)

Probability of rain, percent	Changes from sunny-day activity
10	___
20	___
30	___
40	___
50	___
60	___
70	___
80	___
90	___

b Invent Think of a situation in which your behavior depends on the probability of a future event. (One possible example is the extent of your weekend socializing, depending on the probability of a test next week in chemistry.) Analyze and describe the way your behavior changes as the probability assessment changes.

Exercise 5

Do you pick up your mail from a mailbox? And, if so, do you ever visit the mailbox when there's no mail, or at least nothing important? How high does the probability of mail need to be before you consider it worthwhile to visit your mailbox?

Exercise 6

Is there a certain route you drive where there is often a traffic jam? On a given day, can you assess the probability that a tie-up will occur along that route? How high does the probability need to be before you will take an alternative but longer route?

Exercise 7

Events such as having your home being destroyed by fire, winning a million-dollar lottery, and suffering ill effects of radiation near a nuclear power plant have very low probability. Yet people insure their homes against fire, buy lottery tickets, and protest against nuclear power plants. Examine—by thinking about it yourself and by talking with other people about their views—the question whether people have any intuitive ("gut-level") understanding of very small probabilities or whether their behavior depends very little on what these probabilities actually are. (Accurate probability estimates require information, say, about type and location of a home and type of lottery; but the following figures are not unreasonable: the probability that a home will be destroyed by fire is less than .001; the probability of winning a million-dollar lottery is less than .000001.)

Exercise 8: Journal

Exercises 3 to 7 were designed to get you to think *probabilistically,* that is, to estimate probabilities for uncertain events in your day-to-day life and, as a result, to gain some control over uncertainty. For 3 days, try to engage in probabilistic thinking about whatever you do or say.

For example, as you walk to the cafeteria, estimate the probability of a long wait in line. If your anthropology instructor will give an unannounced quiz next week, what is the probability of its being on Monday? On Wednesday? On Friday? If a numerical estimate is hard to provide, give a verbal description: is the probability high or low?

Write up the results of these 3 days of probabilistic thinking in your journal. What aspects of this type of thinking are useful enough to develop into habits for the future?

Note The Appendix contains complete answers for Exercises 1 and 2.

A MATHEMATICAL FRAMEWORK FOR PROBABILITY

A basic concept in the formal study of probability is the sample space. In an uncertain situation, an exhaustive (that is, complete) list of outcomes which are mutually exclusive (no two can occur at the same time) is called a *sample space.* A sample space is often denoted by the letter S, possibly with a numerical subscript.

For example, if our concern is Saturday's weather, one possible sample space S is the following list of two events:

S Rain occurs; rain doesn't occur.

A given situation may allow many different sample spaces. If a 1980 penny and a 1981 penny are tossed, three of the possible sample spaces (S_1, S_2, and S_3) are as follows:

S_1 The two coins turn up alike.
The two coins turn up different.

S_2 The two coins both show heads.
The two coins both show tails.
The two coins both show one head and one tail.

S_3 The 1980 penny shows a head; the 1981 penny shows a head.
The 1980 penny shows a head; the 1981 penny shows a tail.
The 1980 penny shows a tail; the 1981 penny shows a head.
The 1980 penny shows a tail; the 1981 penny shows a tail.

The events listed in a sample space are called *elementary events.* Any combination of elementary events is referred to as an *event.* For example, in S_3 (above) there are four elementary events—for which we can use the abbreviations HH, HT, TH, and TT. We can let E_1 designate the event of tossing at least one head. Symbolically, we have

$$E_1 = \{HH, HT, TH\}^*$$

The event E_2, tossing twice as many heads as tails, contains no elementary events. In symbols, we would write

$$E_2 = \{ \quad \}$$

or

$$E_2 = \varnothing\dagger$$

Once a sample space has been established, then probability values may be assigned to the events. For an event E, the symbol p_E will designate the numerical probability value associated with E. The expression p_E is commonly read as "probability of event E" or simply "*p* of E."

Probability assignments must obey three rules. These are discussed below.

Probability Rule 1

For any event E,

$$0 \leq p_E \leq 1$$

*Braces { } are used to enclose lists of objects to indicate that all the members of the list form one *set* of objects. Thus the expression $E_1 = \{HH, HT, TH\}$ means that E_1 is identified with the set of outcomes including HH, HT, and TH.

†∅ is a common mathematical abbreviation for a collection or list with no members; this symbol is often called the *empty set.*

This rule summarizes the following pair of conditions:

1 The least that an event can occur is never (none of the time). Thus $p_E \geq 0$ percent or $p_E \geq 0$.
2 The most that an event can occur is always (all of the time). Thus $p_E \leq 100$ percent or $p_E \leq 1$.

Probability Rule 2

If $E_1, E_2, \ldots, E_k$ are the elementary events in the sample space S, then

$$p_{E_1} + p_{E_2} + \cdots + p_{E_k} = p_S = 1$$

Since the sample space is an exhaustive list of elementary events, one of them is certain to occur. Thus the total probability distributed among the elementary events must be 100 percent, or 1.

Probability rules 1 and 2 establish basic limitations of probability measures. A third probability rule ensures consistency when probabilities are assigned to events that are combinations of elementary events.

Probability Rule 3

If $A_1, A_2, \ldots, A_k$ are any mutually exclusive events, then

$$p_{A_1 \text{ or } A_1 \cdots A_k} = p_{A_1} + p_{A_2} + \cdots + p_{A_k} \qquad (1)$$

To illustrate this rule, consider the toss of a die. Suppose we have

$$S = \{1, 2, 3, 4, 5, 6\}$$

and

$$\begin{aligned} p_1 &= 1/6 \\ p_2 &= 1/6 \\ p_3 &= 1/6 \\ p_4 &= 1/6 \\ p_5 &= 1/6 \\ p_6 &= 1/6 \end{aligned}$$

To calculate the probability that a toss will result in an even number, proceed thus:

$$\begin{aligned} p_{\text{even}} &= p_{2 \text{ or } 4 \text{ or } 6} \\ &= p_2 + p_4 + p_6 \\ &= 1/6 + 1/6 + 1/6 \\ &= 3/6 \end{aligned}$$

For an event E, let not E designate the event that E does not occur. For example, when tossing a die, if E denotes the event that the toss is even, then not E denotes the event that the toss is odd. Since all the elementary events of a sample space S are either in an event E or in its complementary event not E, then

$$1 = p_S = p_{\text{E or not E}}$$

Since E and not E are mutually exclusive events, then

$$p_{\text{E or not E}} = p_E + p_{\text{not E}}$$

Thus

$$p_E + p_{\text{not E}} = 1 \tag{2}$$

Formula 2 is useful in deducing new probability values from given ones. For example, when a die is tossed, using formula 2, from $p_5 = 1/6$ we know that

$$1/6 + p_{\text{not 5}} = 1$$

and thus that

$$p_{\text{not 5}} = 1 - 1/6 = 5/6$$

Probability assignments may be visualized on the scale shown in Figure 15-1. You may be surprised to see, in Figure 15-1, that .5 is labeled as the probability value that designates maximum uncertainty. The following comparison illustrates why this is so.

When we toss a fair coin, hoping for a head, the chances are

$$p_H = .5$$
$$p_T = .5$$

FIGURE 15-1
Scale for probability assignments

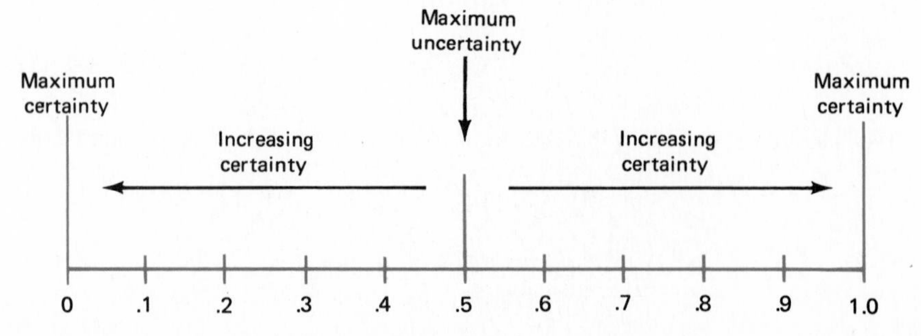

Neither outcome is more likely than the other—a situation of maximum uncertainty. On the other hand, when we toss a balanced die, hoping for a 5, the chances are

$$p_5 = 1/6$$
$$p_{\text{not } 5} = 5/6$$

In this case, one outcome is considerably more likely than the other; thus this situation involves less uncertainty than the coin toss.

You may now be wondering how probability measures fit into the types of measurement scales discussed in Chapter 5. If we treat "likelihood of occurrence" as an attribute associated with events, then probability numbers are measurements of this attribute. Probability measures constitute a scale that is like a ratio scale in that an event with no likelihood of occurrence has measure number 0. However, instead of a fixed unit of measure (such as centimeters or grams) probability measures have a fixed *span* of measure (the interval from 0 to 1) with no uniform unit but, instead, with subdivisions created as needed. Because of their special nature, probability measures are studied as a topic entirely separate from the general types of measurement scales discussed in Chapter 5.

The following exercises offer opportunities to become fluent in the terminology and rules presented so far. The next section of the chapter will investigate how to estimate probabilities for particular events.

Exercise 9

Defend the assertion in the text that maximum uncertainty about whether an event will occur is reflected by a probability assignment of ½. Include in your argument several examples of uncertain situations and probability assignments. Explain why a probability value near 0 indicates more certainty than a probability value near ½.

Exercise 10

a Simpson, Thompson, and Wilson are all vying for promotion to general manager. Give a sample space that includes the possible outcomes for the promotion.

b Marshall, Nelson, and Oberdorfer are Republicans who are running for the local offices of mayor, council member, and district justice, respectively. Each is running against a Democratic opponent. Give a sample space that includes the possible outcomes for the election.

Exercise 11

Suppose that a card is drawn at random from a well-shuffled standard 52-card deck. One possible sample space that describes possible outcomes for this draw is

$$S_1 = \{\text{spade, heart, diamond, club}\}$$

A second possible sample space is

$$S_2 = \{\text{red, black}\}$$

Give two additional sample spaces that also list possible outcomes for this uncertain situation.

Exercise 12

For each of the following uncertain situations give at least two different sample spaces that list all possible outcomes.

a A family is selected at random from a large group of families with three children. What are the sexes of the children?

b An unprepared student takes a five-question true-false quiz and guesses at each answer. How well does he or she do on the quiz?

c A white die and a black die are tossed. What is the outcome of this pair of tosses?

Exercise 13

An American roulette wheel has 38 compartments. Two, labeled 0 and 00, are colored green. The remaining ones, numbered 1 through 36, include 18 red and 18 black. A ball is spun in the direction opposite to the motion of the wheel and bets are made on the number or color where the ball may come to rest. Here, one sample space is the list of all the labels of compartments where the ball may stop: 0, 00, 1, 2, . . . , 36.

Consider a fair wheel in which all the compartments are equally likely; for such a wheel it is reasonable to assign a probability of $1/38$ to the elementary event of a ball landing in any particular compartment.

a Verify that the given probability assignments satisfy probability rules 1 and 2.

b Use formulas 1 and 2, as needed, to answer the following questions:

What is the probability that the ball will come to rest in a green compartment?
What is the probability that it will come to rest in a nongreen compartment?
What is the probability that it will come to rest in a black compartment?

Exercise 14

Peter announces that he has a 90 percent probability of getting an A on tomorrow's history quiz, a 50 percent probability of getting a B, and a 30 percent probability of getting a C or lower. Criticize Peter's assignment of probabilities.

Exercise 15

The course grades given at Uncertain University are A (excellent), B (good), C (satisfactory), and NC (no credit). A student guide to courses estimates, for each course, the probability that a student who registers for it will receive a certain grade. Discuss whether the following probability assignments are valid estimates of what could happen when a student enrolls in a course. (Base your discussion on probability rules 1 and 2.)

a $p_A = .3; p_B = .6; p_C = .1; p_{NC} = .2$
b $p_A = -.2; p_B = .8; p_C = .2, p_{NC} = .2$
c $p_A = 0; p_B = .7; p_C = .2; p_{NC} = .1$
d $p_A = .75; p_B = .05; p_C = .15; p_{NC} = .02.$

Exercise 16

The customer complaint department at X-Mart has never received more than four complaints and requests for exchanges in 1 hour. On the basis of the department's experience, the manager of the department has estimated the following probabilities for complaints per hour:

$$p_{>4} = 0$$
$$p_4 = .05$$
$$p_3 = .20$$
$$p_2 = .30$$
$$p_1 = .20$$

If these probability estimates are accurate, what must be the probability that an hour will be free of complaints?

Exercise 17

A study of air traffic patterns at a metropolitan airport has yielded the following estimates of probabilities for the numbers of aircraft waiting to land:

Number of aircraft waiting	Probability
0	.1
1	.2
2	.4
3	.2
More than 3	.1

a Find the probability that at least three aircraft are waiting to land.
b Find the probability that at most two aircraft are waiting to land.
c Find the probability that more than two aircraft are waiting to land.

Exercise 18

Making consistent probability assignments depends first of all on listing (at least mentally) a sample space—that is, an exhaustive list of mutually exclusive outcomes. When this is not done, inconsistencies can easily result. The following situations illustrate this. In each of them apply probability rules 1 to 3, as needed, to show that the probability assignment is not valid.

a Kerry has been observing a gambling game in which a six-sided die is rolled. It seems obvious to him that the die is unbalanced and, as a result of his observation, Kerry has estimated the following probabilities:

$$p_{1,\ 2,\ \text{or}\ 3} = .6$$
$$p_{1\ \text{or}\ 2} = .4$$
$$p_{3\ \text{or}\ 5} = .1$$

Comment on these values. (Use the information to compute p_3 and p_5.)

b In advance of a public opinion poll assessing the relationship between attitudes toward increased government spending for defense (DF) and attitudes toward increased government spending for space exploration (SE), the pollsters estimated the following probabilities of a favorable attitude on the part of a respondent:

$$p_{\text{DF}} = .6$$
$$p_{\text{SE}} = .3$$
$$p_{\text{DF and SE}} = .4$$

Comment on these probability assessments.

If 50 million people say a foolish thing, it is still a foolish thing.

Anatole France

c After additional training in probability the pollsters in *b* changed their probability estimates to the following:

$$p_{DF} = .3$$
$$p_{SE} = .6$$
$$p_{DF,\ SE,\ or\ both} = .95$$

Comment on these new estimates.

Exercise 19

A pile of 20 identical paper slips each contain a different one of the counting numbers from 1 to 20. Mr. Beck explains to his mathematics class the following procedure for determining whether the class will have a quiz on probability tomorrow. "Place each of the numbered slips in one of these two bags," Mr. Beck said. "Then I will blindfold one of you and have you select a number from one of the bags. If the number selected is even, we'll have a quiz, but if it's odd, no quiz."

How would you put the slips into the bags to have the best chance of drawing an odd number when blindfolded? (See Figure 15-2.)

FIGURE 15-2

Exercise 20

(This problem is based on an example used by Jacques Bernoulli (1655–1705), the author of *Ars Conjectandi,* the first significant book on probability. Bernoulli used the example to illustrate when one may *not* add probabilities.)

Two criminals, Flogg and Mugg, have been sentenced to death. In a surprising act of clemency, the executioner offers them a gamble. Each will get to throw a die. If the two dice turn up equal, both prisoners will go free. Otherwise, the one who gets the smaller number will be executed, and the other will go free.

a What is the probability that Flogg's life will be spared?

b What is the probability that Mugg's life will be spared?

c What is the probability that someone's life will be spared?

d Why can't the answer to *c* be found from the answers to *a* and *b* by adding probabilities?

Exercise 21

People often give answers like "possibly" and "maybe" when asked questions about which they are unsure. Make a list of responses that you commonly use to indicate various degrees of uncertainty. Tell what probability value you associate with each of them.

Compare your probabilities with those of others. Do you use about the same values for the same word? What difficulties arise when people have different values?

Exercise 22

One of my friends is amused by my efforts to "think probabilistically." She will allow that one might harmlessly assign a probability to an event like getting a traffic citation if I park illegally but thinks it improper if I assign a probability to the event that last night's date will telephone me today and calls it absolutely out of bounds for me to try to estimate a probability for the event that I will be alive when my youngest child graduates from college. Her view is that some events are either too personal or too important to have numbers assigned to them. My view is that I want to understand the events and situations of my life, and I am willing to use every tool at my disposal (even probabilities) to accomplish this.

Think about the disagreement that my friend and I have over probabilities. Try to offer some points in support of her view, then some points in support of my view. What is your own view?

Note The Appendix contains complete answers for Exercises 16 and 17*a*. It contains partial solutions or hints for Exercises 10*a* and *b*, 12*a*, 18*a*, 19, and 20.

> The theory of probability is at bottom nothing but common sense reduced to calculus.
>
> Pierre-Simon de Laplace

DETERMINING PROBABILITY ASSIGNMENTS

As has already been noted, the use of probability values to predict future events is subjective: we lack complete evidence about what will happen. Nevertheless, probability values need not be completely arbitrary. We can gather enough objective evidence to make probability assignments that satisfy both mathematical rules and common sense.

Consider, for example, the problem of estimating the probabilities for the several possible outcomes when two coins are tossed. Let us use the sample space

$$S = \{\text{two heads, one of each, two tails}\}$$

TABLE 15-1

ELEMENTARY EVENT	PROBABILITY ASSESSMENTS TOM	DICK	HARRY
E_1: two heads	.3	⅓	¼
E_2: one of each	.3	⅓	½
E_3: two tails	.3	⅓	¼

Suppose that Tom, Dick, and Harry provide the probability assignments shown in Table 15-1. Tom's probability assessments do not satisfy probability rule 2; the sum of the probabilities for the elementary events is unequal to 1. Both Dick's and Harry's probability assessments satisfy probability rules 1 and 2 and thus are mathematically legitimate. But which, if either, is a reasonable description of what actually will occur?

Approach 1

One approach to the problem of deciding on reasonable probability assignments is to illustrate the possibilities with a tree diagram. If both coins are balanced—so that, when each is tossed, a head and a tail are equally likely—then the numbered tree in Figure 15-3 suggests that Harry's assessment is more reasonable.

The tree diagram in Figure 15-3 points out what we may fail to see without it—that the coins may produce event E_2 (one of each) in two different ways, whereas event E_1 or event E_3 can occur in only one way.

The labels on the tree diagram assume symmetry in our coins. If that assumption is valid, then Harry's probability assessments are reasonable and Tom's are not. However, in certain situations we may not be satisfied with the assumption of symmetry; a particular coin may be bent or may show obvious favoritism for one side.

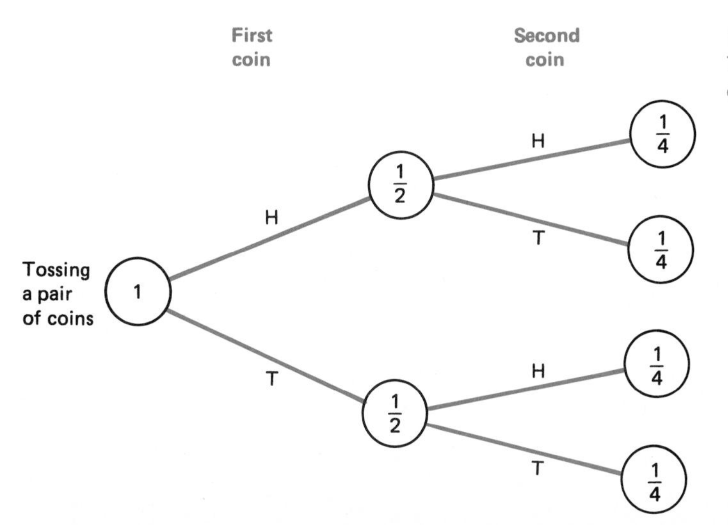

FIGURE 15-3
Tree diagram for two-coin-toss experiment

TABLE 15-2
Experimental results from 40 tosses of two coins

TOSS NUMBER	OUTCOME		
	E_1: TWO HEADS	E_2: ONE OF EACH	E_3: TWO TAILS
1		x	
2	x		
3		x	
4		x	
5			x
6		x	
7		x	
8	x		
9	x		
10	x		
11			x
12		x	
13			x
14	x		
15			x
16		x	
17	x		
18			x
19	x		
20	x		
21		x	
22		x	
23		x	
24	x		
25		x	
26		x	
27		x	
28	x		
29		x	
30			x
31		x	
32		x	
33		x	
34		x	
35		x	
36	x		
37		x	
38			x
39		x	
40	x		
Total	12	21	7

Approach 2

Another approach to the problem of deciding on reasonable probability assignments is experimentation. Seated at my desk, I dug two pennies out of my pocket. I tossed the pair 40 times. Table 15-2 shows what happened.

In summarizing the results of probability experiments a new term, *relative frequency,* is useful. Suppose that an experiment is performed n times. After each trial we observe whether some event E of interest to us has occurred. After completing the n trials, let n_E denote the total number of times E has occurred. The ratio

$$\frac{n_E}{n}$$

is called the *relative frequency* of the occurrence of E in the n trials.

The relative frequency of an event E tells the proportion of times E has occurred in a segment of past experience. If we believe that the future will be like the past, we may use the relative frequency of E as a guide in setting a value for p_E.

On the basis of the results in Table 15-2, we have:

$$\text{Relative frequency of } E_1 = \frac{n_{E_1}}{40} = \frac{12}{40} = .3$$

$$\text{Relative frequency of } E_2 = \frac{n_{E_2}}{40} = \frac{21}{40} = .525$$

$$\text{Relative frequency of } E_3 = \frac{n_{E_3}}{40} = \frac{7}{40} = .175$$

How should we interpret these results? We might use them to make the following probability assessments:

$$p_{E_1} = .3$$

$$p_{E_2} = .525$$

$$p_{E_3} = .175$$

Or we might use them as a basis for arguing that Harry's probability assessments are more reasonable than Dick's. Or we might decide that we'd like more information—and go on to experiment further.

The value of continued experimentation is summarized in the *law of large numbers* (often nicknamed the *law of averages*). One version of this law is: If the number n of trials of an experiment is increased, then the relative frequency n_E/n will tend to stabilize at the true probability for the event E.

To conclude this example, we should make a choice of probability assignments: (1) those given by Harry and illustrated in the tree diagram, *or* (2) those obtained using experimentation and relative frequency. Which would be better?

Most people prefer to use the simple probability assessments given by the tree diagram. However, these describe exactly balanced coins. An actual pair of coins may include a coin that is bent or unbalanced, and in that case experimentation may be the only way to obtain a good probability estimate. But experimentation with one pair of coins cannot be assumed to yield results valid for a second pair.

If important decisions rest on probability measures, then it is important to gain and use as much information as possible in determining the probability values. Experimentation is one way of gaining information. Analysis of data about the past is another way. Life insurance companies, for example, use past data to obtain relative frequencies as probabilities to be used in setting rates. Their procedure resembles the following calculation. Suppose that 1 year ago there were 1 million 20-year-old males living in the United States. Now, 1 year later, 998,000 of these males are still alive. We may think of these data as not unlike an experiment with $n = 1{,}000{,}000$ trials. The event E that a 20-year-old male is alive 1 year later has relative frequency

$$\frac{n_{\mathrm{E}}}{n} = \frac{998{,}000}{1{,}000{,}000} = .998$$

The event not E (a 20-year-old male has died within the year) has relative frequency

$$\frac{n_{\mathrm{not\ E}}}{n} = \frac{2000}{1{,}000{,}000} = .002$$

Summary

Through the example of a two-coin toss, we have considered two ways to gain information useful in making probability estimates for the outcomes of an uncertain situation:

1 Display the possible outcomes using a tree diagram. Use information about symmetry and equal likelihood as a guide in estimating probabilities.

2 Perform repeated trials of an experiment like that whose outcome is under scrutiny. Use the relative frequency of each outcome as a guide in estimating probabilities.

Sometimes actual historical data are treated in the same manner as the repeated trials of an experiment, as in the case of probability estimates by life insurance companies. (Simulation, using random digits—as in Chapter 4—also provides a way to obtain experimental relative frequencies.)

Exercise 23

Shown in Figure 15-4 are four probability assignments for the possible outcomes when a die is tossed. The only way to know which, if any, is correct for a given die is to test by actual tossing. However, some of the assignments are not legitimate, because they do not satisfy the mathematical rules for probabilities. Which are legitimate and which are not—and why?

FIGURE 15-4

Die face	Probability assignment 1	Probability assignment 2	Probability assignment 3	Probability assignment 4
1	$\frac{1}{7}$	$\frac{1}{6}$	$\frac{1}{3}$	$\frac{1}{4}$
 2	$\frac{1}{7}$	$\frac{1}{6}$	$\frac{1}{6}$	$\frac{1}{4}$
 3	$\frac{1}{7}$	$\frac{1}{6}$	0	$\frac{1}{4}$
 4	$\frac{1}{7}$	$\frac{1}{6}$	$\frac{1}{3}$	$\frac{1}{4}$
 5	$\frac{1}{7}$	$\frac{1}{6}$	$\frac{1}{6}$	$\frac{1}{4}$
 6	$\frac{1}{7}$	$\frac{1}{6}$	0	$\frac{1}{4}$

Exercise 24

Possibly with another student to help you, take three coins and toss them 300 times. For each toss, record the number of heads.

a Determine the relative frequency of each of the following outcomes: 0 heads, 1 head, 2 heads, 3 heads.

b On the basis of the results of your tosses, does it seem reasonable to use the following probability assignments for heads?

$$p_0 = 1/4$$

$$p_1 = 1/4$$

$$p_2 = 1/4$$

$$p_3 = 1/4$$

If not, what probability assignments *are* reasonable? Explain.

c Draw a numbered tree diagram to display the outcomes of tossing three fair coins. What values does the tree diagram suggest as reasonable for the probabilities of 0 heads, 1 head, 2 heads, 3 heads? Compare these values with the relative frequencies found in *a;* how well do they agree?

TABLE 15-3

SUM	NUMBER OF TIMES SUM WAS OBTAINED
2	28
3	26
4	30
5	22
6	25
7	28
8	31
9	26
10	25
11	30
12	29

Exercise 25

a After tossing a pair of dice 300 times, Marlene reported the results shown in Table 15-3. Marilyn accused Marlene of making up the values rather than actually carrying out the experiment. Supply reasons to back up Marilyn's accusation: why are Marlene's results unreasonable?

b When two balanced dice are tossed, what are *reasonable* probability values for the 11 possible sums of 2, 3, 4, . . . , 12? Defend your probability values; why are they reasonable?

Exercise 26

a Suppose that families in a certain society have children until they have a boy or until they have five children, whichever comes first. Draw a numbered tree diagram to show the possible outcomes. (Suppose that each birth is equally likely to produce a boy or a girl.)

b Suppose that a family is chosen at random from those families that have completed childbearing in this society. On the basis of your tree diagram, what probabilities would you assign to each of the following events?

E_1: The family has one child.
E_2: The family has two children.
E_3: The family has three children.
E_4: The family has four children.
E_5: The family has five children, including a boy.
E_6: The family has five children, all girls.

What is the probability that the family has at least one boy? What is the probability that the family has at least one girl?

c In Exercise 19 in Chapter 4, completion of Table 4-20 required 100 trials of an experiment, using random digits, that simulated childbearing in the society described in *a,* above. Compare the proportions (relative frequencies) that you obtained in Table 4-20 for events $E_1, \ldots, E_6$ with the probabilities based on your tree diagram. Explain any differences.

d Suppose it is important to have accurate probability values. What steps would you take to check whether your probability estimates are reliable?

Exercise 27

(This exercise is similar to Exercise 23 in Chapter 4.) Use tree diagrams to obtain estimates of the probabilities requested.

a Carol and Fred want four children—two boys and two girls. If they have four children, what is the probability that the four will include two of each sex?

b Dr. Kidney is a surgeon who specializes in a risky operation on critically ill patients. The success rate for this operation is 50 percent—that is, 50 percent of patients who undergo it survive. Two such operations are scheduled for next week. What is the probability that both patients will survive? If three such operations are scheduled for next month, what is the probability that all three patients will survive? If the success rate climbs to 70 percent, what are the answers to these questions?

c (Hard.) A certain lottery sells tickets which, when a covering substance is rubbed away, reveal a hidden letter. The letter may be N, O, or W. Of the tickets, 60 percent have the letter N, 30 percent have O, and 10 percent have W. Winning requires possession of an N, an O, and a W—spelling NOW. What is the probability that you will win with only three cards?

Exercise 28

License plates in the state of New Sylvania include all possible combinations of one letter followed by three digits. Suppose that a car registered in that state is selected at random (so that any car is as likely to be selected as any other). Supply relative frequencies to answer the following questions:

a What is the probability that the license begins with the letter B?
b What is the probability that the license ends with the digit 7?
c What is the probability that the license begins with a vowel?
d What is the probability that the license ends with an odd digit?
e What is the probability that the license has all digits odd?

Exercise 29

A club is selling grapefruit to raise money. A shipment has just arrived, containing the following proportions of types: 40 percent pink seedless, 30 percent white seedless, 20 percent pink with seeds, 10 percent white with seeds. If a grapefruit is selected at random from the shipment, what is the probability that it is

a Seedless? **b** White? **c** Pink and seedless? **d** Pink or seedless?

Exercise 30

The student body at Anaconda College is half male and half female. Ten percent of all male students and 15 percent of all female students major in mathematics and computer science. One student is selected at random from the student body.

a What is the probability that this student is a male majoring in mathematics and computer science?

b What is the probability that this student is a female who is *not* majoring in mathematics and computer science?

c What is the probability that the student is a female or is majoring in mathematics and computer science?

Exercise 31

On a certain slot machine there are three reels, with the digits 0, 1, 2, and 3 and a bumblebee on each reel. When a quarter is inserted and the lever is pulled, each of the three reels spins independently and comes to rest showing one of the four digits or the bumblebee. What is the probability that

a The bumblebee shows on all three reels?

b The bumblebee shows on exactly one reel?

c A bumblebee shows on exactly two reels?

d No bumblebees show?

e A three-digit sequence (either increasing or decreasing) shows?

f One bumblebee shows and the other two reels show odd integers?

Exercise 32

Harry was completely unprepared for an unannounced five-question true-false quiz in economics. He reported that he guessed on every question. Suppose that guessing is interpreted to mean that Harry had probability ½ of answering any question correctly.

Visualize the different possible outcomes of the quiz using a tree diagram. What is the probability that Harry obtained a score of 60 percent or better by guessing?

Exercise 33

Two ambulances are kept in readiness for emergencies. Because the vehicles need maintenance and are sometimes out on calls, each ambulance is able to respond to emergencies about 90 percent of the time. The availability of one ambulance does not depend on or affect the availability of the other.

a Represent the possible combinations of ambulance availability with a tree diagram.

b In the event of a catastrophe, what is the probability that both ambulances will be available?

c What is the probability that neither will be available?

d If an ambulance is needed for an emergency, what is the probability that one (at least) will be available?

Exercise 34

The failure rate for the guided control systems in a certain type of missile is 1 in 1000. Suppose a duplicate but completely independent control system is installed in each missile so that if the first fails, the second can take over. Represent the possible combinations of control-system function and failure with a tree diagram. What is the probability of failure in this system with duplicate backup control? What is the probability that the compound system can be relied upon?

Exercise 35

When the original Pennsylvania lottery was established in 1974, it worked approximately as follows. Each lottery ticket showed a six-digit number. A winning number was drawn each week, and every six-digit number had the same chance of being chosen. (Numbers such as 004391, with leading digits zero, were included in the draw.) A ticket holder won

\$50,000	If the winning number was a perfect match with the ticket number.
\$2000	If the winning number matched the ticket number in the first five or the last five of the six digits. (For example, if the winning number was 004391, then a ticket of the form *x*04391 or of the form 00439*x*—where *x* can be replaced by any digit—would win \$2000.)
\$200	If the winning number matched the ticket number in the first four or the last four digits.

\$40	If the winning number matched the ticket number in the first three or the last three digits.

a Use the multiplication rule to calculate the total number of possible six-digit numbers.

b When a winning ticket is drawn, how many \$50,000 winners can there be? How many \$2000 winners? How many \$200 winners? How many \$40 winners?

c In this lottery, what is the probability of winning the \$50,000 prize? What is the probability of winning \$2000? Of winning \$200? Of winning \$40? Of winning nothing?

Exercise 36

An insurance company sells a term life insurance policy that pays \$10,000 if the insured dies within the next 5 years. Rates depend on the probability that a given person will remain alive year after year. Consider the case of a 21-year-old woman; the probability of her death during one of the next 5 years can be estimated from the mortality table below. The table gives the number of women from a group of 10,000 who were tracked for the years following their twenty-first birthday.

Number who died before twenty-second birthday	5
Number who died before twenty-third birthday	5
Number who died before twenty-fourth birthday	10
Number who died before twenty-fifth birthday	15
Number who died before twenty-sixth birthday	20

a How many of the 10,000 women that were observed lived to age 26?

b Use the information above to estimate probabilities for the following events:

E_1: A 21-year-old woman lives to age 22.
E_2: A 21-year-old woman lives to age 23.
E_3: A 21-year-old woman lives to age 24.
E_4: A 21-year-old woman lives to age 25.
E_5: A 21-year-old woman lives beyond age 25.

Exercise 37

Spring thaws often cause rivers to overflow their banks, but such events are uncertain rather than predictable. Historical records show that there have been major floods along the Susquehanna River three times during the last 100 years (in 1889, in 1936, and in 1972). How would you use this information to estimate a probability value for the statement, "The Susquehanna River will flood next year"? For the statement, "The Susquehanna River will flood during the next 50 years"?

Exercise 38

a A coin is tossed repeatedly. Use a tree diagram to display the possibilities and to obtain estimates of the following probabilities:

The probability that the first toss yields a head.
The probability that a head doesn't turn up until the second toss.
The probability that a head doesn't turn up until the third toss.
The probability that a head doesn't turn up until the fourth toss.
Etc.

Is there a limit to the number of tosses needed to get the first head?

b Get a friend to help. Toss a coin until a head appears; record the number of tosses required; do this 100 times.

What proportion of the 100 trials required just one toss?
Just two tosses?
Just three tosses?
Just four tosses?
Etc.

What is the maximum number of tosses needed?

c Compare the results from *a* and *b*. How well do they agree? Give possible reasons why they differ.

Exercise 39: The Gambler's Fallacy

All eyes were on the green baize dice table, waiting for George's next throw. George's girlfriend, Gertie, put her hand on his shoulder. "I don't think you should chance it, George," she said. "Take the pot and quit. I believe you've used up your luck. You've thrown four 7s in a row, and that means your chance of throwing another 7 is considerably less than it was before." George cashed in his chips and walked away from the table with his winnings. Possibly that was a wise decision; however, it was based on incorrect reasoning. What error did Gertie make?

Exercise 40

a Tom, a gambler, knows that red and black are equally likely to occur on each spin of the roulette wheel. After watching five consecutive reds, he bets heavily on black at the next spin. When asked why, he explains that black is "due" because of the "law of averages." What's wrong with this reasoning?

b After learning that red and black are still equally likely after five reds on the roulette wheel, Tom moves to a card game. The first five cards dealt to him are red. Remembering his lesson about the roulette wheel, he assumes that the next card dealt to him is equally likely to be red or black. Is he right or wrong? Explain.

Exercise 41

You have been enticed into a gambling game. To play the game you toss a fair coin a number of times. You are allowed to choose either 7 or 70 tosses. Suppose that you will win if the relative frequency of heads is between ⅓ and ⅔. In this case, would you choose 7 tosses or 70?

Exercise 42

Which of the following statements is more likely to be true? Discuss the difference between them.

Statement 1: In many tosses of a fair coin the fraction of heads will be close to ½.

Statement 2: In many tosses of a fair coin the number of heads will be close to ½ the number of tosses.

Exercise 43

Sometimes the "law of averages" is interpreted to mean, "If extremes of one type occur, extremes of the opposite type will occur to balance things." Criticize this interpretation. What does the "law of averages" mean? Illustrate with examples.

Exercise 44: Journal

What is a decision you face that depends on uncertain events in the future? Estimate probability values for these uncertain possibilities. Discuss how the probabilities enter into your decision making.

Exercise 45: Testing Probability Values

Consider the following uses of probabilities:

Professor X, your mathematics instructor, says, "On the basis of your work in Algebra 1, I estimate that you have an 80 percent probability of getting an A when you take Algebra 2."

Dr. Y, a surgeon, says, "The operation is risky, but the probability that it will be a complete success is 60 percent."

Mr. Z, a football sportscaster, predicts that Oklahoma will beat Texas with a probability of 90 percent.

When probabilities are used in ways like these, we may wonder how much faith we are to place in the probability values. In each case we suppose that the probability measures the degree of certainty or belief, and that it is based on considerable experience. But how accurate and reliable is the estimate? One method that can be useful in testing probability values is given below.

Suppose that we want to test the sportscaster's estimate of 90 percent as the probability that Oklahoma will beat Texas. We wonder if he really means 90 percent or if he has used that value haphazardly. As a test we set up a thought experiment; the sportscaster is asked to choose one of the following chances to win a marvelous prize:

A: He will win the prize if Oklahoma beats Texas.

B: He will win the prize if a marble drawn at random from a box—containing 90 white and 10 black marbles—is white.

We ask the sportscaster which chance he would prefer, A or B. If he is indifferent between the two ways of getting the prize, then we are convinced that he really thinks that 90 percent is the probability of an Oklahoma victory. If he prefers A to B, then his 90 percent probability estimate is lower than his actual belief. On the other hand, if he prefers B to A, then his 90 percent probability estimate is higher than his actual belief.

a Select a course that one of your good friends is taking this semester and talk with your friend about the course to gain information to estimate the probability that he or she will get an A. Test your probability value by devising a thought experiment similar to the one described above. Keep revising the number of black and white marbles until you obtain a pair of choices between which you are indifferent.

b Invent; Journal Take an uncertain future situation that interests you and develop probability estimates for the various outcomes. Test them using thought experiments.

Exercise 46

Pierre-Simon de Laplace (1749–1827) was a prolific French mathematician whose primary field of work was celestial mechanics. In the quotation on page 258, Laplace was using the word *calculus* to mean "calculation." From your study, do you agree with Laplace that probability is nothing but common sense reduced to calculation? Offer reasons and examples in support of your view.

Exercise 47

Exercise 39 provided an example of a reasoning pattern often called the *gambler's fallacy:* "If I have been losing for a while, now it's time to win." A related fallacy is called the *postmortem fallacy:* "If I have been losing for a while, I must be doing something wrong."

a Explain why these reasoning patterns about chance events need not be correct.

b Sometimes the gambler's fallacy and the postmortem fallacy do not lead to incorrect conclusions. For each fallacy, give an instance in which this is so. How can we recognize cases in which these fallacious reasoning patterns should *not* be applied?

Exercise 48

Probability and statistics have often been used (and occasionally misused) in legal cases.* Many recent cases, in which both sides have used testimony from "mathematical experts," are related to discrimination—in education, in jury selection, in employment, in pension benefits, etc.

*For an interesting introduction to the use of probability and statistics in the courtroom, see Mary Gray, "Statistics and the Law," *Mathematics Magazine,* vol. 56, no. 2, March 1983, pp. 67–81.

Probably the best-known case of misuse of probability—and one that has caused considerable controversy about the responsibility of the courts not to allow experts to overwhelm naive jurors—is *People v. Collins:*

> After an elderly person was mugged in an alley of a Los Angeles suburb, a couple were seen running from the alley. They were described as a black man with a beard and a mustache and a blond woman with her hair in a ponytail. According to witnesses, they drove off in a partly yellow car.
>
> Malcolm and Janet Collins were later arrested. He was black, and although he was clean-shaven when arrested, there was evidence that he had recently had a beard and mustache. She was blond and customarily wore her hair in a ponytail. They had a partly yellow Lincoln.
>
> At their trial the prosecutor called a professor of mathematics as a witness. "Just suppose," the prosecutor said, "that we could assign conservative probabilities to the characteristics noted by the witnesses":
>
> | | |
> |---|---|
> | Partly yellow automobile | 1/10 |
> | Man with mustache | 1/4 |
> | Woman with ponytail | 1/10 |
> | Woman with blond hair | 1/3 |
> | Black man with beard | 1/10 |
> | Interracial couple in car | 1/1000 |
>
> "Now," he continued, "is it not true that if we have independent events we find the probability that they all occur by multiplying the individual probabilities?" The poor professor tried to protest that these probabilities were far from independent but was directed: "Just answer the question." "Yes," he had to admit; "one does multiply." "And the product of these probabilities is 1/12,000,000, is it not?" No arguing with that, and the witness was dismissed, with the dazed defense befuddled by all this high-powered mathematics.
>
> Later in the trial the prosecutor claimed that inasmuch as there were 12 million people in the Los Angeles metropolitan area, and his probability estimates were conservative, the chances of there being another couple like the Collinses in the area must be 1 in 1 billion and thus he had "mathematical proof" of their guilt. Unfortunately the jury convicted on little evidence other than the misuse of statistics.
>
> On Malcolm Collins's appeal, the California Supreme Court reversed the conviction, but only after he had spent considerable time in prison.*

This chapter did not discuss the term *independent.* Two events are said to be *independent* if the occurrence or nonoccurrence of one has no effect on the probability of the other. If I toss a coin and you roll a die, the event of my getting a tail and your getting a four are independent; the occurrence of one event has

---

*Gray, "Statistics and the Law." The excerpt has been edited slightly for the purposes of this text and is used with permission.

no effect on the probability of the other. However, if you and I are each vying for first place in a race, the event of your winning and the event of my winning are not independent; if one occurs, the other is prevented.

a One reason for rejecting the prosecutor's argument is that the characteristics listed are not independent; thus it is incorrect to multiply their probabilities. Examine the list and select a pair for which you can give a good argument that they are not independent.

b Even if we ignore the improper multiplication of probabilities, the prosecutor's argument is weak. Given what you know about probability, explain why you, as a juror, would be skeptical of the information given. What questions would you ask before accepting the prosecutor's argument?

Exercise 49: Test Yourself

Develop a list of new terms and key ideas presented in this chapter. Supply an example of each item on your list.

Review especially the discussion at the beginning of the chapter concerning the two aspects of probability in interpretation 1 and interpretation 2. Examine some of your uses of probability values and note which interpretation or interpretation you have made in each case. In what proportion of your uses is interpretation 2 appropriate? Observe others' uses of probability values; do they use interpretation 2 in cases when it is incorrect to do so? Why?

Note The Appendix contains complete answers for Exercises 23, 28*a,* 31*a* and *d,* 37, and 38*c*. It contains partial solutions or hints for Exercises 24*b,* 30*a,* 34, 35, 41, and 48*b*.

CHAPTER 16

EXPECTED VALUE

Man is not the sum of what he has but the totality of what he does not have, of what he might have.

—Jean Paul Sartre

GAMBLING GAMES

When we toss a fair coin once, we cannot predict, before the toss, whether a head or a tail will turn up. However, if the coin is tossed 100 times, we are able to make a prediction: the number of heads will be about 50.*

This chapter will consider how probabilities can be used to make predictions about groups of events—to see what can be expected to happen "on the average." To begin with, you are asked to play three "gambling games." These will give background experience for the discussion.

To play these games you need three coins—pennies will do—and a pencil and paper. Play each game 40 times. Record the results in a chart like the one shown in Figure 16-1. In the third column use positive numbers for winnings and negative numbers for losses. After 40 plays, total the winnings and losses and also find their mean.

*Advanced probability theory enables us to make the following statement: There is about a 95 percent probability that the number of heads will lie between 40 and 60.

FIGURE 16-1
Form for recording results of gambling games

Game ___ : Summary of results		
Toss number	Outcome of toss (How many heads? tails?)	Amount won (lost)
1		
2		
3		
• • •		
40		
		Total amount won (lost) =
		Mean amount won (lost) =

Stop and predict. Before you actually play any of the games, read the instructions for all three and attempt to predict what will happen overall in 40 plays of each game. Do you expect your total winnings to be positive (a gain) or negative (a loss)? Rank the games in order. Which would you most want to play if given a choice? Which would you prefer least?

Each play of each game requires you to toss your three coins and observe the result:

Game A If the coin toss shows all tails, you win $3. Otherwise you lose $1.

Game B If the coin toss shows 2 heads and 1 tail or 2 tails and 1 head, you win $1. Otherwise you lose $2.

Game C If the coin toss shows more tails than heads, you win $1. Otherwise you lose $1.

Play these games before you read on. After you have played them, take time to *compare.* For which game were your overall winnings (and average winnings) greatest? Least? How do the results of actual play compare with your predictions?

EXPECTED VALUE

The amount won in games A, B, and C in the preceding section depends on the unpredictable outcomes of coin tosses. Despite this, it is possible to predict what will happen "on the average" when these games are played. This prediction is called *expected value*—an idea that will be introduced through examination of another gambling game.

This game involves tossing a balanced die. If the die turns up 3, you win $1; if it turns up any other number, you lose 50 cents. What value do you expect to win?

This question is hard to answer. From the statement of the problem, we can say that you expect either to win $1 or to lose 50 cents, $0.50. Using probabilities, we can say more: There is a ⅙ probability of winning $1, and there is a ⅚ probability of losing $0.50. If the probabilities are treated as *weights,* then we can compute a weighted average amount won. This amount is

$$\frac{1}{6}(\$1) + \frac{5}{6}(-\$0.50) = \frac{\$1}{6} + \frac{(-\$2.5)}{6} = \frac{-\$1.5}{6} = -\$0.25$$

The answer (− $0.25) predicts an average loss of $0.25 per game if you play a large number of games.

For example, if you tossed the die 600 times, you would expect about 100 tosses to turn up 3 and about 500 tosses to turn up 1, 2, 4, 5, or 6. Your total winnings from the tosses of 3 would be $100; your total losses from the other tosses would be $250. Altogether you would have a net loss of $150 with a mean loss of

$$\frac{\$150}{600} \quad \text{or } \$0.25 \text{ per game}$$

The weighted average is an important tool in evaluating uncertain situations where probabilities can be assigned to outcomes. In such cases it has the special name *expected value,* since it measures what we "expect" (on the average) when an uncertain future event takes place. The expected value can be calculated whenever the outcomes listed in a sample space have numerical values. Its precise definition follows.

Suppose that we have an uncertain situation S with possible outcomes having values

$$s_1, s_2, \ldots, s_n$$

Suppose that these outcomes have probabilities

$$p_1, p_2, \ldots, p_n$$

Then the *expected value* of this uncertain situation is given by

$$\text{EV}_\text{S} = p_1 s_1 + p_2 s_2 + \cdots + p_n s_n$$

(The expression EV_S is read "expected value of S.")

The gambling games that you played earlier in the chapter may be analyzed using expected value. To obtain the probability assignments needed for the calculations, let us use the numbered tree diagram in Figure 16-2.

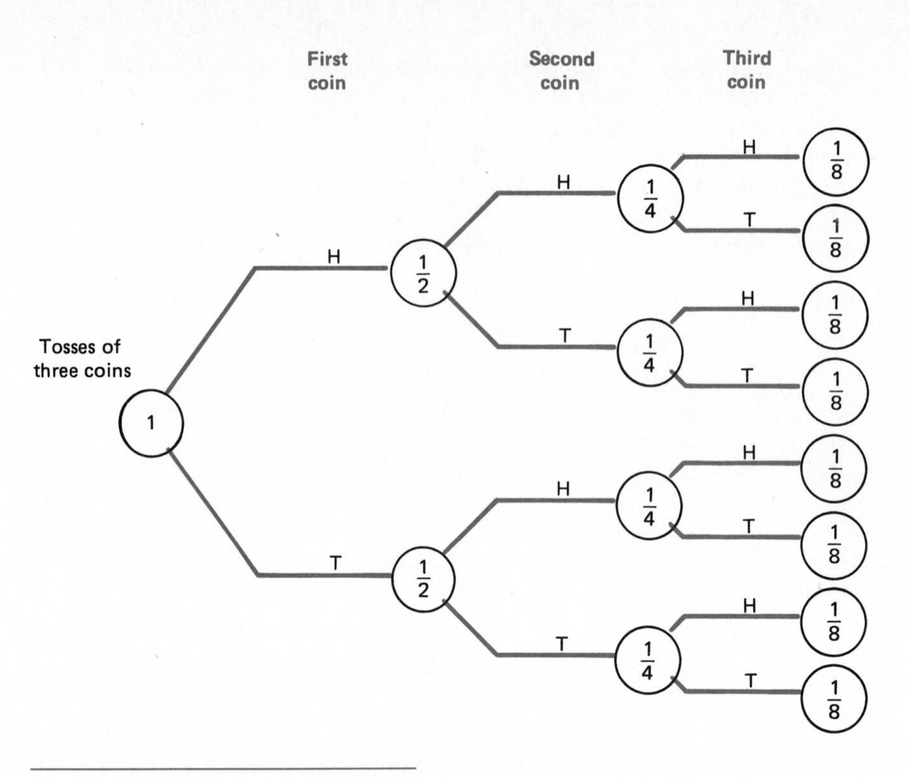

FIGURE 16-2
Numbered tree diagram for three-coin toss

Expected Value for Game A

The possible outcomes of game A have values

$$s_1 = \$3 \quad \text{(for TTT)}$$

$$s_2 = -\$1 \text{ (for any result other than TTT)}$$

Their probabilities are

$$p_1 = \frac{1}{8}$$

$$p_2 = \frac{7}{8}$$

Using these values, we calculate the expected value for game A:

$$\text{EV}_A = \frac{1}{8}(\$3) + \frac{7}{8}(-\$1) = \frac{\$3}{8} + \frac{-\$7}{8} = \frac{-\$4}{8} = -\$0.50$$

Thus, if the probabilities are accurate, when we play game A a large number of times, we can expect to lose (on the average) \$0.50 per game.

Look back. Compare this expected value with the mean value of your winnings when you played game A 40 times. Are the two values close? If they are different, how can the difference be explained?

Expected Value for Game B

The possible outcomes of game B have values

$$s_1 = \$1 \text{ (for a 2-1 combination)}$$
$$s_2 = -\$2 \text{ (for any other toss)}$$

Their probabilities are

$$p_1 = \frac{6}{8}$$
$$p_2 = \frac{2}{8}$$

Thus the expected value for game B is

$$EV_B = \frac{6}{8}(\$1) + \frac{2}{8}(-\$2) = \frac{\$2}{8} = \$0.25$$

That is, repeated play of game B will yield an average gain of $0.25 per game.

Expected Value for Game C

The possible outcomes of game C have values

$$s_1 = \$1 \text{ (for more tails than heads)}$$
$$s_2 = -\$1 \text{ (for any other toss)}$$

Their probabilities are

$$p_1 = \frac{4}{8}$$
$$p_2 = \frac{4}{8}$$

The expected value of game C is

$$EV_C = \frac{4}{8}(\$1) + \frac{4}{8}(-\$1) = \$0$$

In a large number of plays of game C, we would expect to break even.

Look back again. Compare the expected values computed above with the guesses you made beforehand. Did you intuitively guess that game B would have the highest expected value and game A the lowest?

Expected value calculations are not limited to situations for which there are only two outcomes—as Example 1 illustrates.

Example 1 ▶

One of the early state lotteries awarded 145 prizes for each 100,000 tickets sold, as follows:

1	$5000
14	$500
30	$50
100	$5

What was the expected value of a ticket in this lottery?

Solution

If winning tickets are drawn at random, then it is reasonable to assign to each ticket a probability of $^1/_{100{,}000}$ of being drawn. Neglecting the purchase price of a ticket, the possible outcomes have values

$$s_1 = \$5000$$
$$s_2 = \$500$$
$$s_3 = \$50$$
$$s_4 = \$5$$
$$s_5 = \$0$$

Their probabilities are

$$p_1 = \frac{1}{100{,}000} = .00001$$
$$p_2 = \frac{14}{100{,}000} = .00014$$
$$p_3 = \frac{30}{100{,}000} = .00030$$
$$p_4 = \frac{100}{100{,}000} = .00100$$
$$p_5 = \frac{99{,}855}{100{,}000} = .99855$$

The expected value of a ticket is

$$\text{EV} = (.00001)(\$5000) + (.00014)(\$500) + (.00030)(\$50) + (.00100)(\$5) + (.99855)(\$0) = \$0.14$$

A typical lottery ticket costs 50 cents. Generally, as in this case, less than half of the revenue from ticket sales is returned in winnings; the rest is used by the state as a supplement to tax revenues.

From the point of view of the state, lotteries are a good way to raise funds. From the point of view of the consumer, lottery tickets are a poor investment: each ticket bought for 50 cents has an expected return of only 14 cents; and if we take the price of the ticket into account, there is an expected loss of 36 cents on each ticket purchased. ◀

Example 2 ▶

Data collected by a certain insurance company suggest that the probability that a 20-year-old man who is alive today will still be alive 1 year from now is .998. What is the expected value of a 1-year policy for a 20-year-old man if the annual premium is $75 and the death benefit is $25,000? Would such a policy be a good buy?

Solution

In this case, the values of the outcomes are

$s_1 = -\$75$ (net gain to a person who pays the premium and lives)

$s_2 = \$24{,}925$ (net gain to the estate of a person who pays the premium and dies)

Their probabilities are

$$p_1 = .998$$
$$p_2 = .002$$

The expected value of the policy is

$$\text{EV} = (.998)(-\$75) + (.002)(\$24925) = (-\$74.85) + (\$49.85) = -\$25.00$$

The negative sign preceding the expected value tells us that the insurance company pays out (on the average) $25 less in benefits per policy than it takes in as revenue.

An individual policyholder does, on the average, lose money from the purchase of insurance. Most people are not, however, dissuaded from buying insurance for this reason. In exchange for their money they purchase an intangible benefit, "peace of mind." ◀

TABLE 16-1
Utility assignments for weather situations

	UTILITY ASSIGNMENTS	
POSSIBLE OUTCOMES	MARK	MARILYN
Carry an umbrella; it rains.	+5	+3
Carry an umbrella; it doesn't rain.	−2	−3
Don't carry an umbrella; it rains.	−5	−1
Don't carry an umbrella; it doesn't rain.	+5	+3

Example 3 ▶

Let us return to a problem situation first considered in Chapter 5, in which Mark and Marilyn are trying to decide whether to carry umbrellas. The utility values assigned by Mark and Marilyn to the various possible outcomes (Chapter 5, Table 5-2) are given in Table 16-1.

Expected value provides a tool to use in deciding whether to carry an umbrella. The following calculations show how it may be used with Mark's utilities.

Suppose that the probability of rain is 40 percent. Should Mark carry an umbrella? If Mark carries an umbrella, the possible outcomes have the values

$$s_1 = 5 \text{ utils (if it rains)}$$
$$s_2 = -2 \text{ utils (if it doesn't rain)}$$

These have probabilities

$$p_1 = .40$$
$$p_2 = .60$$

The expected value of carrying an umbrella is

$$EV_{yes} = (.4)(5) + (.6)(-2) = 0.8 \text{ utils}$$

If Mark does not carry an umbrella, the possible outcome values are

$$s_1 = -5 \text{ utils (if it rains)}$$
$$s_2 = +5 \text{ utils (if it doesn't rain)}$$

The probabilities again are

$$p_1 = .40$$
$$p_2 = .60$$

The expected value of not carrying an umbrella is

$$EV_{no} = (.4)(-5) + (.6)(5) = 1.0 \text{ utils}$$

Since the expected satisfaction value of not carrying an umbrella is higher than the expected value of carrying one, *on the average* Mark will get the most satisfaction by not carrying an umbrella when the probability of rain is 40 percent. (Exercise 8, below, will ask you to make additional expected value calculations for this situation.) ◀

Example 4 ▶

Donald has two choices of routes to and from work. One route, along back roads, takes a predictable 40 minutes every time. The other route is shorter and more direct but may involve lengthy traffic tie-ups. Specifically, the shorter route takes as long as 1 hour about one-third of the time; it takes about 30 minutes another one-third of the time; the rest of the time it takes only 20 minutes. Which route is the better choice if Donald wants to minimize overall driving time?

Solution

If the given fractions are interpreted as probabilities, the expected length of time for the direct route can be calculated as follows:

$$EV = \frac{1}{3}(60) + \frac{1}{3}(30) + \frac{1}{3}(20) \approx 37 \text{ minutes}$$

Thus, on the average, the direct route will save 3 minutes per trip over the winding route. (Of course the reliability of this figure depends on the reliability of the probability and time estimates.) Overall, Donald will spend less time driving if he takes the direct route.

However, many people would not be willing to use expected or "average" time as a decision criterion in this type of situation. The fact that the direct route may unexpectedly require 1 hour may cause difficulties that outweigh the average time savings—with the result that the longer, more predictable route is preferred. ◀

The exercises below provide practice with expected value calculations and illustrate a variety of situations in which expected value can be useful.

Exercise 1

a Compare the expected values calculated earlier for games A, B, and C with your own average winnings when you played these games. Are there substantial differences? If so, give a plausible explanation for the differences.

b Select one of the games and play it 100 times. Is the average amount won per game closer to the calculated expected value than when you played only 40 times? How does this relate to the "law of averages"?

Exercise 2

Martin has just gotten married and is considering buying a $10,000 life insurance policy. Martin is 21 years old and, on the basis of mortality tables, the probability that he will be alive 1 year from now is .998. One agent quotes a premium of $50 for a 1-year term policy. What is the expected value of this insurance purchase? What should Martin do?

Exercise 3

Middlecreek Construction Company is trying to decide whether to bid on a particular contract valued at $100,000. It estimates that it has an 80 percent chance of receiving the contract but will have to spend $10,000 in consultants' fees to prepare the bid. What is the expected net gain (or loss) to Middlecreek if it submits a bid on this contract?

Exercise 4

In former days—when state-run lotteries tended to avoid gimmicks—the New York State Lottery awarded the following prizes for each million tickets sold:

1 prize of $50,000
9 prizes of $5000
90 prizes of $500
900 prizes of $50

Ignoring the cost of a ticket, what was the expected value of a ticket in this lottery? If a ticket costs 50 cents, what is a ticket purchaser's expected gain or loss per ticket? What percent of the total lottery income did New York State keep?

Exercise 5

a For a family with four children, what is the expected number of girls?

b If an operation with a mortality rate of 50 percent is performed on four patients, what is the expected number of survivors?

Exercise 6

a If a single balanced die is tossed, what is the expected value of the number of spots that turns up?

b If a pair of balanced dice is tossed, what is the expected value of the sum?

Exercise 7

In earlier chapters, many problems were considered in which an expected value could have been computed. Return to the exercises listed below and compute the requested expected values.

a Chapter 12, Exercise 9. Use Myron's past sales record as a basis for obtaining relative frequency estimates of probabilities. What is Myron's expected weekly car sales?

b Chapter 12, Exercise 12*b*. What is the expected claim amount for Paywell Insurance Company?

c Chapter 15, Exercise 16. Find the expected number of complaints per hour at X-Mart.

d Chapter 15, Exercise 17. Find the expected number of aircraft waiting to land.

e Chapter 15, Exercise 35. What was the expected value of a ticket in the Pennsylvania lottery?

f Chapter 15, Exercise 36. Ignoring the premium, what is the expected value of the $10,000 life insurance policy? Suppose that the premium for the 5-year policy is one of the following values: $25, $50, $100. Which of these values is most likely to be the premium? Explain why.

Exercise 8

Return to Example 3 in this chapter.

a If the probability of rain is 50 percent, which decision—to carry an umbrella or not to carry one—offers Mark the highest expected utility?

b For Marilyn, which decision—to carry an umbrella or not to carry one—offers the highest expected utility if the probability of rain is 40 percent? 50 percent? 60 percent? 70 percent?

c Provide your own utility values for the four situations of Table 16-1. How high does the probability of rain need to be before you get more satisfaction from carrying an umbrella than from leaving it at home?

Exercise 9: Journal

a Make a list of types of situations where calculating an expected value could be useful in making a decision. Identify the situations on your list in which you would actually use expected value in making your decision.

b Have you ever made mental estimates of expected values when faced with decisions such as carrying or not carrying an umbrella? Might you use expected value in such situations in the future? Why or why not?

Exercise 10

Jean, a tennis pro, has a contract to give daily lessons at a remote country club. She will waste $10 in transportation costs if she reports for the club to give lessons and it rains. She will make $65 ($75 minus transportation expenses) if she reports to the club and it doesn't rain. Use expected value calculations as a basis for answering the following questions:

a If the probability of rain is 70 percent, what should she do?

b If the probability of rain is 90 percent, what should she do?

Exercise 11

Freezing temperatures endanger the citrus crops in Florida and California. Frank, a farmer, can protect his crop by burning smudge pots at night.

If he burns smudge pots, he will be able to sell his crop for a net profit of $28,000, provided a freeze does develop. If he does nothing, he will either lose $6000 in planting costs if there is a freeze, or he will gain a profit of $16,000 if there is no freeze. If he burns smudge pots and there is no freeze his profit will be only $15,000, since the cost of burning smudge pots is $1000.

a If the weather forecaster says that the probability of a freeze is 40 percent, on the basis of expected value calculations should the farmer use smudge pots?

b If the probability of a freeze is only 10 percent, on the basis of expected value calculations, should the farmer use smudge pots?

c If you were this farmer, would you base your decision on expected value? Explain.

TABLE 16-2

	PROBABILITY	
NUMBER OF ERRORS	OPERATOR A	OPERATOR B
1	0	.1
2	.3	.3
3	.4	.3
4	.3	.2
5	0	.1

TABLE 16-3

PROPOSED PRODUCT	ESTIMATED ANNUAL SALES	PROBABILITY
X	$600,000	.6
	400,000	.3
	200,000	.1
Y	$600,000	.4
	300,000	.4
	200,000	.2
Z	$700,000	.1
	500,000	.3
	400,000	.6

Exercise 12

Ed, the personnel manager of the Process Data Company, must select one of two keypunch operators for an important job. On the basis of observation, he has arrived at the probability estimates shown in Table 16-2 for errors made each hour by the two operators. He wants to determine who is the better keypuncher—that is, who makes fewer errors.

a Compute the expected number of errors per hour for each operator; which operator does this process indicate is better?

b How accurate a measure of error is the expected value in this case? Would you use it to choose an operator? If not, how would you choose?

Exercise 13

The Ramco Manufacturing Company has sufficient capital on hand to invest in new machinery to produce one and only one new product. It is considering three products—X, Y, and Z—and plans to decide which to manufacture on the basis of expected sales. Sales probabilities are estimated in Table 16-3 (page 288). Calculate the expected sales for each product to decide which of them Ramco should market. Discuss whether expected-value calculations provide a reasonable basis for this decision.

TABLE 16-4

	PROBABILITY	
SELLING PRICE	WITH REALTOR	WITHOUT REALTOR
$70,000	.1	0
67,000	.4	.2
65,000	.3	.5
60,000	.2	.3

Exercise 14

Julia and Tony are trying to decide whether to try to sell their house themselves or to list it with a realtor. The realtor has many contacts with prospective buyers and therefore is more likely to be able to sell the house at a higher price; but Julia and Tony must pay the realtor 5 percent of the selling price. Julia and Tony assign probabilities to the various outcomes as shown in Table 16-4. Calculate the expected value of using and not using the realtor. Which offers the higher expected value?

Exercise 15

When asked why she always takes the 15-mile, 25-minute route to and from work instead of the 10-mile, 15-minute route, Paula said that occasionally there is a bad traffic jam on the narrow bridge across the Monongahela and, in such cases, the "short" route can take as long as 40 minutes. When pressed for further information, Paula revealed that the delays occur around 10 percent of the time.

Calculate the expected length of time required for the short route. Discuss whether it is sensible for Paula to avoid that route.

Exercise 16

Operator-assisted long-distance calls cost roughly 1½ times as much as calls dialed directly by the caller. Suppose that your business requires you to make a large number of brief long-distance calls. Which policy is less expensive overall—to dial direct or to ask the operator's assistance in placing a person-to-person call—if the probability that your callee will be in is 25 percent? 50 percent? 75 percent?

TABLE 16-5

VALUE OF PARCEL, DOLLARS	INSURANCE COST, DOLLARS
0-20.00	0.45
20.01-50.00	0.85
50.01-100.00	1.25
100.01-150.00	1.70
150.01-200.00	2.05
200.01-300.00	3.45
300.01-400.00	4.70

Exercise 17

a Hypothetical parcel post insurance rates are given in Table 16-5. If you mail a $40 gift to your parents and if the probability of loss or damage is 2 percent, should you insure? Explain why or why not; include the expected value of an insurance purchase in your explanation.

b Twinkle Lamp Company uses parcel post to ship hundreds of lamps (most of them in the $20-$50 range) to purchasers. If the probability of loss or damage is 2 percent, should the parcels be insured? Explain.

Exercise 18

Steve likes to buy tools on sale, but sometimes he can't resist a bargain and buys something he really doesn't need. Suppose that a tool with a regular price of $100 is selling for 25 percent off. Steve decides to buy it because he feels that there is a 75 percent probability that he'll use it. He reasons, "Since there's a 75 percent chance that I'll use it, then it's worth 75 percent of the full price."

Discuss Steve's reasoning. Does it make sense? Can it be extended to other situations? Does it suggest a rule that you might use as a guide for buying sale items? Develop a possible rule that answers the question, "When is a bargain price really a bargain?" Invent examples to illustrate and test your rule.

Exercise 19

An investor has the choice of investing $25,000 in stocks or in bonds. A broker helps determine a matrix of annual profits based on the possibilities of prosperity and recession during the next year; this is shown in Table 16-6.

a If the likelihood of prosperity is 50 percent, which type of investment leads to a higher expected profit?

b What if the likelihood of prosperity is 20 percent? 30 percent? 40 percent? Greater than 50 percent?

TABLE 16-6

PROSPERITY	RECESSION
$10,000	$ 500
5,000	3500

TABLE 16-7

NUMBER OF COPIES SOLD	PERCENT OF DAYS WITH THESE SALES
50,000	10
45,000	10
40,000	40
35,000	30
30,000	10

TABLE 16-8
Net profits in dollars from sales of *Daily News*

	NUMBER OF COPIES IN DEMAND				
NUMBER OF COPIES ORDERED	50,000	45,000	40,000	35,000	30,000
50,000	5000	3750	2500	1250	0
45,000	4500	4500	3250*	2000	750
40,000	4000	4000	4000	2750	1500
35,000	3500	3500	3500	3500	2250
30,000	3000	3000	3000	3000	3000

**Note:* The following calculation illustrates how the values in this table were obtained. Suppose that 45,000 copies of the *Daily News* were ordered but only 40,000 were sold. Newswatch, Inc., will make a profit of $0.10 × 40,000 = $4000 on the copies sold; however, it will lose $0.15 × 5000 = $750 on the unsold copies. The net profit in this case is $4000 − $750 = $3250. Other values in the table are obtained in a similar fashion.

Exercise 20

Newswatch, Inc., buys copies of the *Daily News* from the publisher for 15 cents and sells them for 25 cents at its numerous city newsstands. Daily sales range from 30,000 to 50,000 copies, and each unsold copy results in a 15-cent loss. Roger Lewis has just replaced Claude Reynolds as Newswatch manager and is considering making a change in the standing order of 50,000 copies daily. Records kept over the past year have revealed the figures shown in Table 16-7. Roger decides to compute the profits for various combinations of supply and demand and formulates the matrix shown in Table 16-8.

Compute the expected profit for each order size, using the percents of days with certain sales as probabilities. What size order should Roger place to maximize the expected net profit for Newswatch?

Exercise 21

Suppose that you are a student at Chester University with only $6 in your pocket. You figure that you need $25 to take a date to a party next weekend. Someone offers to bet you any amount on the outcome of the Pitt-Chester football game. You would bet on Chester; the probability that Chester will win is ⅙. Analyze this decision situation. What would you do and why?

Exercise 22: Invent; Journal

a One situation in which an expected-value calculation might be of interest to you is predicting a grade on a test or in a course.

For example, if Kathy can estimate possible scores and their probabilities for a forthcoming mathematics test as follows,

90	.4
80	.4
70	.2

then her expected score can be calculated as

$$EV = (.4)(90) + (.4)(80) + (.2)(70) = 82$$

Furthermore, if Kathy can estimate probabilities for final grades as follows,

4 (A)	.5
3 (B)	.4
2 (C)	.1
1 (D)	0

then her expected final grade is

$$EV = (.5)(4) + (.4)(3) + (.1)(2) + (0)(1) = 3.4$$

Carefully consider these expected-value calculations. What meaning can you attach to each of the calculated values?

b Consider the courses you are enrolled in and the tests that are coming. Estimate your probability of receiving various grades and compute an expected grade.

After your tests and courses are completed, return to these calculations and compare. How close was your expected grade to your actual grade?

c A variety of situations have been considered in which expected value may be used. Examine your own personal situations, looking for applications. For at least two of the situations that you find, make probability estimates and calculate an expected value. Assess whether the calculation provides useful information.

Exercise 23: Journal—Thinking Probabilistically

My cousin George was all set to buy a Honda. He had read all sorts of consumer reports and was impressed with the economy, durability, and service record of the Honda. But the Friday night before the Saturday on which he was to close the deal, George was at a party and talking about his planned purchase when his friend Barry volunteered the following information. "My brother-in-law just got rid of his Honda. It was a real lemon, always in the repair shop." Barry's story turned George around. The next day he went shopping and visited every local showroom but Honda's.

a Discuss whether George's change of heart was justified.

b Suppose that among George's collection of information about cars, there are these facts: fewer than 1 percent of Hondas require major repairs during the first 60,000 miles of operation, and no other comparably priced (or less expensive) make of car comes close to matching that record. In this case, argue that his change of heart was unjustified. Invent a plausible reason why George changed his mind despite the facts of the matter.

c Think of at least one specific instance in your own experience in which knowledge of one or more particular examples distorted your perception of a probability (just as George's knowledge of Barry's brother-in-law's lemon seems to have distorted his view of the Honda).

d Think of a specific instance in your own experience in which hope or fear caused you to behave as if a very small probability were larger than its actual value.

e Consider a particular event which you repeat again and again and for which the length of time it takes is out of your control. The event might be commuting (Will there be a traffic jam?), eating lunch in the cafeteria (How long a line will there be?), or doing laundry (How long will I have to wait for a dryer?). For a number of repetitions of the event, keep track of how long it takes; try for at least 20 repetitions, if possible. Use these times as a basis for estimating probabilities that the event takes various lengths of times and compute the mean (or expected) length of time. Compare the data you have with the *feelings* you have about the time that the repeated event takes. (For example, does the unpleasantness of a long wait make it seem to have occurred more often that it actually did occur?)

Monitor carefully your probability estimates and how they compare both with your feelings about events and with actual relative frequencies of occurrence. Strive to have all these agree.

Exercise 24: Test Yourself

Develop a list of new terms and key ideas presented in this chapter. Supply an example of each item on your list.

Note The Appendix contains complete answers for Exercises 2, 3, 5*a* and *b,* 7*a* and *e,* 8*a*, and 15. It contains partial solutions or hints for Exercises 7*d,* 8*b,* 14, 16, 18, and 20.

CHAPTER 17

DECISION RULES: DECIDING HOW TO DECIDE

Always do right. This will gratify some people and astonish the rest.

—Mark Twain

INTRODUCTION

The examples in Chapter 16 showed situations in which expected value serves as a method for evaluating decision alternatives. This chapter will consider expected value along with four other decision-making rules: *maximax, maximin, most likely value,* and *minimax regret.* These different rules fit different attitudes and goals of a decision maker; each is suitable for some—but not all—decision situations.

DECIDING WHAT TO STUDY

Roberta faced a problem. It was Friday, and her history professor had just announced a quiz for Monday. The quiz would consist of a single essay question requiring discussion and analysis of one of four topics announced in Friday's class. But over the weekend Roberta's parents were planning to celebrate their silver wedding anniversary, and Roberta knew that she wouldn't have time to study until her return to campus on Sunday evening.

Although she had been regular in her class attendance and an attentive listener, the announced quiz caught Roberta three chapters behind in her reading; she despaired of being well prepared for Monday.

As she hurriedly packed for her trip home, Roberta picked up her history text. "I'll glance through it during the bus ride to Allentown," she thought. Later, in the bus, she did open the book and started leafing through Chapters 7, 8, and 9; but the length of these chapters was dismaying. "I'll probably have time to study only one of the chapters carefully," Roberta figured. Turning the pages, she started to mark those sections that seemed to pertain to the four possible topics for the quiz. But, "Which chapter should I concentrate on?" she wondered.

Last week in her mathematics class Roberta had learned about matrices; she decided to fiddle with her problem using a matrix. She began with the matrix shown in Table 17-1.

After skimming the chapters, Roberta estimated how many points (out of a possible 25) she might score on Monday's quiz for each chapter-question combination. Table 17-2 shows her completed choice matrix.

Roberta felt pleased that she had thought of using a matrix, and she examined it optimistically for dominance—but she found none. Just as she was wondering what to do next, someone sat down next to her and introduced himself as Peter, a business administration major at a nearby university—also on his way home for the weekend.

Roberta's matrix caught Peter's eye. "That looks to me like a decision matrix," he observed.

"Yes," said Roberta, "I need to decide which history chapter to study to prepare for a quiz on Monday."

After Roberta had explained the details, Peter told her that he could show her *five* different methods for deciding which chapter to study. Roberta challenged him to do it.

TABLE 17-1
Roberta's choice matrix

	POSSIBLE QUIZ TOPICS			
CHOICES	A	B	C	D
Chapter 7				
Chapter 8				
Chapter 9				

TABLE 17-2
Roberta's choice matrix; entries are possible quiz scores

	POSSIBLE QUIZ TOPICS			
CHOICES	A	B	C	D
Chapter 7	16	10	10	18
Chapter 8	8	16	5	16
Chapter 9	10	5	15	20

Maximax

"Are you an optimist?" Peter inquired. "Maybe," responded Roberta. "If so," continued Peter, "you might like to use the *maximax* decision rule":

> Identify the best possible outcome for each of your choices. Then select that choice which allows you to achieve the maximum of these best possible outcomes.

"Let's see," replied Roberta. "If I study Chapter 7, the best possible outcome is 18 points—when my professor chooses topic D. If I study Chapter 8, the best possible outcome is 16 points—when the professor chooses topic B or topic D. On the other hand, if I study Chapter 9, the best outcome is 20 points—when the professor chooses topic D. So my optimistic choice is to study Chapter 9, hoping for the maximax value of 20 points. OK, that's one way of deciding; what are the others?"

Maximin

"Well," Peter said, "a pessimist would prefer the *maximin* decision rule":

> Identify the worst possible outcome for each of your choices and then select the choice that achieves the maximum of these worst possible outcomes.

"This time you're suggesting that I avoid the worst instead of seeking the best outcome," said Roberta. "The worst outcome if I study Chapter 7 is 10 points; the worst outcome if I study Chapter 8 is 5 points; the worst outcome if I study Chapter 9 is 5 points. The *best* of these *worst* outcomes is 10—so I should study Chapter 7."

"Yes," agreed Peter, "Chapter 7 is the choice selected by the maximin rule."

Most Likely Value

"Actually," Roberta said, "I think the professor favors topic B. I'd say that he's twice as likely to ask a question on topic B as on any of the other topics."

"You're getting close to another of my five decision rules—the *most-likely-value* rule," said Peter:

> Identify the alternative most likely to occur; then choose an action which maximizes the most likely value."

"Does this rule mean that I should study Chapter 8?" Roberta asked.

"Exactly," Peter said. "Since topic B is most likely, focus only on the outcomes if topic B is selected. Make the choice which has the highest value among those outcomes. Since 16 is the highest point value in column B of your matrix, the most-likely-value rule directs you to study Chapter 8."

Expected Value

But Roberta was worried about focusing only on one topic. "Professor Houseman might ask one of the other questions," she speculated. Perhaps I should be somewhat prepared for all of them."

"In that case, you may find the *expected-value* decision rule preferable," Peter suggested. "Using the probabilities that you have already suggested,

$$p_A = 1/5$$
$$p_B = 2/5$$
$$p_C = 1/5$$
$$p_D = 1/5$$

you can compute the expected value of each of your choices. Then you can choose to study the chapter that offers you the highest expected or 'average' value."

Roberta carried out the following calculations:

$$EV_7 = 1/5(16) + 2/5(10) + 1/5(10) + 1/5(18) = 64/5 = 12.8$$
$$EV_8 = 1/5(8) + 2/5(16) + 1/5(5) + 1/5(16) = 61/5 = 12.2$$
$$EV_9 = 1/5(10) + 2/5(5) + 1/5(15) + 1/5(20) = 55/5 = 11$$

"My expected-value calculations choose Chapter 7 as best 'on the average.' This decision rule seems most reasonable to me; I think I'll spend Sunday evening studying Chapter 7," Roberta concluded.

Minimax Regret

"Don't be so hasty," warned Peter. "I still have one more decision rule to show you. This last one is a bit complicated, so listen carefully. It's called the *minimax-regret* decision rule; it's useful in situations where the decision maker is likely to feel substantial regret if a poor decision is made. A decision maker who is tortured by remorse over mistakes would do well to consider the minimax-regret decision rule."

"Sometimes I'm that way," confided Roberta. "Sometimes I worry a lot after a poor decision. I'd like to hear about this 'low regret' rule."

"Suppose you choose to study Chapter 7," said Peter. "If Professor Houseman gives a question on topic A, you'll have no regret, since 16 points is the most you expect to get on topic A no matter what you study. On the other hand, if you have studied Chapter 7 and the professor gives a question on topic B, you will have some regret. You will get only 10 points, but you could have had 16 points if you had studied Chapter 8. This difference (16 − 10) of 6 points is the amount of your regret. That is, the choice to study Chapter 7 leads to a regret of 6 points if topic B is chosen for the quiz."

TABLE 17-3
Roberta's regret matrix

	POSSIBLE QUIZ TOPICS			
CHOICES	A	B	C	D
Chapter 7	0	6	5	2
Chapter 8	8	0	10	4
Chapter 9	6	11	0	0

"I see," said Roberta. "And if I study Chapter 7 and topic C is given on the quiz, then I have a regret of 5—since I could have obtained 15 points instead of 10 by studying Chapter 9 instead."

Roberta and Peter went on to calculate additional regret values. Since there is a regret value for each combination of chapter and quiz topic, Roberta made a second matrix—shown in Table 17-3—to keep track of the regret values. Each entry in the "regret matrix" gives the number of points lost when a particular chapter is studied and a particular quiz topic is chosen. There is a zero entry in the regret matrix corresponding to the maximum value for each column of the choice matrix. All entries in the regret matrix are found by subtracting a given entry in the choice matrix from the maximum entry in its column.

"Now what do I do?" asked Roberta, as she examined Table 17-3.

"Find the maximum regret for each choice," replied Peter. "Then select the choice with the smallest maximum regret."

Roberta examined her matrix for the following values:

Maximum regret from studying Chapter 7: 6 points
Maximum regret from studying Chapter 8: 10 points
Maximum regret from studying Chapter 9: 11 points

"If I want to minimize maximum regret, then I should study Chapter 7," Roberta observed. "This is the same as the expected-value decision," she observed.

"Those two rules don't always recommend the same decision," Peter warned.

As the bus bumped along the highway, Roberta looked over the notes she had scribbled on a paper in her lap. Peter had indeed made good on his promise to show her five decision rules. She reviewed the results:

Maximax had selected Chapter 9.
Maximin had selected Chapter 7.
Most likely value had selected Chapter 8.
Expected value had selected Chapter 7.
Minimax regret had selected Chapter 7.

As she climbed off the bus at Allentown, Roberta offered profuse thanks to Peter. She had decided that when she returned to campus Sunday evening she would devote her limited study time to Chapter 7.

WHICH RULE SHOULD I USE?

The discussion of Roberta's problem introduced five different rules that can be used as guides for decision making.

When the probabilities of the various outcomes are known—as in Roberta's case—it is reasonable to use this information and apply the expected-value decision rule. However, if the probability values are only vague estimates, then the expected-value rule can yield only a vague estimate of the average outcome.

When we know that one outcome is much more likely than the others, then the most-likely-value rule is worth serious consideration.

When information about probabilities is unavailable, we can use the maximax, maximin, or minimax-regret rule—depending on the circumstances and on the attitude we have toward the decision (are we optimistic, pessimistic, or worried about regrets?).

No firm guidelines can be given concerning which rule (if any) to apply. The expected-value rule is considered superior when probabilities are known and when a decision is to be repeated again and again; however, for one-time-only decisions, many decision makers prefer other methods. A decision maker should experiment with the methods in practice situations like the exercises below and then, in a real and important decision, choose the method which experience has shown will best fit the decision maker's value system and advance his or her goals.

Exercise 1

Suppose that Roberta gets back to campus early on Sunday, in time to study two of the three chapters. Under these revised circumstances, the matrix shown in Table 17-4 (page 300) gives the various possible point scores.

a Which pair of chapters does the maximax decision rule select for Roberta's study time?

b Which pair of chapters does the maximin rule select?

c If topic B is more likely to be chosen for the quiz, which pair of chapters does the most-likely-value rule select?

d If $p_A = 1/5$, $p_B = 2/5$, $p_C = 1/5$, and $p_D = 1/5$ are the probabilities for the various topics, calculate the expected value of each study choice. Which has the highest expected value?

e Find the regret matrix for this decision situation. Which pair of chapters does the minimax-regret rule select?

f On the basis of your analysis in *a–e* above, what chapters would you advise Roberta to study? Explain why.

TABLE 17-4

	POSSIBLE QUIZ TOPICS			
CHOICES	A	B	C	D
Chapters 7 and 8	20	22	12	22
Chapters 7 and 9	16	15	20	25
Chapters 8 and 9	22	16	18	24

Exercise 2

In Exercise 20 in Chapter 16, Roger Lewis, the new manager of Newswatch, Inc., had to decide, how many copies of the *Daily News* to order. Table 17-5 (repeated from Table 16-8) gives the profit amounts for each combination of order quantity and sales quantity. The sales history of the past year is:

On 10 percent of the days, 50,000 copies were sold.
On 10 percent of the days, 45,000 copies were sold.
On 40 percent of the days, 40,000 copies were sold.
On 30 percent of the days, 35,000 copies were sold.
On 10 percent of the days, 30,000 copies were sold.

a Verify that use of the maximax rule results in the decision to order 50,000 copies—the decision made by Claude, Roger's optimistic predecessor.

b What size order would be placed if Roger made use of the maximin rule?

c If the most-likely-value rule is used, what size order will be placed?

d What is the optimum size order if the expected-value rule is used?

e If it is particularly important to Roger to avoid days with low profits, he should consider the minimax-regret rule. What order size is recommended by this rule?

f On the basis of your analysis above, what numbers of copies of the *Daily News* do you think Roger should order? Justify your decision.

TABLE 17-5
Net profits in dollars from sales of *Daily News*

	NUMBER OF COPIES IN DEMAND				
NUMBER OF COPIES ORDERED	50,000	45,000	40,000	35,000	30,000
50,000	5000	3750	2500	1250	0
45,000	4500	4500	3250	2000	750
40,000	4000	4000	4000	2750	1500
35,000	3500	3500	3500	3500	2250
30,000	3000	3000	3000	3000	3000

Exercise 3

Harriet, a hospital administrator, must award a contract for reseeding the hospital lawn. She can give the contract to Groner Construction Company, which has agreed to do the job for $1500 if the weather is good over the next month but will charge $2400 if the weather is bad. Or she can give the job to Yingst Construction Company, which will do the job for a flat fee of $2000 regardless of the weather. Meteorological records indicate that there is a 40 percent chance of bad weather for the month.

a Which of the five decision rules do you think best suits this situation? Explain why. Which construction company does your recommended rule choose?

b Determine the choices made by the four decision rules you did *not* use in *a.* Discuss whether you would consider using one of them instead of the rule you did choose.

Exercise 4

For each of the five decision rules, develop a statement that summarizes how to use it.

Exercise 5

The weather forecaster is predicting a 60 percent chance of a snowstorm today. Duane is trying to decide whether to wear heavy boots. He assigns the following utility values to the several possible outcomes:

Wear boots; it snows.	4
Wear boots; it doesn't snow.	−8
Don't wear boots; it snows.	−2
Don't wear boots; it doesn't snow.	4

What is Duane's best choice, to wear boots or not to wear them? Explain.

TABLE 17-6

	OUTCOMES		
CHOICES	Hard quiz	Easy quiz	No quiz
Study hard	8	5	2
Skim notes	6	8	6
Watch television	1	4	10

Exercise 6

It is Thursday night, and you must decide whether to study hard for tomorrow morning's biology class, to just skim through your class notes, or to watch television. The matrix in Table 17-6 shows the utility value you attach to each possible outcome.

a What will you do Thursday night if you are an optimist?
b What will you do Thursday night if you are a pessimist?
c What will you do Thursday night if you worry about what might have been?
d Suppose you assess the probabilities of the possible outcomes as:

$$p_{\text{no quiz}} = .1$$
$$p_{\text{easy quiz}} = .4$$
$$p_{\text{hard quiz}} = .5$$

What choice does the expected-value rule recommend?
e What choice does the most-likely-value rule recommend?

f **Journal** Modify the details of this situation until it closely resembles a study decision of your own. Explain what decision rule you would use, and why.

Exercise 7

A Coast Guard officer is called upon to organize a search for a missing boat. There are three standard search procedures, and the effectiveness of each depends on weather and sea conditions. The weather during the search will be either clear and calm, foggy, or clear with high seas. The matrix in Table 17-7 gives a rating that indicates the effectiveness of each procedure under each type of weather condition. The higher the number, the more effective the procedure in finding the lost boat.

a Which decision rule do you think the officer should use? Justify your recommendation.

TABLE 17-7

CHOICES	OUTCOMES		
	Clear and calm	Foggy	High seas
Procedure 1	9	3	6
Procedure 2	6	5	5
Procedure 3	7	4	5

b Suppose that the officer knows from the weather bureau that the probability of a clear and calm day is 1/6, the probability of a clear day with high seas is 2/6, and the probability of a foggy day is 3/6. Determine the resulting decision when each of the five decision rules is applied. Do these results help confirm the wisdom of your recommendation in *a*? Discuss.

Exercise 8

Elwood Kull is trying to decide which of three crops he should plant on his 100-acre farm in Lancaster County, Pennsylvania. The three crops he is considering are oats, peas, and beans. The profit from each is dependent on the rainfall during the growing season, which may be substantial (more than 5 inches per month), moderate (3 inches to 5 inches per month), or light (less than 3 inches per month). Using his records from the past and price predictions for the current year, Kull has developed profit estimates for each combination of crop and rainfall, as shown in Table 17-8 (page 304).

a Apply the maximax, maximin, and minimax-regret decision rules to Kull's situation. Which crop does each rule recommend?

b If moderate rainfall is more likely than either light or substantial rainfall, then the most-likely-value decision rule can be used. Which crop does this rule select?

c If substantial, moderate, and light rainfall are all equally likely, which crop offers the highest expected profit?

d The average rainfalls for the past 20 years are as shown in Table 17-9. Use these data to obtain relative-frequency estimates of the probabilities of substantial, moderate, and light rainfall.

e Use the probabilities from *d* to compute the expected profit for each crop. Which crop is the best one to plant "on the average"?

f Examine the results of *a, b, c,* and *e.* Which crop would you recommend? Explain why.

g Kull will face similar decisions next year, the year after that, and so on. What advice would you give him about how to make these future decisions?

TABLE 17-8
Average profit, dollars

CHOICES	RAINFALL		
	Substantial	Moderate	Light
Oats	18,000	20,000	16,000
Peas	20,000	20,000	15,000
Beans	15,000	25,000	18,000

TABLE 17-9

YEAR	AVERAGE RAINFALL PER MONTH FROM MAY TO SEPTEMBER, INCHES
1980	2.1
1979	7.3
1978	3.8
1977	3.2
1976	4.7
1975	4.8
1974	4.1
1973	2.1
1972	5.3
1971	3.9
1970	1.1
1969	3.9
1968	6.3
1967	4.2
1966	6.5
1965	2.4
1964	3.6
1963	4.4
1962	3.1
1961	1.2

Exercise 9*

Mahlon Fredricks, Republican party leader in a large midwestern city, is trying to decide whom to support in the primary campaign for mayor. He can support either Janice Edwards or Norton Simmons, or he can avoid a commitment and simply declare himself neutral in the primary. Fredricks's primary concern about whom to support rests on the possible numbers of patronage jobs that will be allocated to him as a result of his position. Suppose that Table 17-10 summarizes the situation for Fredricks.

*Adapted from C. A. Lave and J. G. March, *An Introduction to Models in the Social Sciences,* Harper and Row, New York, 1975, pp. 154–155.

TABLE 17-10

CHOICES	OUTCOMES	
	Edwards wins	Simmons wins
I am neutral	10	12
I support Edwards	28	0
I support Simmons	5	20

a What will Fredricks do if he wants to maximize expected value and if the probability that Edwards will win is .4?

b What will Fredricks do if he uses the most-likely-value decision rule?

c What will Fredricks do if he uses the maximax decision rule?

d What will Fredricks do if he uses the maximin decision rule?

e What will Fredricks do if he uses the minimax-regret decision rule?

f If you were Janice Edwards and you knew of Fredricks's decision problem, what could you do to improve your position and gain his support?

g Which decision rule seems most suitable to use in this situation? Explain why you think so.

Exercise 10

A jury has spent many hours listening to the arguments of the prosecutor and the defense attorney in the case of a defendant who allegedly shot and killed an elderly man while robbing his home. The defendant has a long record of convictions for armed robbery, but the evidence presented has not completely convinced the jury of the defendant's guilt in this case. One particular juror, William Markovis, is about 90 percent certain of the defendant's guilt. The testimony and arguments have just ended, and Markovis is collecting his thoughts about the case as the jury prepares to deliberate. To aid his decision process, Markovis has sketched the matrix shown in Table 17-11.

a Note Markovis's utility values shown in the matrix. What situation or situations does he prefer most? Which would he prefer in this case, to decide that an innocent defendant is guilty or to decide that a guilty defendant is innocent? Is this an unusual attitude for a juror? Discuss.

b Analyze Markovis's decision problem using the five decision rules. Given his utility values and his probability estimate of guilt, what decision do you think he will make?

c Which of the five decision rules do you see as most suitable for a jury member? Explain.

TABLE 17-11

CHOICES	ACTUAL SITUATION	
	Guilty	Not guilty
Guilty	100	−20
Not guilty	−50	100

TABLE 17-12

BULLDOGS' CHOICES	CRUSADERS' DEFENSES		
	C	B	A
Play 1	0	12	8
Play 2	0	−20	20
Play 3	15	5	−20
Play 4	2	5	3

Exercise 11*

The Bloomsburg Bulldogs have arrived at the moment they've been waiting and preparing for. It is the fourth quarter of the Susquehanna Valley Regional Football Conference championships, and victory over the Clarion Crusaders is within reach. The coaches have always been thorough in their analysis, and this time is no exception. The team has four plays from which to choose:

Play 1: Pass short
Play 2: Pass long
Play 3: Power sweep
Play 4: Quick slant

The opposing team has three basic defensive formations—known as *crash* (C), *bamm* (B), and *alakazamm* (A). The yardage that the Bulldogs can gain on any given play depends on the play chosen and on the defensive formation used against it. Study of statistics on the Crusaders has provided the Bulldogs' coaches with information concerning their play patterns. The Crusaders use defense C about 20 percent of the time, defense B about 30 percent of the time, and defense A about 50 percent of the time. Furthermore, the Bulldog coaches have come up with estimates of the number of yards they should be able to gain from each combination of plays. This information is summarized in Table 17-12. For example, if the Bulldogs choose play 3 while the Crusaders use defense A, the Bulldogs can expect to lose 20 yards.

a If the Bulldogs are on the 2-yard line and want to be sure to score on the next play, which play should they use?

b If the Bulldogs have the ball at midfield and want to try for the greatest possible yardage gain, which play should they use?

c If the Bulldogs want to use the play which will offer the largest expected, or average, gain which will they choose?

d If the Bulldogs' coaches are concerned about second-guessing by fans and sportscasters who say, "If only the team had used play . . . ," which play will they choose?

Exercise 12: Journal

For each of the five decision rules describe at least one situation in which that decision rule would be best. Which of these rules do you use most often? Which rule do you observe other people using most often?

*Adapted from Lave and March, *An Introduction to Models in the Social Sciences*, pp. 138–139.

Exercise 13

Consider the quotation from Mark Twain at the beginning of the chapter. How would his admonition, "Always do right," be as a decision rule?

Exercise 14: Journal

List at least 10 "minor" decisions (such as which route to take to work, which brand of toothpaste to buy, and whether to buy a yearbook). For each, state which of the five decision rules (if any) you would apply and why.

Exercise 15

A person facing a decision might follow one of these procedures:

Select one of the five decision rules as most suitable for the decision situation. Apply the rule and follow its recommendation.

Apply all five decision rules and see what choice each rule recommends. Select whichever choice is recommended by the greatest number of rules.

Discuss these procedures. Which do you think is better? Explain why you think so. Formulate a procedure that you prefer to either of these.

Exercise 16: Invent

In Chapter 7, as in this chapter, matrices were used to display decision information. However, in Chapter 7 weighted averages were calculated and used instead of decision rules. In Chapter 7, moreover, choices were evaluated according to their ability to satisfy certain objectives (such as low cost); here, choices have been evaluated according to their ability to produce certain outcomes (such as points on a quiz). Whether you use the decision methods of Chapter 7 or those of this chapter will depend largely on how you view the decision you are facing.

Invent details for an example in which a decision maker is trying to decide how to spend a vacation. He or she might be trying to decide whether to visit family, to work on a term paper, to return to a part-time job, to travel, or to relax at a favorite beach. Formulate the decision in two ways, first using the ideas of Chapter 7 and second the ideas of this chapter. Evaluate the results; does one approach work better for you? Look for reasons why.

Exercise 17: Journal—Good Decisions Are Made Ahead of Time

"Many decisions must be made at times of emotional stress, when it is difficult to weigh the alternatives objectively. In such situations it is useful to have well-thought-out decision rules to fall back on. Decision makers who have determined—on the basis of both analysis and experience—that a certain decision rule suits a certain type of situation can draw on that knowledge to make a good decision under stress even if they are then incapable of careful thought."

To what extent do you agree with this argument? Discuss in your journal how you can make use of these ideas.

Exercise 18: Journal—Difficult Decisions Are Made When We Get Tired of Thinking about Them

Despite the existence of decision-making rules, many people are reluctant to trust decisions to rules. A man I know described his decision process this way:

> I think about the decision for a while and consider the advantages and disadvantages of each alternative. Then when I get tired of thinking, that means it's time to decide—and I pick the choice that looks best at that time.

a Consider this attitude toward decision making. Is it reasonable that fatigue should determine when a decision is made? What relevance does the importance of the decision have?

b How do emotions and fatigue interact with objective analysis in your decision process? What role do you see for the five decision rules in your own decision situations? What reservations do you have about their use?

Exercise 19: Test Yourself

Develop a list of new terms and key ideas presented in this chapter. Supply an example of each item on your list.

Note The Appendix contains complete answers for Exercises 1 and 8*a–e*. It contains partial solutions or hints for Exercises 2*f*, 3*a*, 8*f*, 9*f* and *g*, 10*c*, and 13.

CHAPTER 18

PROBABILITIES AND ESTIMATION

I shall never believe that God plays dice with the world.

—Albert Einstein

AN ALMANAC SURVEY

In Table 18-1 you will find a survey consisting of 10 difficult questions. These are all examples of "almanac questions." Each has a specific known answer that can be looked up in any almanac or book of facts. For most of the questions you will have little or no idea of the correct answer. You will be asked to supply a range within which you think the answer lies.

This is not an assignment for which you should look up the answers; the answers are provided below. Instead of looking up the answers, since the correct values are unknown to you, you should specify the *range of your uncertainty* about them.

TABLE 18-1
An almanac survey

	1% CHANCE OF BEING LESS THAN	25% CHANCE OF BEING LESS THAN	50% CHANCE OF BEING LESS THAN	75% CHANCE OF BEING LESS THAN	99% CHANCE OF BEING LESS THAN
1 The total number of points scored in the 1949 Orange Bowl game	____	____	____	____	____
2 The total area (in square miles) of Nebraska	____	____	____	____	____
3 The height (in feet) of New York City's World Trade Center	____	____	____	____	____
4 The United States population in 1900	____	____	____	____	____
5 The number of 42-gallon barrels of gasoline produced in the United States in 1970	____	____	____	____	____
6 The average amount of water discharged (in cubic feet per second) from the Susquehanna River from 1941 to 1970	____	____	____	____	____
7 The total number of American casualties in World War I (April 6, 1917–November 11, 1918)	____	____	____	____	____
8 The airline distance (in miles) from New York to Moscow	____	____	____	____	____
9 The year in which Mark Twain's *Tom Sawyer* was first published	____	____	____	____	____
10 The number of tornadoes that occurred in the United States during 1940	____	____	____	____	____

For example:

- Item to be estimated: The 1971 census population of Calcutta, India.
- Task: Specify values, based on your own knowledge, to fill in each of the blanks below.

The requested item has a 1 percent chance of being less than ____.
The requested item has a 25 percent chance of being less than ____.
The requested item has a 50 percent chance of being less than ____.
The requested item has a 75 percent chance of being less than ____.
The requested item has a 99 percent chance of being less than ____.

Suppose that, after some thought, we fill in the blanks as follows:

1 percent chance of being less than 10,000
25 percent chance of being less than 100,000
50 percent chance of being less than 1 million
75 percent chance of being less than 5 million
99 percent chance of being less than 10 million

These values say that we are almost certain that the population was at least 10,000 and less than 10 million. We think there is a 50-50 chance that the population was between 100,000 and 5 million. The estimates we have given are spread over a wide range, and the width of the range shows our degree of uncertainty; if we knew more about the population of Calcutta, the range would be narrower.

Now it's your turn. For each of the 10 questions in Table 18-1, estimate values as was done for the population of Calcutta. To repeat: For most of the questions, you will have little or no idea of the correct answer. Use a wide range of values to reflect your uncertainty. Answer quickly; spend no more than 1 minute on each question.

EVALUATING THE SURVEY RESULTS

The correct answers to the survey items are given in Table 18-2 (page 312). Call your estimates *satisfactory* if the correct value lies between your 1 percent and 99 percent values. These two values are given with a confidence of 98 percent that the true value lies between them; they form the end points of what may be called a *98 percent confidence interval.*

Consider first the answer to the sample question. The 1971 census population of Calcutta was 7,031,382.* The 50 percent value—which we might have hoped would be close to the actual value—is far away from the correct answer (Figure 18-1). Nevertheless, our answer is satisfactory, since the actual population value is captured within the 98 percent confidence interval from 10,000 to 10 million.

Now check your answers to the survey questions. What is your score—that is, how many of your answers were satisfactory? If your estimates adequately reflect your lack of certainty about the correct values, then your score will be 90 percent or 100 percent. If your score was less than 90 percent, you probably lack experience in estimation. And you are not alone—when almanac surveys are given, most people's scores consistently fall below 60 percent.†

It is natural, when we estimate, to want a narrow range for our estimate. A narrow range *is* desirable—but only if it is based on knowledge rather than wild guessing. For example:

> Stuart and Malcolm are overheard trying to recall the distance of the moon from the earth.
>
> Stuart says, "I'm 98 percent confident that the moon is between 450,000 and 500,000 miles away from the earth."
>
> Malcolm says, "I'm 98 percent confident that the moon is between 100,000 and 300,000 miles away from the earth."

***The World Almanac and Book of Facts,* Newspaper Enterprise Association, New York, 1979.

†Richard Fallon, "Subjective Assessment of Uncertainty," Rand Corporation, Santa Monica, Calif., P-5581, 1976.

TABLE 18-2
Answers to survey questions

1 The 1949 Orange Bowl game resulted in a total score of 69 points (Texas 41, Georgia 28).

2 The area of Nebraska is 72,237 square miles.

3 The height of New York City's World Trade Center is 110 stories or 1350 feet.

4 The United States population in 1900 was 75,994,575.

5 The number of 42-gallon barrels of gasoline produced in the United States in 1970 was 2,135,838,000.

6 The average amount of water discharged from the Susquehanna River from 1941 to 1970 was 37,190,000 cubic feet per second.

7 The total number of American casualties in World War I was 370,710.

8 The airline distance from New York to Moscow is 4683 miles.

9 The year in which Mark Twain's *Tom Sawyer* was first published was 1876.

10 The number of tornadoes in the United States in 1940 was 124.

FIGURE 18-1
Population estimates for Calcutta

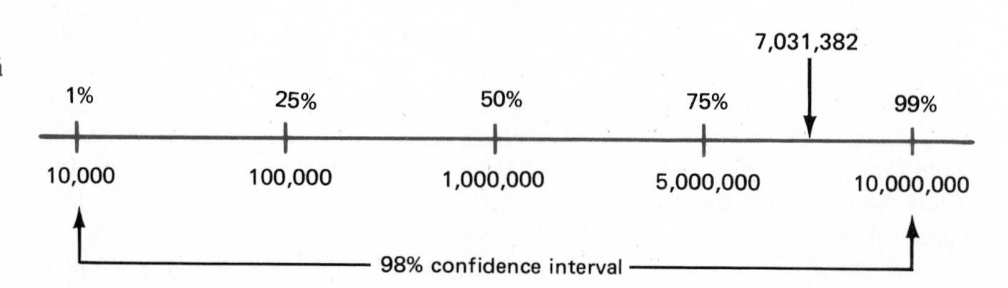

Stuart's estimate is more appealing than Malcolm's because of its narrower range; it offers more precision. But it is unsatisfactory; the mean distance between the moon and the earth is about 239,000* miles—and Stuart's confidence interval does not even come close to including this value. Malcolm, on the other hand, has clearly shown his uncertainty by using a wide confidence interval and thus has given a satisfactory estimate.

When we estimate using a 98 percent level of confidence, then, if our estimates accurately reflect the extent of our knowledge, 98 percent of our estimates will lie within our confidence intervals.

*The actual distance between the earth and its moon varies between about 227,000 and 254,000 miles, with a mean distance of about 239,000 miles.

CONFIDENCE INTERVALS IN STATISTICS

Truth is the
cry of all
but the game
of the few.

George Berkeley,
Bishop of Cloyne

One of the principal activities in the science of statistics is to take data about a particular phenomenon and use them to produce a statement about that phenomenon which is as nearly true as possible. "Nearness to truth" is expressed in terms of confidence intervals. Attempts to determine the true weight of a chunk of metal illustrate this:

> NB-10 is a chunk of chrome-steel alloy owned by the National Bureau of Standards and weighing approximately 10 grams (the weight of two nickels). More precisely, the weight of NB-10 is about 405 micrograms (millionths of a gram) beneath 10 grams. The *exact* weight of NB-10 is *unknowable.* The degree to which we know the exact weight can be described in terms of confidence intervals.
>
> The weight of NB-10 is measured weekly. The mean of a large group (actually 100 in number) of these measurements is 404.6 micrograms below 10 grams, and the standard deviation is 0.6. A 95 percent confidence interval is obtained by going about 2 standard deviations in either direction from the mean. Thus NB-10 weighs somewhere between 403.4 and 405.8 micrograms below 10 grams with confidence 95 percent.
>
> The 95 percent confidence may be interpreted thus. If 100 other investigators set out to determine how much NB-10 weighs and if they use the same careful procedures and equipment that the bureau uses, each taking 100 measurements, then (about) 95 of them should get intervals that cover the stated weight of NB-10 and the other 5 should miss.
>
> Having 100 investigators weigh NB-10 is an experiment that was simulated with a computer, leading to the results shown in Figure 18-2. To get started, the computer program was told that NB-10 weighed 405 micrograms below 10 grams. The program then simulated 100 measurements by each of 100 investigators. Each horizontal line in Figure 18-2 shows the spread of the 95 percent confidence interval (2 standard deviations on either side of the mean of 100 measurements) for a particular investigator.
>
> The confidence level of 95 percent says that about 95 percent of the intervals should cover the stated weight of 405 grams (indicated by the vertical line). In fact, 96 do.*

Exercise 1 on page 315 is an experiment with confidence intervals. In Exercise 2 you will be asked to give the weight of NB-10 in terms of a confidence interval.

---

*Figure 18-2 and the information concerning NB-10 are reproduced (with permission) from the following excellent introductory statistics text: David Freedman, Robert Pisani, and Roger Purves, *Statistics,* Norton, New York, 1978, pp. 401, 402.

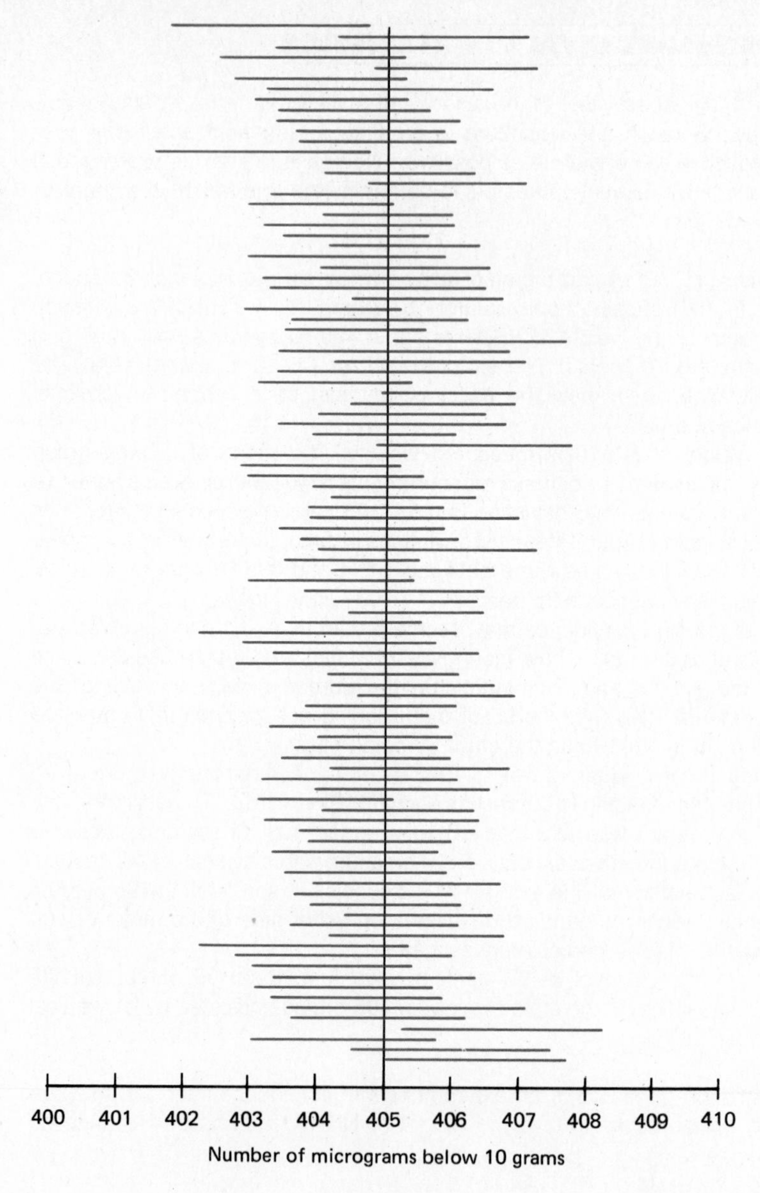

FIGURE 18-2
A computer simulation of 100 confidence intervals, each with confidence level 95%. The intervals have different centers and lengths, due to chance variation; 96 out of 100 cover the exact weight, represented by the vertical line.

Exercise 1: Class Group Activity—Experimentation with Confidence Intervals

This exercise requires 100 repetitions of an experiment. Each experiment requires 100 tosses of a coin. Although the activity is "simple," it is not "easy" unless the tosses are shared by a group. (For a discussion of the distinction between "simple" and "easy," see Exercise 4 in Chapter 9.)

At the beginning of Chapter 15 it was noted that when a fair coin is tossed 100 times there is about a 95 percent probability that the number of heads will be between 40 and 60. The interval from 40 to 60 is thus a 95 percent confidence interval. This exercise describes a procedure through which a 95 percent confidence interval can be examined.

Suppose that American pennies are made to be fair—that is, that the true probability of a head when any penny is tossed is ½. Here is the experiment:

> Toss a penny 100 times. Record the number of heads. Repeat this activity for a total of 100 times. (Thus, all together you will perform 10,000 tosses, in groups of 100.)

Find the interval centered at the number of heads and extending a distance of 10 on either side of it.

If the assumption is correct that pennies are fair and that the true probability of a head is ½, then in the experiments performed by you and your classmates, 95 percent of the intervals found will include 50 (which is ½ of the number of tosses) and 5 percent of them will not.

Exercise 2

What is the weight of NB-10? Answer in terms of a confidence interval.

Exercise 3: Journal

Ability to estimate can be a valuable problem-solving tool, but (as the almanac survey showed) to be good at estimation requires practice.

List a variety of situations that offer you the opportunity to make and check estimates. Select one situation and estimate using a 90 percent confidence interval. Check your estimates against the actual data to see whether you are achieving a 90 percent success rate—that is, check that about 90 percent of your estimates lie within your confidence intervals.

Possible quantities to estimate include:

A score (yours or someone else's) on an upcoming test
A final grade in a course you are now taking
The distance to be traveled on a trip that you're planning
Your weekly expenditures
The total cost of groceries bought on one visit to the market
The age, height, or weight of a stranger
The time that a certain activity will require

Exercise 4: Journal

Exercise 21 in Chapter 15 asks for probability estimates to replace terms like *possibly* and *maybe.* Because of our uncertainty over the meanings that others attach to such terms, does it make better sense to develop interval estimates than to pin them down to a single value? Return to that exercise and supply interval estimates for the terms listed there. Survey at least 10 people to learn their probability estimates for each term in your list. How many of their estimates fall within your intervals?

Exercise 5: Journal

Many people believe that the universe is governed by a predetermined plan, with nothing left to chance. These people believe that probabilities have been invented by humans to describe events whose causes are not fully understood, but that there are nonprobabilistic explanations for all events. Others think that probabilities are actually part of the workings of the universe, that chance has a role in the course of events. The quotation from Einstein at the beginning of this chapter gives his view on this issue.

Think and write about your own view of the role of probability in the creation and operation of the universe. Contrast it with the role of probability in your day-to-day life. Is it possible to make use of probabilistic thinking in everyday life even if you are unsure about whether some events actually occur by chance (rather than according to a plan) or chance is just a human device for describing events whose causes are unknown?

Exercise 6: Test Yourself

Estimation has been the central topic of this chapter. Write a summary of the key ideas concerning estimation that have been presented; then state some guidelines for yourself concerning how you can make use of estimation in your future problem solving.

Note The Appendix contains partial solutions or hints for Exercises 3 and 4.

CHAPTER 19

REDUCING THE EFFECTS OF INTERRUPTIONS

Work expands so as to fill the time available for its completion.

—Northcote Parkinson ("Parkinson's law")

INTRODUCTION

Gwynne is a resident counselor in a dormitory. Her study time is interrupted frequently by knocks on her door and questions from students who live around her. Yet Gwynne always has her class assignments completed on time. When asked how she manages to get her work done despite a lot of interruptions, Gwynne gives a simple rule of thumb:

> After I estimate how much time I think an assignment will take, then I double my estimate. I then schedule the double length of time in order to get everything done.

The ideas of this chapter first appeared in: JoAnne Simpson Growney, "Planning for Interruptions," *Mathematics Magazine,* vol. 55, no. 4, September 1982, pp. 213–219. And later in: JoAnne Growney, "Controlling the Effects of Interruptions," UMAP Module 657, *UMAP Journal,* vol. 5, no. 3, 1984, pp. 301–336. Both sources go beyond the content presented here and include derivation of formula 1.

Gwynne is not unique in her need to schedule more time for a task than she really thinks it will take in order to allow for interruptions. Many people—students, teachers, managers, parents—spend much of their time in situations in which tasks cannot be completed without interruptions. Some even find themselves with projects that *never* get completed, because interruptions break their concentration and cause so many setbacks that it is continually necessary to start over.

Often, as in Gwynne's case, dealing with interruptions is a necessary part of one's job or one's responsibilities, so that locking the door on interruptions is not a reasonable alternative. But neither does it seem reasonable to allow, as Gwynne does, *twice* as much time as the job really needs. This chapter will consider another way to deal with setbacks caused by interruptions: organization. A person who is willing to invest some time in organizing tasks can significantly reduce the time wasted because of interruptions.

SIMULATION OF INTERRUPTIONS

Let us begin with a simple example of a task beset by interruptions that can be illustrated easily by simulation. The example is definitely make-believe but provides a model for later, realistic examples.

Example 1: Paul ▶

Paul has just become aware of flab around his waist and other signs of physical unfitness. He has resolved to return to being trim and fit. Being a sensible fellow, Paul consults his physician, who recommends a daily stint of 5 uninterrupted minutes on a treadmill. If Paul should be interrupted during the 5 minutes, then he must start over and complete 5 additional minutes without interruption.

Sounds simple! However, Paul has another difficulty—a chronic skin disorder. Randomly, but on the average of once every 5 minutes, Paul has to stop what he is doing to scratch an itch. No matter how hard he scratches at any given time, he cannot prevent the random and frequent reoccurrence of this annoyance.

How does Paul fare as he tries to carry out his fitness program?

Simulation

If Paul's itches occur randomly but on the average of once every 5 minutes, then we may describe this by saying that he has probability $p = 1/5$ of an interruption during any given 1 minute.

We can simulate Paul's experience on the treadmill using a table of random digits. (Recall Table 4-11 in Chapter 4.) Starting at a random point in the table we can read digits and interpret each digit as 1 minute on the treadmill—with digits 1, 2, 3, 4, 5, 6, 7, and 8 denoting minutes in which no itch interrupts and digits 9 and 0 denoting interrupted minutes. A simulation of a 5-minute uninterrupted jog is completed when we encounter five digits from 1, 2, 3, 4, 5, 6, 7, and 8 in a row.

Row 1	82698	26610	90511	08055	80364	70233	91451	34528	30357	27456
	93680	27051	67692	57437	08779	81065	50586	20621	28296	43353
	45153	17985	74725	08526	09220	89778	59814	02387	78112	16035
	65055	40547	20834	50243	23998	59708	12313	89349	25103	43682
	80863	76681	73173	48970	91202	81344	89446	60285	12653	95567
Row 6	65704	35329	80233	67505	22518	58994	63968	79316	53447	65610
	16862	82356	69963	61171	96043	56593	73637	82198	51634	71363
	76048	34462	57543	98743	80838	42517	42094	98970	07496	22223
	92003	32221	39595	99113	43596	90842	87684	80098	54888	32782
	74244	90661	80795	20305	92055	54532	99534	34660	41569	88305
Row 11	38128	35924	55245	97971	52694	92422	15875	18971	20058	78333
	33729	56998	99535	52712	21558	36734	24131	95807	80922	85010
	63971	68875	13322	07349	73991	41072	31419	29611	10297	85465
	57653	56330	22804	71402	62635	33217	85828	69039	77095	57063
	36395	30423	96224	53481	23420	44921	30883	56083	32038	63699
Row 16	90543	52660	09346	76795	89783	87944	92379	34576	18055	67418
	58133	19098	70130	16092	43843	80508	96387	42270	35335	18264
	57487	88972	50914	65331	87902	42601	85407	19867	77391	48159
	77128	23219	48346	02047	63984	66444	83317	40167	39020	00798
	13964	87042	24341	25448	30779	30472	92064	71532	47311	33061

TABLE 19-1
1000 random digits

A table of 1000 random digits (rows 1–20 of Table 4-11) appears here as Table 19-1. If we start our simulation with the first digit of the first row of Table 19-1, we have the following results:

Digit	Interpretation
8	First minute of jogging completed
2	Second minute of jogging completed
6	Third minute of jogging completed
9	Stop to scratch; start over
8	First minute of jogging completed
2	Second minute of jogging completed
6	Third minute of jogging completed
6	Fourth minute of jogging completed
1	Fifth minute of jogging completed

In this case, the effort to complete 5 uninterrupted minutes of jogging required 9 minutes all together.

Of course, one simulation experiment provides only a sample of what can happen. To determine what can be expected on the average we need to perform many simulations and find the mean of the lengths of time that individual simulations require.

Suppose that we decide to perform 25 repetitions of the simulation. Table 19-2 indicates the blocks of random digits that constitute the simulations; each group of five digits enclosed in a box designates 5 uninterrupted minutes.

Table 19-3 shows the number of minutes needed to complete each experiment. Although the lengths of time required vary from 5 to 28 minutes, the mean length of time for the 25 experiments is approximately 11.7 minutes. ◀

Row 1	82698	26610	90511	03055	80364	70233	91451	34528	30357	27456
	93680	27051	67692	57437	08779	81065	50586	20621	28296	43353
	45153	17985	74725	08526	09220	89778	59814	02387	78112	16035
	65055	40547	20834	50243	23998	59708	12313	89349	25103	43682
	80863	76681	73173	48970	91202	81344	89446	60285	12563	95567
Row 6	65704	35329	80233	67505	22518	58994	63968	79316	53447	65610
	16862	82356	69963	61171	96043	56593	73637	82198	51634	71363
	76048	34462	57543	98743	80838	42517	42094	98970	07496	22223
	92003	32221	39595	99113	43596	90842	87684	80098	54888	32782
	74244	90661	80795	20305	92055	54532	99534	34660	41569	88305
Row 11	38128	35924	55245	97971	52694	92422	15875	18971	20058	78333
	33729	56998	99535	52712	21558	36734	24131	95807	80922	85010
	63971	68875	13322	07349	73991	41072	31419	29611	10297	85465
	57653	56330	22804	71402	62635	33217	85828	69039	77095	57063
	36395	30423	96224	53481	23420	44921	30883	56083	32038	63699
Row 16	90543	52660	09346	76795	89783	87944	92379	34576	18055	67418
	58133	19098	70130	16092	43843	80508	96387	42270	35335	18264
	57587	88972	50914	65331	87902	42601	85407	19867	77391	48159
	77128	23219	48346	02047	63984	6644	83317	40167	39020	00798
	13964	87042	24341	25448	30779	30472	92064	71532	47311	33061

TABLE 19-2
25 random-digit simulations of Paul's jogging effort

TABLE 19-3
Lengths of 25 experiments simulating Paul's jogging effort

EXPERIMENT	TOTAL NUMBER OF MINUTES NEEDED TO INCLUDE 5 UNINTERRUPTED MINUTES
1	9
2	27
3	5
4	6
5	16
6	6
7	23
8	7
9	5
10	9
11	28
12	5
13	26
14	12
15	15
16	8
17	5
18	5
19	12
20	13
21	9
22	8
23	8
24	7
25	18

Mean length of time for 25 simulation experiments

$= \frac{292}{25} \approx 11.7$ minutes

Exercise 1

Continuing from the point in Table 19-2 where we finished our twenty-fifth simulation, perform 25 more simulations of Paul's jogging effort. Record your results in a continuation of Table 19-3. Find the mean length of time for the 50 experiments.

MORE INTERRUPTION PROBLEMS—AND A FORMULA

The example of Paul's jogging was unrealistic and somewhat silly, but it can serve as a springboard to discussion of some problems that are more realistic and more important.

The problems that will now be considered have the following characteristics:

- A person has a certain task to complete, and the task requires a definite number U of uninterrupted time units.
- The person is subject to interruptions during the task; the probability of interruption during any given time unit is p.
- Interruptions are independent; that is, occurrence of one interruption has no effect on the probability of subsequent interruptions.
- Interruptions are devastating: that is, every time an interruption occurs the person must start the task over.

Paul's jogging problem satisfies these four characteristics; and as students we encounter many realistic problems that fit them at least roughly:

Tasks such as writing papers and solving complex mathematics problems require periods of uninterrupted concentration.

Often we must perform our tasks at times and in places where there is a likelihood of random interruption—music, visitors, telephone calls, even fatigue and hunger. If we can estimate an average number i of time units between interruptions, then the probability of p of interruption during any given time unit is $p = 1/i$.

Independence of interruptions is hard to determine. If interruptions are not at all subject to our control, it may be reasonable to suppose that they are independent.

Sometimes when our concentration on a task is interrupted, our progress is not wholly devastated; we don't need to return all the way to the start after the interruption. Often, though, interruptions are accompanied by a buildup in frustration and fatigue that is almost equivalent to the effect of starting over.

In situations of the sort just described, in which a task requires U uninterrupted time units and interruptions that are independent and devastating occur randomly with probability p, then the expected (or average) length of time t required to complete the task is given by the following formula:*

$$t = \frac{1}{p}\left[\frac{1}{(1-p)^{U}} - 1\right] \tag{1}$$

In Example 1 above, we have

$$U = 5 \text{ minutes}$$
$$p = 1/5 = .2$$

Substitution of these values in formula 1 leads to

$$\begin{aligned} t &= \frac{1}{p}\left[\frac{1}{(1-p)^{U}} - 1\right] \\ &= \frac{1}{.2}\left[\frac{1}{(.8)^{5}} - 1\right] \\ &= 5\left[\frac{1}{.32768} - 1\right] \\ &\approx 5\,[3.052 - 1] \\ &\approx 10.3 \text{ minutes} \end{aligned}$$

Although the logical derivation of formula 1 is not provided, the simulation completed in Exercise 1 above provides a check on the reasonableness of the result. The average for these 50 simulation experiments is just under 10.5 minutes—a result that is close to the answer obtained using the formula.

Several exercises below provide practice with calculations using formula 1. The next section of this chapter will show how the formula can be useful in evaluating different scheduling options.

Exercise 2

a Read problems *b, c,* and *d* before solving any of them. Note their similarity and, before using formula 1 to calculate values for *t, guess* an answer to each.

b Max has a job to do that requires 10 uninterrupted minutes. When an interruption occurs, he must start over. Max's interruptions occur randomly

*Details of the derivation of formula 1 involve far more extensive background in probability theory than this text has provided. The interested reader may find them in the references cited on page 318.

and independently and, on the average, once every 5 minutes. (Thus the probability of an interruption during any given minute is $p = 1/5 = .2$.) What is the expected length of time that Max's job will require?

c Marsha has a job to do that requires 10 uninterrupted minutes. When an interruption occurs, she must start over. Marsha's interruptions occur randomly and independently and, on the average, once every 10 minutes. What is the expected length of time that Marsha's job will require?

d Melanie has a job to do that requires 10 uninterrupted minutes. When an interruption occurs, she must start over. Melanie's interruptions occur randomly and independently and, on the average, once every 20 minutes. What is the expected length of time that Melanie's job will require?

e Compare the answers to *b, c,* and *d* with each other and with your guesses. Are the results surprising? Why is Max's expected time so long?

Exercise 3

Mike and Michelle have a talent for building houses of cards. (A *house of cards* is a structure built from ordinary playing cards, stood on edge, using only each other for support.) This weekend they want to break their record of a 10-story house. But they have an interruption problem—their younger brother Mitch keeps breaking their concentration or creating a breeze or vibration that brings the house down. Mike and Michelle estimate that they will need at least 1 uninterrupted hour (60 minutes) to build a house of cards that is higher than 10 stories.

a If Mitch can be induced to watch television, Mike and Michelle estimate that he will cause interruptions randomly and independently and, on the average, about once every 15 minutes. With this interruption pattern, what is the expected length of time that it will take for Mike and Michelle to build a record-breaking house of cards?

b If the weather is nice on Saturday, Mitch will spend most of his time outside. In this case he will interrupt only, on the average, once every 1 hour. With this interruption pattern, what is the expected length of time that it will take for Mike and Michelle to build their record-breaking house of cards?

c On the basis of the answers you obtained in *a* and *b,* what advice would you give to Mike and Michelle?

Exercise 4

The formula for t (total time that a task requires because of startovers due to interruptions) depends on two quantities:

U = number of uninterrupted time units required
p = probability of interruption during any given time unit

To become familiar with exactly how changes in U and p affect t, it is helpful to perform a series of related calculations. To guide you, selected values of U and p are supplied in Table 19-4, parts *a, b,* and *c* (page 326).

a Use formula 1 to calculate the values of t in Table 19-4*a*.
b Use formula 1 to calculate t in Table 19-4*b*.
c Use formula 1 to calculate t in Table 19-4*c*.

Items *d–f* are questions that will help you analyze table values. Use Table 19-4 to answer them.

d For a fixed value of U, how are the values of t and p related?

e Suppose that Wayne is someone who has an "interruption tolerance" of 50 percent; that is, he is unwilling to have t exceed U by more than 50 percent of U.

Should Wayne attempt to complete a 5-minute uninterruptible task if the probability of his being interrupted during any given minute is $p = .25$? What if $p = .2$? If $p = .1$? If $p = .05$? Should he attempt to complete a 10-minute uninterruptible task when $p = .2$? When $p = .1$? When $p = .05$?

f The formula $i = 1/p$ gives the expected length of the interval of time i between interruptions. Observe that in Table 19-4*a* when $U = i$, then $t \approx 2U$. Can the same observation be made in parts *b* and *c*?

Might most people intuitively expect that when $U = i$, then t will have a value close to U? Supply a verbal (rather than mathematical) explanation to convince such people that their intuitive expectation is incorrect.

g If the probability of interruption in any given 1 minute is $p = .25$, use values from Tables 19-4*a* and *b* to answer the following question. Which will take longer: three 5-minute uninterruptible tasks or one 10-minute uninterruptible task?

h If the probability of interruption in any given 1 minute is $p = .1$, which will take longer: three 10-minute uninterruptible tasks or one 20-minute uninterruptible task?

i Provide an explanation for the "surprising" answers to *g* and *h*. Invent similar questions and determine the answers. Assess your calculations and comparisons to develop a feel for the relationships between p, U, and t.

Note The Appendix contains complete answers for Exercises 2*a* and *b*, 3*a*, and 4*b, e,* and *g*. It contains partial solutions or hints for Exercises 2*e* and 4*f*.

a	U	5 MIN	5 MIN	5 MIN	5 MIN	5 MIN	5 MIN	5 MIN
	p	1/100 = .01	1/20 = .05	1/10 = .1	1/5 = .2	1/4 = .25	1/3 ≈ .33	1/2 = .5
	t							
b	U	10 MIN	10 MIN	10 MIN	10 MIN	10 MIN	10 MIN	10 MIN
	p	1/100 = .01	1/20 = .05	1/10 = .1	1/5 = .2	1/4 = .25	1/3 ≈ .33	1/2 = .5
	t							
c	U	20 MIN	20 MIN	20 MIN	20 MIN	20 MIN	20 MIN	20 MIN
	p	1/100 = .01	1/20 = .05	1/10 = .1	1/5 = .2	1/4 = .25	1/3 ≈ .33	1/2 = .5
	t							

TABLE 19-4

ORGANIZING TASKS TO REDUCE THE EFFECT OF INTERRUPTIONS

Sometimes we are reluctant to spend time needed to organize the details of a task. For example, when an English composition is due, we want simply to write it and get it done rather than first taking time to organize our thoughts with an outline and then taking more time for the writing.

Waiting to start a task until we have first taken time to organize it is an approach that many of us resist. We question whether taking extra time just to get organized can possibly save time overall. Perhaps calculations you have already done convince you of the usefulness of organization as a timesaving strategy. Example 2 is designed to convince you further; it describes a situation for which this approach is definitely beneficial.

Example 2: Jim and Jean ▶

Two political science students, Jim and Jean, have each been assigned to write a brief paper that analyzes and compares different voting methods. Each estimates that the rough draft will take 1 hour of intense concentration—that is, it must be done without interruptions. Any interruption will be so devastating that starting over will be required.

For both Jim and Jean interruptions are independent and occur randomly—on the average of once every 20 minutes.

Jim

Jim started writing his rough draft at 6 P.M. yesterday. Under the given conditions, what is the expected time at which he finished?

Analysis

The uninterrupted time required is

$$U = 60 \text{ minutes}$$

The likelihood of interruption during any given minute is

$$p = \frac{1}{20} = .05$$

Substitution of these values in formula 1 gives

$$t = \frac{1}{.05}\left[\frac{1}{(1 - .05)^{60}} - 1\right] \approx 414 \text{ minutes}$$

Thus Jim's rough draft required an estimated 7 hours. If his interruptions were brief and did not add significantly to this time, his expected completion time was around 1 A.M. today. (It is important to note that the formula for t measures only time on task and does not include time actually spent dealing with lengthy interruptions. We keep track only of time spent working on the desired task and do not include time spent on interruptions.)

Jean

Jean is astounded when a sleepy-eyed Jim tells her how long it has taken him to complete his rough draft. Because of his experience, Jean decides to try a different approach. Jean believes that with 15 minutes of intense concentration she can outline her paper and divide it into four sections, each of which will take 15 minutes to write. If she is interrupted during any 15-minute section, she will need to start over and spend 15 more minutes on that portion of her task. However, once her outline or any one of the four subsections is complete, no interruption will affect it.

Jean thus has 15-minute tasks to complete. How long will this take her? If she begins at 6 P.M. tonight, when can she expect to finish?

Analysis

The uninterrupted time required is

$$U = 15 \text{ minutes}$$

The likelihood of interruption during any given minute is

$$p = \frac{1}{20} = .05$$

Substitution of these values into formula 1 gives

$$t = \frac{1}{.05}\left[\frac{1}{(1 - .05)^{15}} - 1\right] \approx 23.2 \text{ minutes}$$

This value for t is the expected time required for Jean to complete one 15-minute task. Since she has set up five such tasks for herself, the total time that Jean's rough draft will require is

$$5t \approx 116 \text{ minutes}$$

Jean's rough draft thus will require almost 2 hours for completion.* If she starts at 6 P.M. today, she can expect to be done around 8 P.M. (or later if the actual time needed to deal with the interruptions is substantial). ◀

It is hard to believe that Jim would actually spend 7 hours on a 1-hour paper. We can picture him, after a while, placing a DO NOT DISTURB sign on his door, locking it, and not permitting any more interruptions. Avoiding interruptions is the easiest way to deal with them. But some people—friends, parents, teachers, managers—have a responsibility to deal with unscheduled as well as scheduled requests for their time. Since they cannot refuse interruptions, they must schedule around them. Their best defense is to do as Jean did—to devote time to organizing their tasks into subtasks of shorter duration, so that the devastating effects of interruptions are less costly.

The following observation can be helpful in estimating the length of time to complete a task that is subject to interruptions. (For background evidence for the observation, refer to the results of Exercise 4.) A task for which the required uninterrupted time U is the same as the expected time between interruptions (which is i or, equivalently, $1/p$) will take an average of about twice as long as it should (2U) because of time wasted by interruptions.

This observation leads to a rule of thumb for deciding whether to invest time to organize tasks into subtasks:

- If you expect, on the average, i time units between interruptions, and if you value your time, then you should organize your tasks into subtasks each requiring a length of time substantially less than i.

In Example 2, Jean, whose expected time between interruptions was 20 minutes, organized her assignment into five 15-minute subtasks. Her method offered substantial time savings over Jim's "do it all at once" method, even though she planned for a total of 75 minutes whereas Jim planned for only 1 hour.

---

*This estimate, which predicts *average* time for completion, is more likely to be close to the actual time for Jean, who has five tasks to complete, than for Jim, who has only one.

The time savings that resulted from the extra time Jean invested in organization illustrates a principle that any person who values time and who is beset with interruptions should seriously consider:

> Time invested in organizing a lengthy and complex task into shorter and simpler subtasks can be well rewarded by a reduction in time wasted because of interruptions.

Although the examples and exercises encourage you to examine the effects of interruptions through calculations using formula 1, in real problems of your own, interruptions may not be independent and devastating—and you may not be able to supply good estimates for U and p. Thus you will be unable to use formula 1 and calculate t.

However, even when the formula cannot be used, these ideas about interruptions can be useful in a general way. Your recognition that you may be able to shorten the overall time needed for certain tasks by organizing them into shorter subtasks can increase your control over your time schedule and make you less vulnerable to the setbacks caused by interruptions.

Exercise 5

Examine the results of Exercise 4; to what extent do they support the rule of thumb stated above? Under what circumstances would you be unwilling to follow that rule?

Exercise 6

a Describe the probable attitude of someone who believes Parkinson's law toward the process described in this chapter for organizing tasks to reduce time wasted because of interruptions.

b How does the remark by Christopher Robin relate to the content of this chapter?

c Discuss Henry Ford's statement, "Nothing is particularly hard if you divide it into small jobs." To what extent do you agree with it? To what extent do you disagree? Illustrate with examples.

"Organization is what you do before you do something, so that when you do it, it's not all mixed up."

Christopher Robin in A. A. Milne's *Winnie the Pooh*

Exercise 7

Recall Gwynne, introduced at the beginning of the chapter. Gwynne is a resident counselor in a dormitory. Because of the responsibilities of her job, Gwynne's study schedule is subject to interruptions. On a particular evening, Gwynne starts at 9 P.M. trying to complete a difficult mathematics problem that

Nothing is particularly hard if you divide it into small jobs.

Henry Ford

she estimates will take 15 minutes of uninterrupted concentration. Any interruption will require her to start over. At this time of evening Gwynne estimates that her interruptions occur randomly and independently and, on average, about once every 10 minutes.

a Under the stated conditions, how long can Gwynne expect to spend on her 15-minute problem?

b Suppose that Gwynne is willing to spend 5 (uninterrupted) minutes organizing some of the problem information; as a result she will be able to complete the problem in three additional uninterrupted 5-minute intervals. If interruptions occur with the pattern given above, how long will this scheme for solving her problem be expected to take?

c Suppose that Gwynne resists organizing her solution method as described in *b*. She says, "Why should I deliberately choose to spend 20 minutes on a 15-minute problem?" Develop a response to her question.

Exercise 8: Parable of the Watchmakers*

Once upon a time, so the story goes, there were two watchmakers—Hora and Tempus. Each produced very fine watches that began to be in great demand. Their workshop telephones rang frequently, bringing orders from new customers.

Both men made watches that consisted of about 1000 parts. Tempus had constructed his so that if several pieces were assembled and he had to put the assembly down—for example, to answer the telephone—all the pieces fell apart and had to be reassembled from scratch. Naturally, the better his customers liked his watches, the more they told their friends—who interrupted his work to place orders for more watches. As a result it became more and more difficult for Tempus to find enough uninterrupted time to finish a watch.

Hora, on the other hand, had designed his watches so that he could put together components of about 10 parts. As with Tempus, if Hora was interrupted while working on a particular assembly, the pieces fell apart and had to be reassembled from scratch. But the 10-part components, once completed, were "stable" and could be set aside for later use as a unit. Ten of these components could later be assembled into a larger stable subsystem. Eventually, the assembly of the 10 larger subsystems constituted the whole watch.

Hora prospered while Tempus became poorer and poorer and finally lost his shop. What was the reason?

*My source for this tale is Herbert A. Simon, *The Sciences of the Artificial,* 2d ed., MIT Press, Cambridge, Mass., 1981. The development of stable subsystems is described there as a critical factor in determining the speed of biological evolution or the evolution of complex social systems.

To answer this question, we need some additional numerical information. Let us use as a time unit the length of time required to add 1 part to the watch assembly. Then, for specificity, let us suppose that the probability of interruptions during any given time unit is $p = 1/100 = .01$. From the statement of the parable, we know that interruptions are devastating; we assume that they are also independent.

a Under the stated conditions, with startovers due to interruptions, how many time units will Tempus need to complete a watch of 1000 parts?

b Using the method of assembling 10-part components, how many time units will Hora need to complete a watch of 1000 parts?

c If a time unit (the time required to add 1 watch part to the assembly) is 5 seconds long, how long can it be expected to take Hora to complete a watch? For Tempus to complete a watch? Why did Hora prosper and Tempus become poorer and poorer?

d Traditionally, a parable has a moral. State a moral that can be learned from this one.

Exercise 9

Jane Jacobs is an assistant vice president for a large financial institution. In addition to coping with the day-to-day responsibilities of her present position, Jane is anxious to show that she is capable of a higher level of leadership. She believes that she has good ideas about the future growth of her institution. But, for her input to receive the serious consideration of the president and her associates, her ideas must be thought through carefully and presented as a written proposal. Jane has been keeping notes of her ideas but estimates that she needs an uninterrupted interval of 2 hours to get the report written. Interruptions—matters needing her immediate attention—come on the average of once every ½ hour.

Instead of using minutes as her unit of time, Jane uses quarter hours. ("In less than 15 minutes, I can't get a thing done," she comments.) Her report will require $U = 8$ consecutive uninterrupted quarter hours. In a given quarter hour she has probability $p = 1/2 = .5$ of interruption. Interruptions for Jane are independent and devastating. When an interruption occurs, it completely diverts her thinking, so that she must rethink her reasoning up to the point of interruption when she returns to work on her report.

a Using the information given above, and formula 1, estimate the number of working hours it will take for Jane to complete her report.

b The result in *a* might be described as "ridiculous." We can hardly expect Jane to persist and invest over 100 hours in writing her report. After a while, we would expect her either to give up or to give her secretary orders to prevent all interruptions until the report is finished.

However, avoiding interruptions could be very undesirable. These so-called "interruptions" may be events that are part of her job responsibility. (The term *interruptions* implies only that the events are unschedulable; it says nothing about whether or not they are necessary or desirable.)

There are other alternatives for Jane to consider. She could structure her report writing in a different way. Suppose that, after thinking about the problem, Jane estimates that she could complete her report in six half hours. She would use the first half hour to organize her notes into an outline with four main parts. The next four half hours would be used to write the four main parts of the report. The final half hour would be used to edit and synthesize the four parts into a unified report. But Jane resists this alternative. "This way would require me to devote 3 hours to a task that should take only 2 hours," she protests.

Calculate an estimate of the length of time it will take for Jane to complete her desired report using this scheme of organization.

c On the basis of the results calculated in *a* and *b* and for time-organization schemes that you invent, develop a recommendation for Jane. What do you advise? Explain why.

d Reconsider *b* using minutes as the time unit. In this case $U = 30$ and $p = 1/30$. Why is the calculated length of time so much less in this case?

Exercise 10

For this exercise you will find it useful to refer back to the results of Exercise 4. In each of situations *a* to *e*, calculate the amount of time that could be saved by reorganizing an uninterruptible 20-minute task (1) into two 10-minute subtasks; (2) into four 5-minute subtasks.

a The probability of interruption during any given 1 minute is $p = .01$.

b $p = .05$

c $p = .10$

d $p = .20$

e $p = .25$

f If the reorganization of a 20-minute task into two 10-minute uninterruptible subtasks will require an extra time of 5 uninterrupted minutes—just to get organized—in which of the situations above would this extra time for organizing be worthwhile?

g If the reorganization of a 20-minute task into four 5-minute uninterruptible subtasks will require two extra uninterrupted 5-minute intervals—just to get organized—in which of situations *a* to *e* would the extra organizational time be a worthwhile investment?

Exercise 11: Invent

Perhaps you know someone like Fred Biddlesmith, who is on academic probation at a nearby college. Last term, as a first-semester freshman, Fred did not do well in his courses. In particular, he failed his introductory statistics course. Fred is a bright young man who is entirely capable of college work. Nevertheless, he has had difficulty completing the tasks he starts. With the discussions and examples of this chapter in mind, invent a description of what may have been Fred's situation. How might he remedy things in the future?

Exercise 12

Kim McNulty was a journalism major in college and wrote for various student publications. She also worked in the college public relations office, providing text and layouts for college publications. Now Kim is married and staying at home caring for two young children. She would like to do freelance writing but says, "I never have big enough blocks of time to get anything done." Develop some suggestions for Kim.

Exercise 13

Consider the activities of faculty members Kline and Lewis last Tuesday morning. Both were in their respective offices preparing materials for their 11 A.M. classes.

Professor Kline's door was open and, as she worked at her desk, students dropped in to chat and to ask questions about their homework assignments. Feeling responsible to help them, Professor Kline pushed aside her preparations and dealt with their questions. The students went away satisfied. At 11 A.M., Professor Kline conducted a disorganized, poorly prepared class.

Professor Lewis, on the other hand, spent Tuesday morning with the office door closed. Instead of responding to knocks on his door, he diligently prepared his lesson. Some of his students were disgruntled about their professor's unavailability to help them. But Professor Lewis' 11 A.M. class was well organized and effective.

a Discuss the conflicting responsibilities that professors Kline and Lewis face.

b Formulate suggestions for each of these professors about how to meet their conflicting responsibilities.

Exercise 14

It is a common belief that since interruptions are unscheduled events, one cannot plan for them. Develop a convincing argument that this is not so. Argue that even though interruptions are unscheduled events, one can make plans that allow for them and minimize their effect.

Exercise 15

Make a list of tasks for which interruptions cause you serious setbacks. For which of these tasks would you be able to reduce the overall time required if you spent time organizing them into shorter subtasks?

Exercise 16: Invent, Journal

Focus on an interruption problem (either real or hypothetical) that interests you. Develop a detailed description of the problem; assess whether the interruptions are independent and devastating; estimate values for U and p and apply formula 1. Consider several ways of dividing the task into subtasks; attempt to discover a best way of organizing and completing the task.

Exercise 17: Test Yourself

Develop a list of new terms and key ideas presented in this chapter. Supply an example of each item on your list.

Note The Appendix contains complete answers for Exercises 7*b*, 8*a* and *b*, 9*d*, and 10*a*, *b*, and *f*. It contains partial solutions or hints for Exercises 6*a* and *c*, 13, 15, and 16.

P A R T S I X

GROUP DECISION PROCESSES

PART CONTENTS

CHAPTER 20

INDIVIDUAL THRESHOLDS AND GROUP BEHAVIOR

Chain Reactions
Threshold Values and Possible Behaviors
Role of Environment in Threshold Analysis
References
In Conclusion

CHAPTER 21

METHODS OF RECOGNIZING CONSENSUS

Introduction
Sequential Pairwise Voting—Avoid It If You Can
Different Voting Methods Lead to Different Decisions—Beware
Five Voting Decision Methods
Criteria for a Voting Decision Method

CHAPTER 22

THE PRISONER'S DILEMMA: A MODEL OF THE CONFLICT BETWEEN INDIVIDUAL AND GROUP BENEFITS

Introduction
What Is "Rational" Decision Making?
Prisoner's Dilemma
Prisoner's Dilemma as a Model for Other Decision Problems
Circumventing the Dilemma
Suggestions for Further Reading
Multiperson Prisoner's Dilemma Situations

CHAPTER 20

INDIVIDUAL THRESHOLDS AND GROUP BEHAVIOR

The dogmas of the quiet past are inadequate to the stormy present.

—Abraham Lincoln

CHAIN REACTIONS

Sometimes it is a puzzle why two apparently similar groups exhibit opposite behaviors. For example, in one college students may organize and stage a protest against a proposed tuition increase; in another, similar college no action is taken. In one of two similar high schools shaving heads may catch on among the boys while in the other this fad never takes hold. One community may develop a strong protest movement preventing construction of a nuclear plant while another, similar community permits such construction. Moreover, such opposite behaviors can occur even when the individuals in one group hold attitudes similar to those of individuals in the other; Example 1, below, provides a simple illustration of how this can happen.

A somewhat different version of this chapter, "I Will If You Will . . . ," has been published as UMAP Module 539 and has appeared in *UMAP Journal,* vol. 11, no. 4, 1981, pp. 81–112.

A key to such situations is this: In many situations that involve choices between opposite behaviors, such as protesting and not protesting or taking up a fad and letting it pass, many individuals are not firmly committed to one choice or the other. Instead, each has a certain threshold *t:* he or she will engage in a certain behavior when at least *t* others do so.

Example 1: An Antinuclear Protest ►

Near to each of two cities, Bordenton and Grand Forks, there are sections of land which are being considered as sites for proposed nuclear power plants. Within each of these cities 100 people have formed an organization called Nuclear Concern. These people are determined to educate themselves about the merits and dangers of nuclear plants and then take appropriate action. The degree of concern among group members varies from person to person. Each group met recently to decide whether to stage a protest at the executive offices of the electric power company that has proposed nuclear construction. The attitudes of group members and the outcomes of their meetings are given below.

Bordenton Nuclear Concern

1 member favors protest even if no one else does.
1 member favors protest if at least 1 other agrees.
1 member favors protest if at least 2 others agree.
1 member favors protest if at least 3 others agree. . . .
1 member favors protest if at least 99 others agree.
The group has decided to stage a protest.

Grand Forks Nuclear Concern

1 member favors protest if at least 1 other agrees.
1 member favors protest if at least 2 others agree.
1 member favors protest if at least 3 others agree. . . .
1 member favors protest if at least 99 others agree.
1 member does not favor protesting no matter how many others do.
The group has decided not to stage a protest.

Examination of the attitudes of the members of the two groups reveals that they are almost identical. However, in the Bordenton group the single person who favored protest regardless of the attitudes of others started a "chain reaction." Once this person's attitude became public at the meeting, the attitudes of others permitted them to agree; thus the decision to protest was born. In the Grand Forks group, however, there was no person who would take the first step. Although many members leaned toward protest, there was no one whose commitment to protest was unconditional; thus the protest never materialized. ◄

Exercise 1

a For Example 1, invent a brief description of how the meeting of the Bordenton Nuclear Concern group might have proceeded toward the decision to stage a protest.

b Invent a description of how things may have gone at the Grand Forks meeting, where the group did not decide to protest.

Exercise 2

a Suppose that Mike Norad—who is committed to protesting nuclear power plants no matter what others do or don't do—moves to Grand Forks and joins the Nuclear Concern group there. Describe what can be expected to happen.

b Observers of Nuclear Concern's activity after Mike Norad's arrival might conclude that he changed a lot of people's minds about protesting. Develop an explanation that would show these observers how the protest could have evolved without anyone's changing an earlier attitude.

Exercise 3

Make a list of pairs of opposite behaviors for which individuals may decide what to do on the basis of how many others are doing likewise. Here are several examples to start your list:

Obeying (or exceeding) the 55-mile-per-hour speed limit
Studying hard (or not) for an upcoming test
Participating in a standing ovation (or remaining seated) after a mediocre concert

Note The Appendix contains partial solutions or hints for Exercises 1, 2*a*, and 3.

THRESHOLD VALUES AND POSSIBLE BEHAVIORS

Example 1 showed how members of two groups could have similar attitudes and yet exhibit different behaviors. That example is useful to keep in mind whenever one observes a swing in social or political behavior. Dramatic shifts in group behavior may take place while many of the individuals involved change their attitudes only slightly, if at all.

Definitions

To make the discussion of the relationship between individual attitudes and group behavior more efficient, we will need some new notation and terminology.

For convenience, when we want to discuss some optional behavior—such as staging a protest, studying for a test, or wearing a bow tie—we can refer to it as *behavior X.*

If a person states, "I will choose behavior X when I will be part of a group of size at least t that does chooses X," then t is called the *threshold value* for that person. (For example, John may say, "I will stay home from the party tonight and study for my mathematics quiz if at least half of my class will be doing so. Otherwise I'll take my chances and go to the party instead." If John's class has 30 students, then John's study threshold value is $t = 15$; or, in terms of percent, we can say $t = 50$ percent.)

When the members of a group are considering behavior X, let us use S_t to denote the total number of members who are satisfied by a threshold of t or less. If we return to Example 1 and restate the information about members' attitudes in terms of t and S_t, the results are as shown in Table 20-1.

Much of the analysis of behavior involving threshold values is based on the *threshold criterion:* individuals' attitudes will support group behavior only if

$$t \leq S_t$$

(that is, when the threshold does not exceed the number satisfied at that threshold).

Examination of Table 20-1 shows that in Grand Fork individual attitudes support only the group behavior "no protest." In Bordenton, on the other hand, individual attitudes permit any number to protest (since $t \leqslant S_t$ always); as members get to know each other's attitudes, we would expect a chain reaction that will produce a unanimous decision to protest.

TABLE 20-1
Threshold information for antinuclear protests

BORDENTON NUCLEAR CONCERN		GRAND FORKS NUCLEAR CONCERN	
t	S_t	t	S_t
0	0	0	0
1	1	1	0
2	2	2	1
3	3	3	2
4	4	.	.
.	.	.	.
.	.	.	.
.	.	98	97
99	99	99	98
100	100	100	99

Example 2 illustrates another application of the threshold criterion—in a paradoxical situation. The paradox is this: most group members favor a certain behavior (attending a "problem session") but their individual threshold values do not support it.

Example 2: An "Optional" Problem Session ▶

A phenomenon familiar to many mathematics and science students is the "dying problem session." In a course where most class time is devoted to introduction of new material, students often request special "problem sessions" for discussion of assigned out-of-class work.

Typically, in such situations, the instructor asks how many class members are interested in a problem session, and most say that they are. The instructor then schedules sessions at a time convenient to as many as possible, and the sessions begin. Since these sessions are not part of the formal course, attendance at them is optional.

More often than not, the problem session lasts only a few weeks. Attendance is good at the first session but decreases steadily after that until finally no one shows up. Why does this occur? Do students really not want the extra sessions after all? Are the sessions poorly conducted or of little value? What factors lead to their seemingly inevitable death?

Let us turn to threshold analysis for a possible explanation and consider a mathematics class with 50 members, 40 of whom want the problem sessions and attend the first session. However, each attendee is conscious of the number of others in attendance. No one wants to be the center of attention; each wants to listen to the questions and discussions of others. Students' attitudes toward attendance are summarized in Table 20-2.

Because the pairs of values in Table 20-2 for t and S_t have $t \leq S_t$ only when $t = S_t = 0$, the only group behavior supported by individual attitudes is for no one to attend.

Initially, of course, the class does not know that zero attendance is the only satisfying group behavior; this is arrived at by trial and error.

THRESHOLD VALUE t	CLASS MEMBERS WITH THRESHOLD t*	CLASS MEMBERS WITH THRESHOLD t OR LESS, S_t
0	0	0
10	0	0
20	10	10
30	10	20
40	10	30
50	10	40
Number of class members who will never attend	10	

*These figures were obtained by means of a survey.

TABLE 20-2
Threshold information for attendance at mathematics problem session

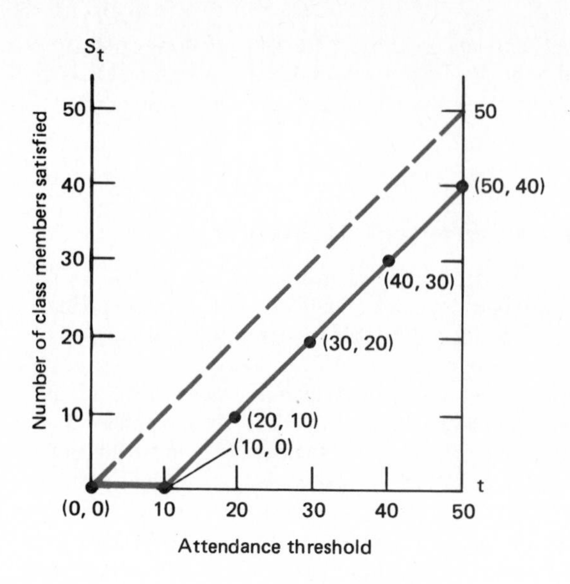

FIGURE 20-1
Graph for threshold information for problem session attendance

Here is one possible description of the "dying" process: Suppose that each person who attends the problem session on a given day uses that day's attendance as an estimate of the number who will attend the next session. As was noted earlier, 40 students attend the first session. However (see Table 20-2), only 30 students are satisfied with a group of 40; thus only those 30 attend the second session. But there are only 20 students who are satisfied with a group of 30; thus the third session attracts only 20 students. Of these 20, only 10 are satisfied with the size of the group, and therefore the fourth sessions has only those 10 students. These 10 students are unwilling to be part of such a small group, and no one shows up for the next session. Paradoxically, although most of the class wanted the sessions, they did not continue. Because individuals' attitudes were not unconditional but depended on the attendance of others, what most of the group "wanted" did not take place; instead, the problem session "died."

A graph can provide a useful summary of the threshold-value analysis. In Figure 20-1, the points that represent the pairs (t, S_t) of Table 20-2 have been located and joined.* Points on or above the dashed line (called the *equilibrium line*) describe pairs for which the threshold criterion $t \leq S_t$ is satisfied. In Figure 20-1 only one point satisfies the threshold criterion, the point (0, 0). This point corresponds to the "dead" problem session with zero attendees. ◀

---

*Points along the segments joining data points have no meaning in this analysis. These segments are simply a visual aid; they connect the data points and make it easier to see the trends.

Interpreting Threshold Information

Our observations suggest an amplification of the threshold criterion. Suppose that at a given time, t people are engaged in (or considering) a certain behavior X. The following can be expected:

1 If $t > S_t$, then the number of people who will do X will tend to decrease until an equilibrium with $t = S_t$ is reached.

2 If $t \leq S_t$, then the number of people who will do X will tend to increase either until the entire group is doing X or until a further increase would result in $t > S_t$.

Exercise 4

a Explain how the threshold criterion and statement 1 above are illustrated by the Grand Forks Nuclear Concern group in Example 1 and the mathematics class in Example 2.

b Explain how the threshold criterion and statement 2 above are illustrated by the Bordenton Nuclear Concern group in Example 1.

Exercise 5

New regulations have been imposed that restrict the use of the piano lounge for group meetings. Two different commuter groups—the "Other Side of 30" group and the "Married Couples" group—have been accustomed to hold weekly meetings there. Both groups have met—with 10 members present in each case—to decide whether to abide by or to violate the restrictions. Their attitudes are summarized in Tables 20-3 and 20-4 (page 344).

a Complete Tables 20-3 and 20-4 by filling in values for S_t.

b Graph the pairs (t, S_t) for the threshold information given in Tables 20-3 and 20-4. Use the format in Figure 20-2; join data points as was done in Figure 20-1.

c Describe the group behavior that is likely to result from the meeting of the "Other Side of 30" group. Will the group violate the restrictions on the use of the piano lounge? Explain how you arrived at your answer.

d What behavior can be expected from the "Married Couples" group? Will they violate the restrictions? Explain.

TABLE 20-3
Threshold information for "Other Side of 30" commuter group

VIOLATE RESTRICTIONS; THRESHOLD VALUE t	GROUP MEMBERS WITH THRESHOLD t	GROUP MEMBERS WITH THRESHOLD t OR LESS, S_t
0	0	
1	2	
2	2	
3	1	
4	0	
5	5	
6	0	
7	0	
8	0	
9	0	
10	0	

TABLE 20-4
Threshold information for "Married Couples" commuter group

VIOLATE RESTRICTIONS; THRESHOLD VALUE t	GROUP MEMBERS WITH THRESHOLD t	GROUP MEMBERS WITH THRESHOLD t OR LESS, S_t
0	0	
1	0	
2	0	
3	2	
4	0	
5	3	
6	5	
7	0	
8	0	
9	0	
10	0	

FIGURE 20-2

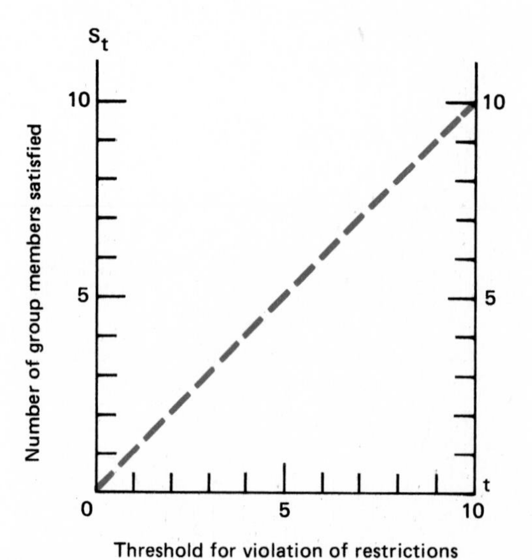

Exercise 6

a Explain why S_t is nondecreasing; that is, as *t* increases, S_t either stays the same or increases.

b How does the nondecreasing nature of S_t show up in graphs of threshold data?

Exercise 7

Class participation is a phenomenon that can be analyzed using threshold values. Suppose that a confidential survey is made of the 50 students in a psychology class, with each member being asked to complete the following statement by supplying a threshold value: "I will participate in class when at least ____ of the class members participate." Results of the survey are given in Table 20-5.

a Complete Table 20-5 by calculating the values of S_t.

b Using the format of Figure 20-1, make a graph of the pairs (t, S_t) from the data of Table 20-5.

c Examine the results of *a* and *b*. What amount of participation do you expect in this psychology class? (This problem will be discussed in the next section of the chapter. For now, you may find it baffling; try it, anyhow. Speculate about what you think will happen and why.)

TABLE 20-5
Threshold information: Participation in psychology class

CLASS PARTICIPATION; THRESHOLD VALUE t	CLASS MEMBERS WITH THRESHOLD t	CLASS MEMBERS WITH THRESHOLD t OR LESS, S_t
0	0	
5	0	
10	5	
15	0	
20	15	
25	10	
30	10	
35	0	
40	0	
45	0	
50	5	
Number of class members who will never participate	5	

Exercise 8

Suppose that in Example 2 the mathematics instructor is unable to find a room large enough to accommodate more than 30 students at a single problem session. Therefore two sessions are scheduled, each at a time suitable to half of the class of 50 students. Under these new circumstances, the students have revised their threshold values to those given in Tables 20-6 and 20-7. The two halves of the class are designated group A and group B.

a Complete Tables 20-6 and 20-7 with values of S_t.

b Using the format of Figure 20-1, make graphs of the pairs (t, S_t) from the data of Tables 20-6 and 20-7.

c Examine the results of *a* and *b*. What do you expect to happen in each of groups A and B?

d Compare the results of this exercise with Example 2. Even if room sizes did not require it, might it be worthwhile to divide the class into two problem-session groups? Discuss the merits of this "divide and conquer" strategy.

TABLE 20-6
Threshold information: Attendance at problem session, group A

ATTENDANCE: THRESHOLD VALUE t	GROUP A MEMBERS WITH THRESHOLD t	GROUP A MEMBERS WITH THRESHOLD t OR LESS, S_t
0	0	
5	0	
10	0	
15	15	
20	10	
25	0	
Number of group A members who will never attend	0	

TABLE 20-7
Threshold information: Attendance at problem session, group B

ATTENDANCE: THRESHOLD VALUE t	GROUP B MEMBERS WITH THRESHOLD t	GROUP B MEMBERS WITH THRESHOLD t OR LESS, S_t
0	0	
5	0	
10	0	
15	5	
20	10	
25	0	
Number of group B members who will never attend	10	

Exercise 9: Journal

Think of groups to which you belong where individual participation in some activity depends on how many others participate. Try to recall, in particular, at least two instances in which group behavior made a large swing (from few participants to many or vice versa). Analyze these situations; was the swing the result of a major shift in attitudes or only a slight shift followed by a chain reaction as thresholds were met?

Note The Appendix contains complete answers for Exercises 6 and 8*a*. It contains partial solutions or hints for Exercises 4*a*, 5*c*, and 8*c* and *d*.

ROLE OF ENVIRONMENT IN THRESHOLD ANALYSIS

In Exercise 7, threshold information was provided to describe the attitudes of psychology students toward class participation. Table 20-8 adds to the information given in the exercise by supplying values for S_t. The discussion of possible class behaviors will be easier in terms of a graph. Pairs (t, S_t) in Table 20-8 have been graphed in Figure 20-3.

When the psychology class first meets, it is likely that the students do not know each other's attitudes toward participation. Some trial-and-error experimentation is likely; students will try participating and see what happens.

Five students in the class will never participate (see Table 20-8); thus the greatest possible number of participants is 45. Table 20-9 presents the various participation levels that might occur during the first week of classes and analyzes the results.

TABLE 20-8
Threshold information:
Psychology class participation

CLASS PARTICIPATION; THRESHOLD VALUE t	CLASS MEMBERS WITH THRESHOLD t	CLASS MEMBERS WITH THRESHOLD t OR LESS, S_t
0	0	0
5	0	0
10	5	5
15	0	5
20	15	20
25	10	30
30	10	40
35	0	40
40	0	40
45	0	40
50	5	45
Number of class members who will never participate	5	

FIGURE 20-3
Graph for threshold information on participation in psychology class

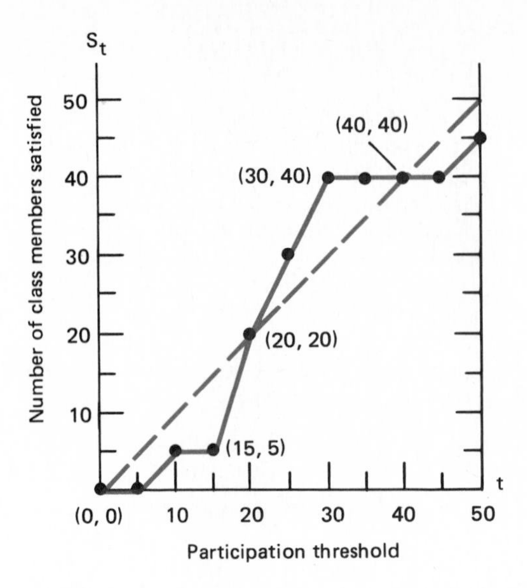

TABLE 20-9
Consequences of various initial participation levels in psychology class

BEHAVIOR OBSERVED DURING WEEK 1	RELEVANT DATA POINT (t, S_t)	ANALYSIS
45 students participate	(45, 40)	When t = 45 students participate, only S_t = 40 students are satisfied and likely to continue. Thus the number participating can be expected to drop to 40.
40 students participate	(40, 40)	When t = 40 students participate, then S_t = 40 students are satisfied and willing to continue. This is an equilibrium situation and can be expected to persist.
35 students participate	(35, 40)	When t = 35 students participate, S_t = 40 students are willing to participate. Thus the number who participate can be expected to climb to 40.
30 students participate	(30, 40)	When t = 30 students participate, S_t = 40 students are willing to participate. Thus the number who participate can be expected to climb to 40.
25 students participate	(25, 30)	When t = 25 students participate, S_t = 30 students are willing to participate. Thus the number who participate can be expected to climb to 30.
20 students participate	(20, 20)	When t = 20 students participate, then S_t = 20 students are satisfied and willing to continue. This is an equilibrium situation, but unstable.
15 students participate	(15, 5)	When t = 15 students participate, only S_t = 5 are satisfied and likely to continue. Thus the number participating can be expected to drop to 5.
10 students participate	(10, 5)	When t = 10 students participate, then only S_t = 5 are satisfied and will continue to participate. Thus the number participating can be expected to drop to 5.
5 students participate	(5, 0)	When t = 5 students participate, no students are satisfied with participation. Thus the number participating can be expected to drop to 0.
0 students participate	(0, 0)	When t = 0 students participate, S_t = 0 students are satisfied. Stated differently, all are satisfied not to participate. This is a stable equilibrium situation and can be expected to continue.

Equilibrium Points

On a graph displaying threshold information, a data point (t, S_t) for which $t = S_t$ will be called an *equilibrium point.* Of course, each equilibrium point lies along the dashed line called the *equilibrium line.* An equilibrium point is described as *stable* if the slope* of each segment connecting it to an adjacent data point is less steep than the slope of the equilibrium line. Stable equilibrium points describe group behaviors which, once achieved, are likely to continue. Unstable equilibrium points describe group behaviors which may continue but which are sensitive to slight changes in the group environment.

In threshold situations (such as that of the psychology class) where group attitudes coalesce to yield more than one equilibrium point, the environment in which the group finds itself determines which equilibrium will be achieved. If, in the psychology class, participation during the first week is strongly encouraged by the instructor, then we might suppose that a large number of students will try participating, and (see the analysis in Table 20-9) an equilibrium with 40 students participating is the probable result.

If the instructor is intimidating, the equilibrium with no students participating, corresponding to the data point (0, 0), is likely. Or, if enough students are highly motivated, despite an intimidating instructor, a 20-student level of participation is possible. However, the steepness of the graph (Figure 20-3) around the data point (20, 20) indicates that many students are making their decisions at thresholds near this level and that slight changes in their perceptions may cause major shifts in group behavior. Such an equilibrium is not readily maintained; that is, 20 is an unstable equilibrium.

Guidelines for Application of Threshold Analysis

To aid in the application of these ideas to other analyses of group behavior, the following guidelines are provided.

Type of situation to which this analysis is applicable A situation in which the members of a group must decide individually whether to engage in a certain behavior X, and for many of the members, the decision depends on how many others do likewise.

Information that is needed to perform the analysis Each individual, unless completely unwilling to engage in behavior X, must supply a threshold value t to complete the following statement: "I will engage in behavior X when at least ____ of the group members are doing X."

Mathematics Find values for S_t, the number of individuals satisfied with a threshold of t or less. Graph the pairs (t, S_t) on an appropriately labeled diagram. Identify equilibrium points.

---

*Recall that the *slope* of a (nonvertical) line segment joining two points (x_1, y_1) and (x_2, y_2) is equal to the ratio $(y_2 - y_1)/(x_2 - x_1)$.

Interpretation Group behavior can stabilize with t people engaged in behavior X if (t, S_t) is an equilibrium point. If there is more than one equilibrium point, then environmental factors must be examined to see which equilibrium is more likely.

Because it takes a while for group members to get to know each other's attitudes and to observe each other's behavior, the group's behavior generally does not begin at an equilibrium point. If at a certain time, t persons in a group are observed engaging in behavior X, then the following trends are likely:

1 If the point (t, S_t) is below the equilibrium line, then group behavior will tend toward the value of the first equilibrium point to the left of (t, S_t). That is, the number engaged in behavior X will decrease until equilibrium is reached.

2 If the point (t, S_t) is above the equilibrium line, then the group behavior will tend toward the value of the first equilibrium point to the right of (t, S_t). That is, the number engaged in behavior X will increase until equilibrium is reached.

3 If the point (t, S_t) is on the equilibrium line, then examination of the stability of the equilibrium is the next step. If the equilibrium point is stable, then group behavior is likely to continue with t persons engaged in behavior X. If the equilibrium point is not stable, future group behavior is uncertain. Slight environmental changes—such as an increase in apathy or enthusiasm—can easily cause behavior shifts. If group members learn of each other's attitudes, group behavior will tend toward whichever equilibrium publicized attitudes support.

REFERENCES

The following paper is a valuable reference for those interested in learning more about threshold analysis:

Mark Granovetter, "Threshold Models of Collective Behavior," *American Journal of Sociology,* vol. 83, no. 2, May, 1978, pp. 1420–1443.

Two points Granovetter makes which are pertinent to this discussion are:

1 Thresholds vary from situation to situation. (A person's participation threshold, for example, is not a single number carried from one class to another; instead, it results from the configuration of costs and benefits of different behaviors in a particular situation.)

2 There are cases in which a small change in the distribution of thresholds can generate a large difference in group behavior. (For example, the addition to Grand Forks Nuclear Concern of one person unconditionally committed to protest (see Exercise 2) causes a change in the whole group's behavior.)

A second and highly readable source of further information about threshold analysis is the following book:

Thomas C. Schelling, *Macromotives and Microbehavior,* Norton, New York, 1978.

In his third chapter, Schelling discusses *critical-mass models* and incorporates much of what has been called *threshold analysis* here. The term *critical mass* has been adopted from nuclear engineering, where it refers to the amount of a radioactive substance necessary to sustain a chain reaction. In the Bordenton Nuclear Concern group, the decision to stage a protest can be described in these terms. The one person who would protest even if no one else would provided a critical mass that satisfied the protest thresholds of others, so that a chain reaction took place. Similarly, the outcome of the mathematics problem session in Example 2 can be translated as follows: the problem session died because there was no critical mass to sustain it.

IN CONCLUSION

Threshold analysis supplies a method of analyzing situations rather than solving problems. In other words, it helps us to understand *how things are* instead of telling us what to do. Despite this limitation, threshold analysis can still be useful in problem solving. (For example, those members of the mathematics class with a strong interest in a problem session might use the threshold analysis as a basis for deciding to adjust their thresholds so that a problem session could survive.)

A final word of caution is in order. When you apply threshold analysis to study new situations—real situations that you seek to understand or improve—you may run into difficulties not encountered in the text and exercises of this chapter. Survey data may be hard to get, hard to summarize neatly, and hard to interpret so that you can make any but very tentative predictions about group behavior. Such difficulties are not unusual; seldom do real problems fit exactly into patterns we have learned. However, a variety of approaches, tried with patience and persistence, ultimately will lead to some insights into confounding situations.

Exercise 10

a Table 20-10 is a completion of Table 20-6 (Exercise 8), and Figure 20-4 provides a graph of data pairs (t, S_t). Assess what can be expected to happen in the group A problem session by completing Table 20-11. Analyze possible attendance levels, following the pattern of Table 20-9.

b If 25 students attend the first meeting of the problem session, how many are likely to show up at the second meeting? Explain.

c If 20 students attend the first meeting of the problem session, how many are likely to show up at the second meeting? Explain.

d If illness and studying for tests causes attendance to drop to 15 for one of the meetings, what do you expect to happen as a consequence of this?

e If attendance at a lecture by a renowned chemist causes 15 members of group A to miss one problem session, what do you expect to happen?

f If unexplained factors cause 15 members of group A to miss one of the meetings, what do you expect to happen as a result of this? (Is this question any different from *e*? Explain.)

TABLE 20-10
Threshold information: Group A's attendance at problem sessions

ATTENDANCE: THRESHOLD VALUE t	GROUP A MEMBERS WITH THRESHOLD t	GROUP A MEMBERS WITH THRESHOLD t OR LESS, S_t
0	0	0
5	0	0
10	0	0
15	15	15
20	10	25
25	0	25
Number of group A members who will never attend	0	

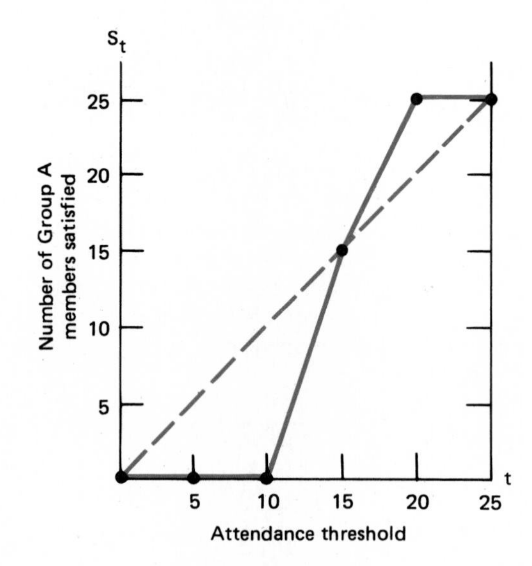

FIGURE 20-4
Graph for threshold information on attendance at problem session: Group A

TABLE 20-11
Consequences of various attendance levels for group A problem session

BEHAVIOR OBSERVED AT A GIVEN SESSION	RELEVANT DATA POINT (t, S_t)	ANALYSIS: WHAT BEHAVIOR IS LIKELY TO FOLLOW?
25 students attend	(25, 25)	
20 students attend	(20, 25)	
15 students attend	(15, 15)	
10 students attend	(10, 0)	
5 students attend	(5, 0)	
0 students attend	(0, 0)	

Exercise 11

Although graphs of threshold data must intersect the equilibrium line—at least at (0, 0) and possibly elsewhere—these intersections need not occur at data points. (For example, see Figure 20-5*a* and *b*.) Since it is stipulated that no meaning is assigned to points on the segments between data points, we face the question how to interpret these crossings.

a Explain why the behavior of the group whose threshold data are summarized in Figure 20-5*a* can be expected to stabilize with 45 members choosing behavior X.

b Explain why the behavior of the group whose threshold data are summarized in Figure 20-5*b* will stabilize with either 0 members or 100 members choosing behavior X.

c Generalize from your analysis of Figure 20-5*a* and *b* to complete the following statements: (1) If a graph of threshold data has a slope less than 1 when it crosses the equilibrium line at a non-data point, then __________. (2) If a graph of threshold data has a slope greater than 1 when it crosses the equilibrium line at a non-data point, then __________.

FIGURE 20-5

S_t

100

80

60

40

20

Number satisfied

(40, 45)

(60, 55)

20 40 60 80 100 t

Threshold value
(for behavior X)

(a)

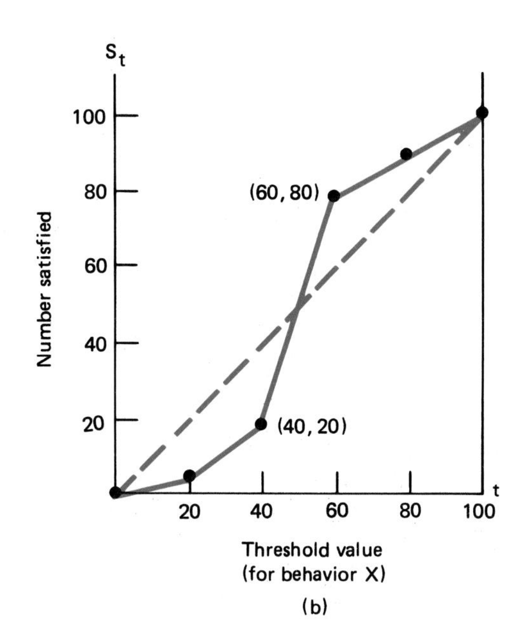

(b)

TABLE 20-12
Threshold information on head shaving

HEAD SHAVING; THRESHOLD VALUE t	MEN WITH THRESHOLD t	MEN WITH THRESHOLD t OR LESS, S_t
0	0	
10	5	
20	15	
30	15	
40	15	
50	15	
60	5	
70	0	
80	0	
90	10	
100	0	
Number who will never shave their heads	20	

Exercise 12

A sociology class was studying how a fad catches on. One student, who had spent her summer at a Jersey shore resort, mentioned a fad that had swept the beach community—young men shaving their heads. As an experiment, the class collected threshold data on head shaving from the 100 men enrolled in that course. The results are given in Table 20-12.

a Complete Table 20-12 by calculating the values of S_t; make a graph of data pairs (t, S_t).
b If, at the time that the class was surveyed to obtain threshold information, none of the 100 men had shaved heads, would you expect any changes to result? Explain why or why not.
c Suppose that 30 of the males in the sociology class have their heads shaved as part of a fraternity initiation. Explain how, as a result of this, 70 of the men could eventually sport shaved heads.

Exercise 13

Several students in Mr. Ogive's statistics class came to him privately to discuss the cheating that was going on among class members. "I don't think people really want to cheat," one student confided, "but the problems are long and hard. A few people started copying solutions, and then others joined them, and now it seems as if almost everyone is cheating—just to pass the course."

Mr. Ogive was dismayed and promised to do what he could to remedy the situation. He thought of gathering threshold data but was afraid that he could

TABLE 20-13
Hypothetical threshold information on cheating

CHEATING; THRESHOLD VALUE t	CLASS MEMBERS WITH THRESHOLD t	CLASS MEMBERS WITH THRESHOLD t OR LESS, S_t
0	0	
10	15	
20	10	
30	10	
40	15	
50	30	
60	10	
70	0	
80	0	
90	0	
100	0	
Number of class members who will never cheat	10	

not get honest data to analyze. Finally, he decided to invent data. Using hypothetical data, he would discuss the problem with his students and hope that they would gain insight from the hypothetical situation which they would transfer to the real situation. The data for a fictitious class of 100 students that Mr. Ogive invented for class discussion are given in Table 20-13. The threshold values are responses to the statement, "I will cheat when at least ___ of the class members cheat."

a Analyze the threshold information in Table 20-13 using the methods that have been discussed. What group behavior does your analysis predict?

b After a class discussion of the threshold information, Mr. Ogive made the following comment: "Often we assume that for group behavior to change from one type to its opposite there must be major changes in the attitudes of group members. This is not so. Small changes in individual attitudes can make a large change in group behavior because of the cumulative effect of such changes." To illustrate his point, he referred to the threshold data in Table 20-14. "Suppose," he said, "that everyone except the confirmed noncheaters has a 10 percent change in attitude. That is, suppose each of the 90 students who might cheat raises his or her cheating threshold by 10 percent."

Apply the methods of threshold analysis to the data of Table 20-14. Compare these results with your analysis of data from Table 20-13. What different behaviors might result? Under what types of circumstances?

c What insights into their own behavior might Mr. Ogive's class gain from consideration of these two sets of cheating data?

d Mr. Ogive's analysis focused on students' attitudes and on what changes the students might make to improve the cheating situation. If, however, students are cheating "just to pass the course," what changes might need to be made by the professor himself to reduce the cheating?

TABLE 20-14
Hypothetical threshold information on cheating: Thresholds raised 10 percent

CHEATING; THRESHOLD VALUE t	CLASS MEMBERS WITH THRESHOLD t	CLASS MEMBERS WITH THRESHOLD t OR LESS, S_t
0	0	
10	0	
20	15	
30	10	
40	10	
50	15	
60	30	
70	10	
80	0	
90	0	
100	0	

Exercise 14*

Consider Bowie, a hypothetical western town in which the typical person's decision whether or not to carry a gun depends on how many others in the town carry guns. In general, the more guns people see worn by others, the more they will be motivated to carry guns themselves. Some may wear guns regardless of what others do; some may refuse to wear guns no matter how many others wear them; but most people will wear guns if enough others do. The meaning of *enough* varies from person to person.

Let us apply threshold analysis to this situation. Suppose that the 5617 adult residents of Bowie have been surveyed with the following results (for convenience, percents are used instead of numbers):

- 90 percent responded: "I prefer a situation in which no (adult) resident of Bowie carries a gun."
- 10 percent responded: "I prefer a situation in which all (adult) residents of Bowie carry guns."
- Responses to the statement "I will carry a gun when at least *t* percent of the (adult) residents of Bowie do so" are summarized in Table 20-15. The percent of residents who said, "I will never carry a gun" is recorded in the last line of Table 20-15.

Apply threshold analysis to the data in Table 20-15 and describe the group behavior that is likely to result in Bowie as a result of the individual attitudes expressed.

*This exercise is adapted from an example suggested by Thomas C. Schelling of Harvard University.

TABLE 20-15
Threshold data for gun carrying

THRESHOLDS, PERCENT t	BOWIE ADULT RESIDENTS WITH GIVEN THRESHOLD, PERCENT
0	0
25	35
50	40
75	0
100	15
Percent of Bowie adults with no gun-carrying threshold	10

TABLE 20-16
Reinterpretation of threshold data on gun carrying

REFUSAL THRESHOLDS, PERCENT t	ADULTS WITH GIVEN THRESHOLD, PERCENT
0	0
25	25
50	0
75	40
100	35
Percent of adults with no refusal threshold	0

Exercise 15

A resident of Bowie, Mr. Disarm, who observes that most of the residents of Bowie prefer that no one carry a gun, becomes concerned about the way the situation had been described. He thinks that people's thresholds for gun carrying were affected by the way the problem was posed. He proposes a different approach: consider instead refusing to carry a gun and determine people's thresholds for this behavior. Disarm takes the data of Table 20-15 and reinterprets it to obtain Table 20-16. Explain how Disarm could have obtained the values in Table 20-16. Apply threshold analysis and describe the group behavior that is likely to result in Bowie as a result of these attitudes. Compare and contrast this analysis with that of Exercise 14.

Exercise 16

Mr. Disarm is chagrined by the results of his efforts at reformulating the gun problem—he was so sure that his new way of looking at the problem would make a difference. But on his way to the paper shredder to destroy the evidence of his failed attempt, he meets Ms. Cooperation, who asks why he is so dismayed and urges him not to give up yet. She reasons thus: Because the original survey data were collected in response to the statement, "I will carry a

TABLE 20-17
New data from gun survey

REFUSAL THRESHOLDS, PERCENT t	ADULTS WITH GIVEN THRESHOLD, PERCENT
0	0
25	25
50	50
75	15
100	0
Percent of adults with no refusal threshold	10

gun when at least t percent of the adult residents of Bowie do so," the survey contained the subtle suggestion that carrying a gun was cooperative behavior. Even when Mr. Disarm had looked at the data from a different point of view, this built-in bias could not be removed.

Together, Cooperation and Disarm decide to gather new data. They ask the 5617 adult residents of Bowie to respond to a new question: "I will *refuse to carry a gun* when at least t percent of the adult residents of Bowie refuse to carry a gun."

As all westerns must, this story has a happy ending. Discover the happy ending by analyzing the threshold data from the second survey, summarized in Table 20-17. Analyze the new data to discover what behavior is likely to result.

Exercise 17: Invent

As summer progresses toward autumn, many communities begin to worry about whether their water supply will last until it is replenished by rainfalls. At such times, these communities may impose restrictions on use of water. Such restrictions are hard to enforce, however. Furthermore, if one person is seen watering the lawn, neighbors who have been scrupulously limiting their own use of water may wonder, "What good does it do for me to conserve water when others don't?" and may then violate the limits themselves.

a Consider the hypothetical community of Watertown, where 1000 families share a common reservoir. Suppose that 900 of the 1000 families support water conservation if everyone else conserves. Invent data which illustrate that, despite 90 percent support for conservation, a situation can result in which no one conserves.

b Suppose that you are the mayor of Watertown and believe that it is essential for the community members to abide by the restrictions. You know that 90 percent of your constituents favor conservation but have observed that their threshold values have a cumulative effect which is preventing conservation. What steps would you take to help the community achieve the conservation that most people want?

Exercise 18: Journal

In what ways can threshold analysis help you understand the collective behavior of groups that interest you? Describe specific examples. Make speculations about the role of individual threshold values in determining a particular group's behavior. Test your speculations by continued observation of the group.

Exercise 19: Journal; Project

Analysis of threshold values is of most interest when we can apply it to gain insights into the behavior of groups to which we ourselves belong. You can gain—and share—insight into a group of which you are a concerned member by conducting your own analysis of threshold values.

Step 1 Choose a group to which you belong and a situation in which members face a choice of engaging (or not engaging) in a certain behavior X. Possible behaviors include attending (or not attending) a particular event; conserving (or not conserving) a valuable resource; supporting (or not supporting) a political candidate or issue; wearing (or not wearing) a particular type of clothing, such as a scarf in cold weather or a helmet when cycling; conforming (or not conforming) to certain rules, such as liquor and drug laws, parking restrictions, and speed limits; joining (or resisting) a particular fad.

Step 2 Survey the group members to learn

a Number of members favoring each behavior (X or not X).

b Threshold values.

Step 3 Tabulate the survey results and construct a graph.

Step 4 Identify the equilibrium points. Describe external circumstances that could lead to each different equilibrium behavior. How likely is a particular equilibrium to be sustained?

Step 5 Think about the results. Consider questions such as:

a Is the group's behavior in accord with what most members want? If not, how can this be changed?

b What would be the probable effects of more (or less) publicity concerning members' thresholds?

c Has the wording of the survey influenced the data gathered? What effects may this have?

d Could the group be divided into two or more subgroups for which the behavior of (at least) one of the subgroups would be markedly different from the behavior of the entire group?

Step 6 Evaluate your methods.

a Is threshold analysis well suited to the situation to which you applied it?

b If you were to start over, what would you do differently to obtain better results?

Exercise 20: Test Yourself

Develop a list of new terms and key ideas presented in this chapter. Supply an example of each item on your list.

Note The Appendix contains complete answers for Exercises 11*a*, *b*, and *c*. It contains partial solutions or hints for Exercises 12*a*, 13*a*, 14, 15, and 17.

CHAPTER 21

METHODS OF RECOGNIZING CONSENSUS

One man with courage makes a majority.

—Andrew Jackson

INTRODUCTION

When we are part of a group that must make a decision—voters electing a senator, council members deciding how to allocate funds, company executives choosing among proposed new products to develop—we often suppose that if we put the matter to a vote, then the alternative that wins the election is the one the group wants most and thus the best selection. But it is not difficult to find examples which show that this idea is naive. An election can produce a result that is not what the group members really wanted.

If an election turns out wrong, apathy, misinformation, or even fraud may be cited as the cause: the voters didn't care enough or know enough to make the right decision, or the election was corrupt. But, as a systematic investigation of voting decision methods will reveal, the fault may lie in the method. The way individual votes are merged into a group decision may not actually reflect the group consensus; thus a careful scrutiny of methods is essential. This chapter examines several voting decision methods, most of them familiar, and proposes guidelines for selecting a method that will effectively coalesce individual opinions into a group consensus. Note that this investigation of voting methods will assume that voters are informed about choices and have already formed their opinions. The focus will be on how these opinions can be effectively translated, through voting, into an election result that is what the group wants most.

Let us begin with three examples of elections whose results did not reflect what the voters really wanted.

Example 1 ▶

In the 1888 presidential election* the Republican candidate, Benjamin Harrison, received 5,444,337 popular votes; the Democratic candidate, Grover Cleveland, received 5,540,050 popular votes. But when the Electoral College convened, Harrison received 233 electoral votes whereas Cleveland received only 168. Thus Cleveland was defeated by a candidate who had received fewer popular votes. ◀

Example 2 ▶

In the 1970 New York senatorial race there were three candidates:

Ottinger (liberal Democrat)
Goodell (liberal Republican)
Buckley (Conservative)

Buckley won the election with 39 percent of the vote, followed by Ottinger with 37 percent and Goodell with 24 percent. For the liberal majority (71 percent of the voters), the election did not produce either of the outcomes they would have preferred. ◀

Example 3 ▶

Last year Donald Hughes was chosen from among his classmates to receive an award as "mathematics student of the year" at Southern High School. The faculty committee in charge of the selection considered four qualified students:

Alice Simpson
Benjamin Walters
Carla Thompson
Donald Hughes

On the basis of their evaluations of these students, the faculty members—Mrs. Peters, Mr. Quincy, and Ms. Roudebush—ranked the students as shown in Table 21-1.

Although at first glance it is hard to decide from Table 21-1 which student should have been selected, it is easy to criticize the choice of Donald. Each teacher's ranking shows a higher preference for Ben than for Donald. How did the committee ever select Donald? (This question will be taken up again in the next section of the chapter, "Sequential Pairwise Voting.") ◀

*_The World Almanac,_ Newspaper Enterprise Association, New York, 1984, p. 292.

CHOICE	MRS. PETERS	MR. QUINCY	MS. ROUDEBUSH
1	Alice	Carla	Ben
2	Ben	Alice	Donald
3	Donald	Ben	Carla
4	Carla	Donald	Alice

TABLE 21-1
Faculty's rankings for mathematics award

Exercise 1

Using the Electoral College method to choose a president is criticized because a candidate may receive a majority of the popular vote and still fail to be elected. (This was illustrated by Example 1.) In fact, a presidential candidate could be elected by the Electoral College while receiving less than 25 percent of the popular vote; this is a rather dismaying consideration, especially in recent times, when only about half the eligible voters go to the polls. To see how the Electoral College can choose a candidate with a low popular vote, let us consider a simplified example.

Suppose that we have an electorate consisting of 9 states, each of which has 1000 voters and 5 electoral votes. As is generally true with the real Electoral College, electoral votes will be cast in a block: the candidate who receives a majority of votes in a state obtains all of its electoral votes. Our election has two candidates, Fitzgerald and Milhous, and the election results are as shown in Table 21-2.

a From the data in Table 21-2, who won the election? What percent of the total popular vote did the electoral winner obtain? What percent of the total popular vote did the electoral loser obtain?

b Use the ideas of this example as a basis for an argument against the Electoral College as a method of selecting a president.

TABLE 21-2

	POPULAR VOTE		ELECTORAL VOTE	
STATE	FITZGERALD	MILHOUS	FITZGERALD	MILHOUS
Arkazona	490	510	0	5
East Virginia	490	510	0	5
Illiana	490	510	0	5
Mashington	490	510	0	5
Michigas	490	510	0	5
North York	1000	0	5	0
Old Jersey	1000	0	5	0
Pennsyltucky	1000	0	5	0
West Dakota	1000	0	5	0
Totals	6450	2550	20	25

TABLE 21-3

CANDIDATE APPROVED	PERCENT OF VOTERS
Buckley (only)	39
Goodell (only)	7
Ottinger (only)	13
Goodell and Ottinger	41

Exercise 2

As in the 1970 senate race in New York (Example 2), a candidate who is preferred by only a minority of the voters can be elected because the majority vote is divided. On way to remedy this situation is a method called *approval voting.** With this method, each voter examines each candidate and decides whether he or she "approves" of that candidate or not. In the election, voters cast "approval votes" for as many candidates as they find satisfactory. The candidate who obtains the largest number of approval votes is the winner.

Suppose that the approval method had been used in the 1970 senate race, with the results shown in Table 21-3.

a What percent of the voters approved of Buckley? Of Goodell? Of Ottinger?

b Who would have won the election if the approval method had been used?

c Discuss possible pros and cons of using the approval method in public elections with more than two candidates.

Note The Appendix contains complete answers for Exercise 1*a* and 2*b.* It contains partial solutions or hints for Exercise 2*c*.

*The following reference includes a rationale for, and a detailed discussion of, approval voting (this method is receiving some consideration for use in presidential primaries, which often include so many candidates that preferences are hard to assess by a one-vote, plurality method): Samuel Merrill, "Approval Voting: A 'Best Buy' for Multicandidate Elections?" *Mathematics Magazine,* vol. 52, no. 2, March 1979, pp. 98–102.

SEQUENTIAL PAIRWISE VOTING—AVOID IT IF YOU CAN

In Example 3, earlier in this chapter, the selection of Donald Hughes for the mathematics award seemed unbelievable. How could he have been chosen when, as Table 21-1 showed, each member of the selection committee preferred Ben to Donald? Here is how it happened:

> When the faculty committee first met to consider students for the award, there were two nominees, Alice Simpson and Benjamin Walters. Mrs. Peters and Mr. Quincy gave strong arguments in favor of Alice, and she became the committee's selection. Later, however, Ms. Roudebush requested a second meeting to consider a nominee overlooked in the first round. She suggested Carla Thompson, a student whom she considered superior to Alice. Mr. Quincy agreed, and the committee decided to recommend Carla instead of Alice—despite Mrs. Peters's negative vote and strong objection. As she continued to ponder the committee decision, Mrs. Peters took a close look at Donald Hughes, who sat in a front-row seat in her advanced algebra class. "In my class, Donald is a better student than Carla," she observed. She convened the committee once again and asked that they consider Donald. Ms. Roudebush agreed with Mrs. Peters that Donald was a better choice than Carla, and so, this time with Mr. Quincy dissenting, Donald became the group's final choice.

In selecting Donald Hughes, the committee members were unaware that they had made a choice which did not reflect their actual preferences. In fact, it wasn't until after the award had been presented to Donald that they realized the truth. Having received a good deal of criticism for choosing Donald instead of Alice, Ben, or Carla, the committee members decided to get together to discuss their selection process. It was at this meeting that they discovered each others' rankings (Table 21-1) and learned that all three of them had preferred Ben to Donald—and began to wonder how they had gone wrong.

A diagram (Figure 21-1) can help to clarify the method that the committee used: *sequential pairwise voting.* The characteristics of sequential pairwise voting are these:

- There are several (more than two) available choices, and these are listed in a specific order called an *agenda.*
- A vote is taken between the first two agenda items, and a winner is selected. This winner is then paired with the third agenda item, and another vote is taken. The process continues, at each step choosing between the winner of the previous election and the next agenda item.

Sequential pairwise voting is an undesirable decision method because of a so-called agenda effect: the order of items on the agenda influences the outcome.

To illustrate the agenda effect, let us consider a variation of the agenda that led to the selection of Donald Hughes. Suppose that the students were considered in the reverse order—first Donald and Carla, then Ben, then Alice. The result of sequential pairwise voting based on this agenda is that Alice is the committee's

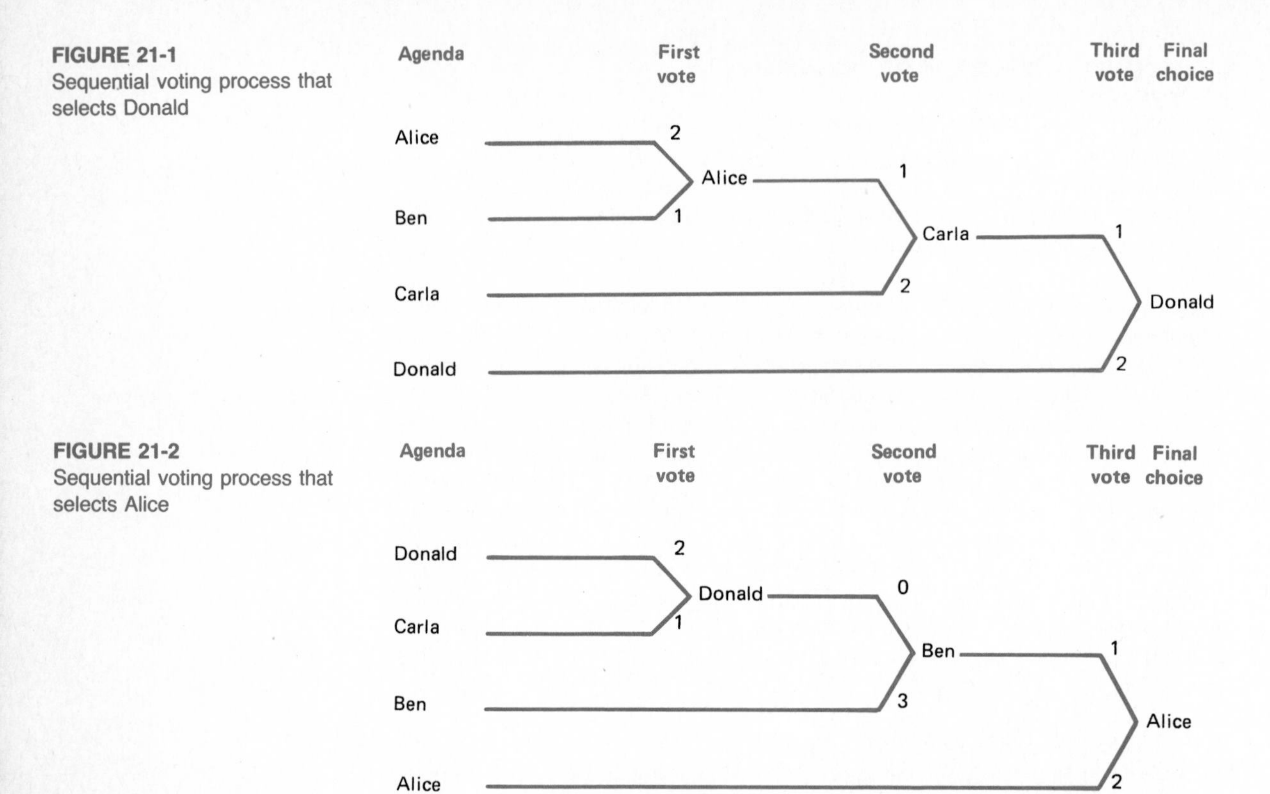

FIGURE 21-1
Sequential voting process that selects Donald

FIGURE 21-2
Sequential voting process that selects Alice

choice. The voting is summarized in Figure 21-2; the votes are derived from the committee members' original rankings of the students shown in Table 21-1.

When a group must choose from among more than two alternatives, if the members vote on them all at once there is a chance that no alternative will gain majority support. Sequential pairwise voting is appealing because alternatives are considered only two at a time, and (except for ties) winners will have a majority vote. However, as the example has shown, this method may distort the group's opinion (Donald was selected when Ben was preferred); thus it is an undesirable voting decision method.

Exercise 3

a Experiment with different agendas (orders) for selecting a "mathematics student of the year" and find an agenda that will lead to the selection of Carla. (Use sequential pairwise voting.)

b Find an agenda that will lead to the selection of Ben.

c Discuss whether the following statement is true: "In sequential pairwise voting the agenda, rather than the voters, makes the decision."

Exercise 4

The following rule of thumb has been proposed as a summary of the agenda effect on sequential pairwise voting:

> Regardless of voters' opinions, the later an alternative's position on the agenda, the better its chances of selection as the winner.

Test this rule by examining the 24 possible agendas that could be used for selecting the "mathematics student of the year." Find the winner for each agenda and examine whether the rule applies.

Exercise 5

Three candidates—Kurtz, Levin, and Mason—are being considered for an important job. Suppose that five people have interviewed these applicants and have ranked them as shown in Table 21-4 (page 368).

a Suppose that the five-person interviewing committee plans to make its decision using sequential pairwise voting. Discuss possible agenda effects.

b If you are interviewer 5 and also chair of the committee, could you propose an agenda that would result in the selection of your preferred applicant (M)?

c If, as committee chair, you wanted to ensure fairness in the decision process, what voting procedure would you propose?

Exercise 6

Individual decision making can also evolve as a sequence of pairwise choices. Exercise 7 in Chapter 7 presented Raymond's choice among jobs in Atlanta, Birmingham, Chicago, and Dallas. Suppose that in Raymond's decision problem the criteria location, salary, and opportunity for advancement are treated as voters. Raymond's problem then resembles the election with three voters shown in Table 21-5 (page 368).

a If Raymond's job offers don't come all at once, he may have to select the better of two before another offer comes along. Suppose that this is the case and the agenda of offers is: first Atlanta and Birmingham, then Chicago, then Dallas. What decision will result? What's wrong with the decision?

b Examine thoroughly the possible agenda effects on Raymond's decision problem. What conclusions can you draw?

c What guidelines can you suggest to avoid detrimental agenda effects in personal decision making?

TABLE 21-4

	INTERVIEWER				
CHOICE	1	2	3	4	5
First	K	K	L	L	M
Second	L	L	M	M	K
Third	M	M	K	K	L

TABLE 21-5

	VOTER		
CHOICE	1	2	3
First	Chicago	Atlanta	Birmingham
Second	Atlanta	Birmingham	Dallas
Third	Birmingham	Dallas	Chicago
Fourth	Dallas	Chicago	Atlanta

Note The Appendix contains partial solutions or hints for Exercises 3*c*, 4, and 5*b*.

DIFFERENT VOTING PROCEDURES LEAD TO DIFFERENT DECISIONS—BEWARE

To provide a background for later discussion of the strengths and weaknesses of various voting decision methods, let us examine a realistic decision problem and try to find a fair way to summarize the group consensus.

A budget decision It's budget time, and the Mount Olympus school board is faced with some hard decisions about how much money to spend and what to spend it on. The high school principal, Mr. Zeus, has presented a strong argument to the board advocating the expenditure of $300,000 to improve athletic facilities. The elementary school principal, Ms. Hera, has provided extensive justification for her request that $300,000 be allocated for the renovation of the elementary school library. The board members have weighed these alternatives along with a third option—holding down taxes by doing neither.

Each of the nine board members has decided on a ranking of the three options, and the board has met to make a collective decision. Mr. Read suggests that their first order of business should be to decide on a method of deciding. Mr. Run greets that suggestion with alarm. "Why should we waste time on that?" he asks. "Let's just make our preferences known and come up with a sensible and democratic decision through open discussion and voting." The other board members agree with Mr. Run, and so Mr. Read does not press his suggestion.

	MEMBER								
CHOICE	MR. THROW	MR. RUN	MR. JUMP	MR. VOLLEY	MRS. BOOK	MR. READ	MRS. STORY	MR. SAVE	MRS. CUT
First	A	A	A	A	L	L	L	H	H
Second	H	H	H	H	H	H	H	L	L
Third	L	L	L	L	A	A	A	A	A

TABLE 21-6
School board "preference profile"

The board members embark on a lively and opinionated discussion, with each in turn stating and defending his or her preferences. Their discussion reveals the preferences that are summarized in Table 21-6. ("A" denotes improving athletic facilities; "L" denotes renovating the library; "H" denotes holding down taxes.)

Stop for a moment and think about the problem yourself before you read on. What decision do *you* think is the consensus of the board? Why?

Each board member, becoming aware of the opinions of the others, may begin to look for a voting decision method that will lead to selection of the alternative that he or she prefers. This possibility—based on the fact that different voting methods can result in different choices—adds a dimension to voting decisions of which most voters are unaware.

We agree with Mr. Read that the first order of business for a decision-making body is selection of a decision method to be used. Otherwise, manipulation can occur; once cognizant of others' opinions, voters may advocate a method that they think is likely to merge individual opinions into the choice they themselves prefer. The three decisions that follow illustrate possibilities for such manipulation.

Voting Decision 1

After lengthy discussion of the spending issue, Mr. Throw calls for a vote. "Let's stop the discussion now and select the alternative that most of us prefer," he says.

"What if no alternative gets a majority vote?" asks Mrs. Story.

"Then we'll call the alternative with the highest number of votes the winner," says Mr. Throw, smiling at Mr. Run, Mr. Jump, and Mr. Volley.

Voting results

A: 4 votes (winner)
H: 2 votes
L: 3 votes

Note Mr. Throw has advocated a voting decision method (called the *plurality* method) that selects the alternative he favors (A).

Voting Decision 2

After lengthy discussion of the spending issue, Mrs. Book calls for a vote. She suggests that if no alternative gets a majority on the first vote, then the alternative with the least votes should be eliminated and a second vote should be taken.

Voting results in round 1

A: 4 votes
H: 2 votes (eliminated)
L: 3 votes

Voting results in round 2

A: 4 votes
L: 5 votes (winner)

Note Mrs. Book has advocated a voting decision method (called the *plurality-with-elimination* method) that selects the alternative she favors (L).

Voting Decision 3

After lengthy discussion of the spending issue, Mrs. Cut calls for a vote. "But," she argues, "in order to evaluate our consensus on a decision about which we all feel so strongly, we should vote for our second choices as well as our first choices. I think that the fairest decision method would be for each of us to give two points to his or her first choice, one point to the second choice, and no points to the third choice. That type of vote will incorporate complete information about our preferences and thus will be the best way of reaching a decision."

Voting results

A: 8 points
H: 11 points (winner)
L: 8 points

Note Mrs. Cut has proposed a voting decision method (called the *Borda count* method) that selects the alternative she favors (H).

The three voting decisions described above illustrate how different voting methods can provide different results as they merge individual opinions.

Below the mechanics of five of the most commonly encountered voting decision methods are discussed; the exercises that follow will provide practice in using them.* In the last section of the chapter, criteria for evaluating voting methods will be presented, and we will see how various methods fare when judged by these criteria.

---

*An extended and elementary discussion of these and other voting methods can be found in: Philip D. Straffin, Jr., *Topics in the Theory of Voting,* Birkhauser, Boston, 1980.

FIVE VOTING DECISION METHODS

As you read the various examples and discussions of voting methods, you may wonder why the familiar "majority rule" is given so little attention:

> If an alternative or candidate gets a majority of the votes cast, then that choice is the group's decision.

As comments throughout the chapter will reveal, the majority rule is a good one and should be used. But when there are more than two alternatives to choose from, it can easily occur that no alternative is chosen by a majority.* To deal with many and varied possibilities, other election methods are needed.

Dictatorship Method

The simplest of all voting methods is *dictatorship.* One voter has the role of dictator and this voter's first choice becomes the winner, regardless of the votes of others. Although a dictatorship usually violates our ideas of fairness, it is sometimes used. In situations where we don't know or care much about possible outcomes, we may select one person and say, "You decide." Or we may select one person as our "representative," empowered by the group to make its decisions.

> Whenever you have an efficient government, you have a dictatorship.
>
> Harry S Truman

Plurality Method

Plurality voting is a simple and widely used decision method. Many primary and general elections in the United States use the plurality rule:

> Each voter votes for one alternative, and the alternative that gets the largest number of votes wins.

If the number n of alternatives is large, plurality voting can result in an extremely weak mandate. An alternative might "win" with just over $1/n$ of the voters favoring it, and with the remainder disliking it intensely.

Plurality-with-Elimination Method

Some analysts of voting decisions argue that a group should stick with its voting process until some alternative (or candidate) obtains the support of a majority. This "eventual majority" is seen as an important part of the psychology of consensus: that is, a voter will support a group decision if (and often only if) he or she has voted for it.

---

*For example, in the presidential elections of 1824, 1860, 1892, 1912, 1948, and 1968, no candidate received a majority of the votes cast.

The plurality-with-elimination method works this way:

> Each voter votes for one alternative, and an alternative is declared the winner if it receives a majority of the votes. If no alternative gets a majority, the alternative with the smallest number of votes is eliminated and voting takes place again. This process continues until a winner is selected.
>
> When a tie occurs between two (or more) alternatives to be eliminated, all should be eliminated.

Borda Count Method

The Borda count method is named for the French politician Jean-Charles de Borda, who in 1781 first proposed its use. To apply this method a voter must first rank all the available alternatives. Once ranked, each alternative is awarded one "Borda point" for every alternative ranked beneath it. Points are then tallied, and the alternative that receives the highest number of Borda points is the winner.

Condorcet Method

The marquis de Condorcet, an eighteenth-century mathematician, philosopher, and political leader was an advocate of pairwise voting. He liked the simplicity of choosing between only two alternatives. But he also recognized that sequential pairwise voting allows the undesirable "agenda effect."

Condorcet proposed that elections be decided by the following means:

> Each pair of alternatives should be considered in its own separate election and the winner determined. If one alternative emerges as winner over all others in these separate two-way contests, then that alternative should win the overall election.

Condorcet's method is appealing: any choice that is preferred to all other choices, taken one at a time, surely deserves to be the winning choice. But the method suffers an important drawback: there may not be such a preferred choice.

To illustrate how these methods work, let us apply them to a hypothetical election involving 17 voters.* For simplicity, the alternatives among which these voters must choose are designated W, X, Y, and Z. Three of the voters rank W first, X second, Y third, and Z last—information that can be summarized concisely as follows:

3 voters

W
X
Y
Z

*The dictatorship method does not require discussion; only the other four methods will be applied here.

3 VOTERS	5 VOTERS	3 VOTERS	4 VOTERS	2 VOTERS
W	Z	W	Y	X
X	X	Y	X	W
Y	Y	X	W	Y
Z	W	Z	Z	Z

TABLE 21-7
Preference information for 17 voters

If other voters' preferences are designated similarly, the complete election information is as shown in Table 21-7.

Before you go on, examine the voters' preferences and determine which alternative you think should win this election.

Applying the Plurality Method

Table 21-7 gives more information than is needed for plurality voting. With this method a voter declares only his or her first choice; from the first line of Table 21-7, the results are:

W: 6 votes (winner)
X: 2 votes
Y: 4 votes
Z: 5 votes

Applying the Plurality-with-Elimination Method

The first-vote results are as above (for plurality voting):

W: 6 votes
X: 2 votes (eliminated for next round)
Y: 4 votes
Z: 5 votes

Table 21-7 gives enough information so that we do not actually need to conduct another election for the second round; we need only reexamine voters' preferences. We eliminate X from consideration and interpret preferences as in Table 21-8.

3 VOTERS	5 VOTERS	3 VOTERS	4 VOTERS	2 VOTERS
W	Z	W	Y	—
—	—	Y	—	W
Y	Y	—	W	Y
Z	W	Z	Z	Z

TABLE 21-8
Preferences after first elimination

TABLE 21-9
Preferences after second elimination

3 VOTERS	5 VOTERS	3 VOTERS	4 VOTERS	2 VOTERS
W	Z	W	—	—
—	—	—	—	W
—	—	—	W	—
Z	W	Z	Z	Z

For the second round of voting, we count only votes for W, Y, and Z, with each voter casting a vote for first choice among those three. Second-round results are as follows:

W: 8 votes
X: eliminated in first round
Y: 4 votes (eliminated for next round)
Z: 5 votes

The information needed to make the final decision (after Y is also eliminated) is given in Table 21-9.

In the third round of voting the results are:

W: 12 votes (winner)
X: eliminated in first round
Y: eliminated in second round
Z: 5 votes

Applying the Borda Count Method

From the information shown in Table 21-7 we see that

W has 6 first-place votes; these yield 18 Borda points.
W has 2 second-place votes; these yield 4 Borda points.
W has 4 third-place votes; these yield 4 Borda points.
W has 5 last-place votes; these yield no Borda points.

In all, W obtains 26 Borda points. Similarly, we can evaluate Borda points for alternatives X, Y, and Z. The complete election results are:

W: 26 Borda points
X: 33 Borda points (winner)
Y: 28 Borda points
Z: 15 Borda points

3 VOTERS	5 VOTERS	3 VOTERS	4 VOTERS	2 VOTERS
W	—	W	—	X
X	X	—	X	W
—	—	X	W	—
—	W	—	—	—

TABLE 21-10 Preferences concerning W and X only

Applying the Condorcet Method

To use this method, we must consider six two-way elections:

W versus X
W versus Y
W versus Z
X versus Y
X versus Z
Y versus Z

Although voters may vote on these six pairings separately, the information needed may be gathered from Table 21-7. For example, to determine the outcome of the two-way election between W and X, we can ignore the information about Y and Z in Table 21-7, focusing only on the information in Table 21-10.

The result of the two-way election W versus X is

W: 6 votes
X: 11 votes (winner)

The other two-way elections have the following results, all obtained in the same way just used for W versus X.

W versus Y
W: 8 votes
Y: 9 votes (winner)

X versus Y
X: 10 votes (winner)
Y: 7 votes

Y versus Z
Y: 12 votes (winner)
Z: 5 votes

W versus Z
W: 12 votes (winner)
Z: 5 votes

X versus Z
X: 12 votes (winner)
Z: 5 votes

Because X wins its two-way contest with each other alternative, X is the winner chosen by the Condorcet method.

A digraph is useful for displaying the results of the various two-way contests. Figure 21-3 shows the outcomes of the six contests evaluated above. An arrow is directed from the winner to the loser.

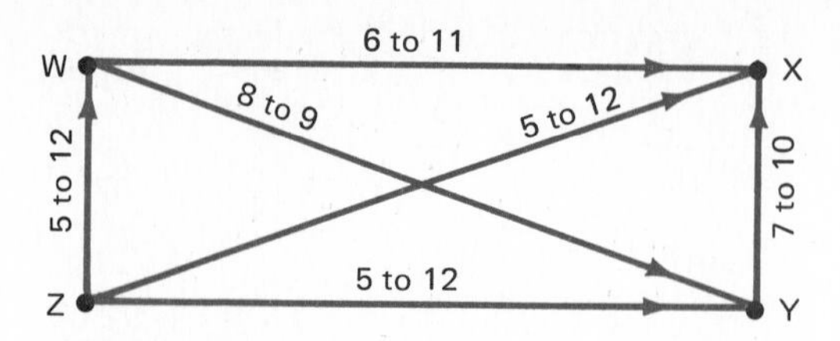

FIGURE 21-3
Digraph of results of six two-way contests

Exercise 7

a Apply the Condorcet voting decision method to preferences of the Mount Olympus school board given in Table 21-6.

b When the example of the Mount Olympus school board was first given, you were asked to state which decision you thought was the consensus of the board. Which of the four methods—plurality, plurality with elimination, Borda count, or Condorcet—led to the decision you chose? Which decision do you now think is the consensus? Explain.

Exercise 8

For the 17-voter example that was used to illustrate how the different voting methods work, plurality and plurality with elimination selected W and the Borda count and Condorcet methods selected X. Is either of these the alternative that you chose as the consensus when the example was first introduced? Which alternative do you now think is the voting group's preferred choice? Explain.

Exercise 9

For each of the eight sets of election information given in Table 21-11, determine the following:

a The alternative that you think is the group's consensus—based on your examination of preferences.

b The alternative selected by the plurality method.

c The alternative selected by the plurality-with-elimination method.

d The alternative selected by the Borda count method.

e The alternative selected by the Condorcet method.

f Compare your choice in *a* with the alternatives selected in *b, c, d,* and *e.* Which method or methods provide results that agree with your original assessment?

TABLE 21-11

ELECTION	4 VOTERS	3 VOTERS	2 VOTERS
1 (9 voters)	B	O	G
	G	G	O
	O	B	B

ELECTION	6 VOTERS	2 VOTERS	5 VOTERS	2 VOTERS
2 (15 voters)	S	T	V	V
	T	V	S	T
	V	S	T	S

ELECTION	5 VOTERS	4 VOTERS
3 (9 voters)	A	B
	B	C
	C	A

ELECTION	2 VOTERS	3 VOTERS	3 VOTERS	3 VOTERS
4 (11 voters)*	F	F	G	H
	G	E	E	E
	H	G	F	G
	E	H	H	F

ELECTION	5 VOTERS	3 VOTERS	3 VOTERS	1 VOTER
5 (12 voters)	J	L	K	L
	K	J	L	K
	L	K	J	J

ELECTION	4 VOTERS	3 VOTERS	2 VOTERS
6 (9 voters)	E	G	F
	F	E	G
	G	F	E

ELECTION	4 VOTERS	3 VOTERS	2 VOTERS
7 (9 voters)	P	Q	R
	Q	R	S
	R	S	P
	S	P	Q

ELECTION	7 VOTERS	6 VOTERS	5 VOTERS	3 VOTERS
8 (21 voters)	A	B	C	D
	D	D	B	C
	B	C	D	A
	C	A	A	B

*When a tie occurs for which alternative to eliminate, eliminate all tied alternatives.

Exercise 10

Exercise 9, which asks you to try all voting methods on each election, is designed to help you see the similarities and differences among the methods. However, in actual elections the voting method should be chosen (and announced) before the voting; to do otherwise allows for unfair manipulation of the voting results.

Review the results of the eight elections in Exercise 9. Select the one voting method that you think would be best to use for all of them. Give the reasons for your choice.

Exercise 11

a Consider election 9*a* in Table 21-12, involving 21 voters. If the plurality-with-elimination method is used, which alternative is selected?

b Suppose that just before the vote in election 9*a* is taken, the final two voters decide that they like P better than Q and change their rankings to those shown in Table 21-13. Under these new circumstances, what alternative is selected when the plurality-with-elimination method is used?

c Compare the results of *a* and *b;* what went wrong?

Exercise 12: Invent

This exercise will seem hard at first, but once you start experimenting, you will find it much easier.

For each of the following, experiment with voter-preference lists and devise an election involving 15 voters who must choose among alternatives A, B, and C.

TABLE 21-12

ELECTION	8 VOTERS	6 VOTERS	5 VOTERS	2 VOTERS
9*a* (21 voters)	P	R	Q	Q
	Q	P	R	P
	R	Q	P	R

TABLE 21-13

ELECTION	8 VOTERS	6 VOTERS	5 VOTERS	2 VOTERS
9*b* (21 voters)	P	R	Q	P
	Q	P	R	Q
	R	Q	P	R

a Invent an election for which the plurality method selects A and the plurality-with-elimination method selects B.

b Invent an election for which a majority of the voters rank A as their first choice but the Borda count method selects B.

c Invent an election for which the plurality winner and the Condorcet winner are different.

d Invent an election for which the plurality-with-elimination winner and the Condorcet winner are different.

e Invent an election for which the Borda count winner and the Condorcet winner are different.

f Invent an election for which the plurality, plurality-with-elimination, and Borda count winners are all different. What is the Condorcet winner in this case?

Note The Appendix contains complete answers for Exercise 9, election 2. It contains partial solutions or hints for Exercise 9, election 3 (question *a*) and election 7 (question *a*); and for Exercise 11.

CRITERIA FOR A VOTING DECISION METHOD

The discussion of voting decision methods began by noting that we are interested in ways of merging individual opinions to express group consensus. We have experimented with methods and examples, and it is now time to evaluate the methods.

If the result of a voting method is to provide an undistorted picture of group opinion, the following criteria seem reasonable.

Pareto criterion* The method should *not* select alternative Y as a winner if every voter prefers alternative X to Y.

Condorcet loser criterion The method should *not* select Y if Y would lose in pairwise contests against every other alternative.

Majority criterion The method *should* select alternative X as a winner if a majority of the voters have X as their first choice.

Consistency criterion The method *should* select X if it originally selected X and if one or more voters change their preferences in a way favorable only to X.

Condorcet winner criterion The method *should* select X if X could win pairwise majority contests against every other alternative.

---

*Vilfredo Pareto (1848–1923) was an Italian economist; his 80-20 rule was introduced in Exercise 14 in Chapter 5.

VOTING DECISION METHOD	CRITERION				
	PARETO	CONDORCET LOSER	MAJORITY	CONSISTENCY	CONDORCET WINNER
Plurality	Yes	No	Yes	Yes	No
Plurality with elimination	Yes	Yes	Yes	No	No
Borda count	Yes	Yes	No	Yes	No
Condorcet	Yes	Yes	Yes	Yes	Yes
Dictatorship	Yes	No	No	Yes	No
Sequential pairwise	No	Yes	Yes	Yes	Yes
Approval	Yes	Yes	No	Yes	No

TABLE 21-14
Criteria satisfied by voting methods

Table 21-14 provides a summary of the criteria satisfied by the various voting methods discussed. According to that summary, the only voting method which satisfies all five of the criteria is the Condorcet method. However, the Condorcet method does not always select a winner—as was illustrated by elections 2, 5, 6, and 7 of Exercise 9.

To counteract this defect of the Condorcet method, the political scientist Duncan Black has proposed a hybrid method. The Black method is this:

> If the Condorcet method produces a winner, that alternative deserves to win the election. If the Condorcet method fails to select a winner, switch to the Borda method.

Black's method satisfies the five criteria and also always* determines a winner. (However, see Exercise 29 for a discussion of a reasonable criterion that is not satisfied by the Black method.)

If we apply our criteria, the Black method stands out as superior to the other methods considered. However, it is not a simple method, and for large elections with voters who are poorly informed about their choices, it may not be best.

For decisions made by small groups, or groups in which each member has a thorough knowledge of the alternatives and a keen interest in the decision, the desirable properties of the Black method make it worth the effort it requires.

Exercise 13

Explain how Example 3, concerning the selection of Donald Hughes as "mathematics student of the year," illustrates that sequential pairwise voting does not satisfy the Pareto criterion.

---

*With any voting method a tie is possible, and a tie is not considered an exception to always determining a winner. Ties for winner are generally decided by a random method—such as tossing a coin or drawing a number.

Exercise 14

Show that the dictatorship method satisfies the Pareto criterion.

Exercise 15

Which election in Exercise 9 shows that the Borda count method doesn't satisfy the majority criterion?

Exercise 16

Explain how the pair of elections 9*a* and 9*b* in Exercise 11 show that the plurality-with-elimination method doesn't satisfy the consistency criterion.

Exercise 17

Select elections from Exercise 9 which show that neither plurality, plurality with elimination, nor the Borda count method satisfies the Condorcet winner criterion.

Exercise 18

Invent an election which shows that the dictatorship method does not satisfy the majority criterion.

Exercise 19

Invent an election which shows that the approval method (Exercise 2) does not satisfy the majority criterion.

Exercise 20

In a certain group there is a majority whose members always agree. These elect one person to make all group decisions. Discuss this dictatorship method. What are its advantages? Its disadvantages?

Exercise 21

Consider a situation in which a committee of three city council members must choose among four alternative ways of spending federal revenue sharing funds. Their preferences are:

Committee member 1	Committee member 2	Committee member 3
Day care center	Community orchestra	Public library
Public library	Day care center	Bicycle paths
Bicycle paths	Public library	Community orchestra
Community orchestra	Bicycle paths	Day care center

a What is the group consensus?

b Provide an argument to justify your answer to *a.*

Exercise 22

Suppose that you are a staff member for a state legislative committee on transportation. Many issues have been presented to the committee as requests for proposed legislation. From these, after much information gathering and discussion, four issues have emerged as having top priority. The transportation committee has been charged with undemocratic decision making in the past, and it has requested staff help in developing a sound decision process. You have been put in charge of this project. What voting decision method would you recommend to this 15-member committee so that it can fairly choose, on the basis of the rankings of its individual members, the highest-priority issue for group endorsement? What reasons can you supply to convince the committee that your recommendation is a good one?

Exercise 23

List at least three situations in which you participate in group decision making by voting. For each, identify the voting method used. Evaluate whether it is the best possible method for the given situation. If not, what method should be used? Why?

Exercise 24*

Once there was a club run by a miser, a prohibitionist, and an alcoholic. The prohibitionist proposed a clubhouse. The alcoholic was in favor, provided it had a bar. The miser was against the clubhouse and even more strongly against equipping it with a bar. Of course, there could be no bar without a clubhouse.

a From the information provided, give the ranked order of each person's preferences for the three possible outcomes:

C: clubhouse without a bar
B: clubhouse with a bar
N: no clubhouse

b Their voting procedure was to decide first whether to build a clubhouse and then, if a clubhouse was desired, to decide whether to equip it with a bar. If the three members vote sincerely (according to their preferences) what outcome will result?

c Sometimes voters are *sophisticated*—that is, they look ahead to the probable election outcome and then vote for something other than their first choice if, by so doing, they can obtain a more desirable final outcome.

Suppose that the alcoholic becomes "sophisticated" enough to look ahead to the outcome determined above in *b.* What can the alcoholic do to obtain a more preferable outcome?

Exercise 25

Suppose that a person is asked to rank the colors red, blue, and green in order of preference and says:

"I like blue better than green."
"I like red better than blue."

We would visualize the ranking as follows:

1: Red
2: Blue
3: Green

And we would deduce a third statement from the two given:

"That person likes red better than green."

---

*Adapted from an example in Robin Farquharson, *Theory of Voting,* Yale University Press, New Haven, Conn., 1969.

However, in group decisions involving three or more preferences, it is possible to find situations for which:

The group prefers B to G.
The group prefers R to B.
But the group does *not* prefer R to G.

Such a situation illustrates the so-called voters' paradox.

Invent an election involving 15 voters each of whom ranks the alternatives R, B, and G and for which—when the alternatives are considered pairwise—R beats B, B beats G, and G beats R.

Exercise 26

(This is a variation of Exercise 6 in Chapter 7.) Suppose that Betsy, who is planning to buy a car, has ranked her top five choices as follows:

Cost	Gas economy	Style and comfort
Chevrolet	Renault	Volkswagen
Renault	Plymouth	Plymouth
Ford	Ford	Ford
Plymouth	Volkswagen	Chevrolet
Volkswagen	Chevrolet	Renault

Using the weights of 30 percent, 50 percent, and 20 percent (given in part *a* of Exercise 6 in Chapter 7) for these criteria, we can model Betsy's decision problem by the following election:

3 voters	5 voters	2 voters
Chevrolet	Renault	Volkswagen
Renault	Plymouth	Plymouth
Ford	Ford	Ford
Plymouth	Volkswagen	Chevrolet
Volkswagen	Chevrolet	Renault

a What voting decision method should Betsy use? Why? Which car does that decision method select?

b Discuss the pros and cons of solving this decision problem by treating it as a voting problem.

Exercise 27: Journal

a List examples of individual decision situations (such as Betsy's in Exercise 26) in which voting methods could be useful. For each example, state which voting method you would use.

b Find a decision that interests you and experiment by treating it (as Betsy did in Exercise 26) as a voting problem. Evaluate this decision process. What are its strengths and weaknesses for personal decisions?

Exercise 28: Small Group Project— A Committee's Decision Problem*

What to do Follow steps 1 to 5:

Step 1: Find four other people who will work with you on this problem.

Step 2: Read the problem and decide which of you will be councilpersons A, B, C, D, and E.

Step 3: As a group, discuss what method or methods you will use to decide on the allocation; select a decision method. For simplicity, suppose that only the allocations shown in Table 21-15 are possible.

Step 4: Make a committee decision. (Where information is missing, ask questions and invent realistic answers.)

Step 5: Write up your decision in the form of a recommendation to the city council. Explain how you arrived at it and why you think it is the best possible.

The problem The Yorkham city council has established a planning committee of five of its members to recommend how to spend $40,000 in state funds specifically targeted for community recreation projects. The council has received three requests totaling $80,000. These have been publicized in local newspapers and public meetings, and each has gained strong support.

Proposal 1: Yorkham Beautification Organization wants $25,000 to create a small park (with benches, trees, and a fountain) in the downtown area. This park would provide a restful spot for shoppers, students, and businesspeople.

Proposal 2: The Yorkham Riders, a bicycle club, proposes construction of bicycle paths. Construction costs about $1000 per mile, and federal matching funds are available. At least a 40-mile system is requested; an 80-mile system would be ideal. Because the city has many narrow streets, paths are needed to relieve congestion and for safety. Most of the cyclists are schoolchildren and university students and professors. A few of Yorkham's elderly people cycle—and many more indicate interest if safe paths are provided.

---

*Adapted from Peter Rice, "Committee Decision Making," Mathematical Association of America, 1976.

TABLE 21-15
Possible allocation

ALLOCATION	PROPOSAL 1	PROPOSAL 2	PROPOSAL 3
1	$25,000	$15,000	$0
2	20,000	20,000	0
3	15,000	25,000	0
4	20,000	0	20,000
5	25,000	0	15,000
6	5000	0	35,000
7	0	25,000	15,000
8	0	5,000	35,000
9	0	20,000	20,000
10	0	40,000	0
11	10,000	10,000	20,000

Proposal 3: The Yorkham Youth Soccer Committee is requesting that the city purchase land and construct a pair of soccer fields. In the last few years, the Yorkham Youth Soccer program has grown to include over 500 young people of both sexes in the age range 7–14. The cost of the fields is estimated at $35,000. At present all Yorkham soccer games are played at out-of-town fields—with much inconvenience to players, coaches, and parents.

The planning committee Although members of the council are concerned with all matters affecting the city, they are particularly interested in those that affect themselves and their constituencies. The information shown in Table 21-16 thus enters into the decision-making process.

The planning committee's assignment Agree on how much money to allocate to each of the three proposals.

Proposal 1: ____________________

Proposal 2: ____________________

Proposal 3: ____________________

TABLE 21-16

COUNCILPERSON	DESCRIPTION	CONSTITUENCY
A	Age 32, parent of young children	Working-class residential area; many families with young children
B	Age 45, parent of teenage children, college professor	Upper-middle-class older residential area
C	Age 58, owner of men's clothing store	Middle- and working-class residential area and downtown business district
D	Age 50, accountant	Middle-class suburban area including student dormitories (students have a vote in local elections)
E	Age 67, retired	Working-class residential area including low-income rental apartments and factories

Exercise 29

Analysis of voting methods was pioneered by Kenneth Arrow,* a Nobel prize-winning economist. Arrow's results were both enlightening and discouraging; he showed that it is possible to give a short list of reasonable criteria for a voting decision method that no possible voting method can satisfy. (Actually, Arrow's results apply only to elections with three or more choices; the majority rule satisfies his criteria in elections with only two choices.) Arrow's criteria require that a voting method V satisfy these conditions:

1: V must always work (that is, it must be able to supply a complete ranking of all choices, possibly including ties) for all possible elections.

2: V must not be a dictatorship.

3: For an election in which all voters prefer X to Y, method V must place X above Y in the final ranking.

4: If, when using V, X is ranked above Y and a revote is taken in which voters may have changed the rankings of other alternatives but each voter who ranked X above Y continues to do so and each voter who ranked Y above X continues to do so, then V must still rank X above Y.

Condition 4 is called the *independence-of-irrelevant-alternatives* criterion and is the most stringent of Arrow's conditions. Arrow argued that the presence of "irrelevant" alternatives should not affect the outcome of an election. (For an example of irrelevant alternatives, refer back to Exercise 11 in Chapter 5. In the host's decision whether to serve coffee or tea, the other five beverages are irrelevant alternatives, yet their different positions in two sets of guests' rankings seemed to provide reasons for different decisions.)

a Explain why the Condorcet method satisfies the independence-of-irrelevant-alternatives criterion.

b Show that the Borda method fails to satisfy independence-of-irrelevant-alternatives by considering election 10 in Table 21-17 (page 388).

Which of alternatives A, B, C, and D is the winner of election 10 when the Borda method is used? If alternative C is eliminated, which of alternatives A, B, and D is the winner by the Borda method?

c Explain why election 10 does not show that the Black method fails to satisfy the independence-of-irrelevant-alternatives criterion. Use election 11 (Table 21-18) to show that the Black method does not satisfy this criterion.

d Pick one of the other voting methods considered and either provide an argument showing that it does satisfy the independence-of-irrelevant-alternatives criterion or invent an example showing that it does not.

*Kenneth J. Arrow, *Social Choice and Individual Values,* Wiley, New York, 1963.

TABLE 21-17

ELECTION	5 VOTERS	4 VOTERS
10 (9 voters)	A	B
	B	C
	C	A
	D	D

TABLE 21-18

ELECTION	4 VOTERS	3 VOTERS	2 VOTERS
11 (9 voters)	E	F	G
	F	G	E
	G	E	F

Exercise 30: Test Yourself

Develop a list of new terms and key ideas presented in this chapter. Supply an example of each item on your list.

Note The Appendix contains a complete answer for Exercise 13. It contains partial solutions or hints for Exercises 21, 22, 23, and 24*a* and *c*.

CHAPTER 22

THE PRISONER'S DILEMMA: A MODEL OF THE CONFLICT BETWEEN INDIVIDUAL AND GROUP BENEFITS

The mutual confidence on which all else depends can be obtained only by an open mind and a brave reliance upon free discussion.

—Learned Hand

INTRODUCTION

Many of the problems confronting society today are difficult to resolve because individual members face a conflict: Shall I do what is best for me as an individual, or shall I do what is best for the group to which I belong? For example:

- An individual manufacturer must choose whether to install expensive pollution controls. For the individual manufacturer, profits will probably be higher if controls are not installed; but for society as a whole, environmental quality will probably be higher if this manufacturer (along with others) controls pollution.
- A family must choose whether to contribute $200 toward the construction of a community recreation area. For the individual family, spending $200 to meet its own needs offers more opportunities for immediate satisfaction than contributing that amount toward community recreation; but for the community as a whole, the availability of swimming, tennis, and picnic areas will be a great asset.

- A nation must choose whether to harvest whales without restraint or to abide by quotas proposed by an international conference. For the individual nation, fishing without restraint may be the most profitable choice; but for the world economy, the preservation of a sizable whale population may be an important resource.

These are examples of situations where individuals or organizations must decide between something that benefits them individually and something that benefits a group to which they belong. Often, individual benefits flow out of the choice that primarily benefits the group, but these may not be as great as those the individual or organization could obtain independently; hence the conflict.

Decisions about whether to cooperate with others in supporting a group activity are often accompanied by strong emotions or moral convictions; thus objective analysis is difficult. To avoid subjective difficulties for the time being, let us first examine such decision conflicts using a model—a story called the *prisoner's dilemma.* It is widely used in biology, economics, philosophy, political science, and psychology as an abstract model that captures the essence of many contemporary decision problems.

WHAT IS "RATIONAL" DECISION MAKING?

In analyzing decision situations, it is helpful to know something about the decision makers. This discussion assumes that decision makers are *rational.* Here a *rational* person or organization is simply one that makes choices with the highest utility value—those that offer the most satisfaction.

PRISONER'S DILEMMA*

A district attorney suspects two persons, recently taken into custody, of together having committed a serious crime. Although the suspects are known "almost certainly" to be guilty of the crime, the district attorney does not have sufficient evidence to convict either one.

To each prisoner, confined alone in a cell and unable to communicate with the other, the district attorney makes the following offer:

> "Tomorrow I will visit you in your cell and give you the opportunity to admit your guilt.
>
> "If you confess and the other prisoner does also, each of you will receive a 7-year jail term.
>
> "If you confess and the other prisoner fails to confess, you will go free in exchange for the evidence you have provided, and the other prisoner will face a 10-year jail term.

*This tale was invented by A. W. Tucker, Stanford University, 1950.

TABLE 22-1
Payoff matrix for prisoner's dilemma

	PRISONER 2	
PRISONER 1	Keep quiet	Rat
Keep quiet	(−3, −3)	(−10, 0)
Rat	(0, −10)	(−7, −7)

"If neither you nor the other prisoner confesses, each of you faces 3-year prison terms as the result of a list of minor charges for which we already have evidence.

"To whichever prisoner I visit second, I will reveal nothing about the results of my first visit."

Each prisoner has a long night during which to think about the district attorney's offer. What should he decide?

To analyze the prisoner's dilemma, it is convenient to summarize the given information using a matrix (Table 22-1). In Table 22-1, rows of the matrix are labeled with the choices for one prisoner, henceforth called *prisoner 1*; and columns are labeled with the (identical) choices for the other, called *prisoner 2*. The matrix entries give pairs of *payoffs*—numerical values (often utilities) that measure the results of decisions. The first number in each pair designates the payoff to prisoner 1, and the second number designates the payoff to prisoner 2. The higher a payoff value, the more a prisoner prefers it.

Prisoner 1's Decision

If we examine the decision from the point of view of prisoner 1 alone, looking only at his or her choices and payoffs, we need focus only on the information included in the matrix shown in Table 22-2.

In Table 22-2, dominance is easy to spot: the second row dominates the first. No matter which column prisoner 2 chooses, prisoner 1's payoff value is higher if he or she chooses to rat.

TABLE 22-2
Payoff matrix for prisoner 1

Keep quiet	−3	−10
Rat	0	−7

Prisoner 2's Decision

Likewise, prisoner 2, looking at the decision only from his or her point of view, will focus only on the information given in the matrix in Table 22-3.

The matrix in Table 22-3 also exhibits dominance: the second column dominates the first. No matter which row prisoner 1 chooses, prisoner 2's payoff value is higher with "rat."

TABLE 22-3
Payoff matrix for prisoner 2

Keep quiet	Rat
−3	0
−10	−7

Putting the Decisions Together

This analysis suggests that when the district attorney sees them tomorrow, both prisoners will rat. That is the "rational" choice—the choice that individual analysis shows to offer the higher payoffs.

As a result of this choice, both prisoners will face 7-year jail terms.

Afterthoughts

If we return to Table 22-1, we see that this outcome (7-year prison terms)—although it probably will please the district attorney—is actually not the best outcome from the selfish point of view of the prisoners. If both had chosen to keep quiet, then each would face only a 3-year term.

Suppose that the prisoners, in their night of thinking about their decisions, discover that they will be better off if both keep quiet. What can we expect them to do?

Despite the fact that *both* prisoners would be better off if *both* chose "keep quiet" instead of "rat," the fact that they are separated makes reasoning as a group impossible. And, if they reason as individuals, each will mimic our earlier analysis and observe that the best decision is "rat."

You may find the prisoner's dilemma confusing; if so, you are becoming aware of the complex and paradoxical nature of this decision problem. Reread the dilemma and the subsequent analysis; then organize your thoughts to support the following observation:

> For each prisoner deciding alone, the dominant decision choice is "rat." However, as a result of making individual "best" choices the prisoners fare worse (7-year terms) than if they both had chosen otherwise.

Later, the prisoner's dilemma will be analyzed further, but first you are to carry out an exercise that will involve you in a similar situation and will provide with some background experience for the discussion.

Exercise 1: Class Exercise

Find a fellow student who will participate with you in this activity. Designate one of you as student 1 and the other as student 2.

This exercise will make use of the matrix shown in Table 22-4. (This matrix is similar to the prisoner's dilemma payoff matrix in Table 22-1; the difference is that entries here are each 5 greater than corresponding entries in Table 22-1. This change introduces positive as well as negative values, but the numerical relationships still capture the decision dilemma.)

To carry out the exercise, each student must choose C or D. The first number in each pair tells the payoff to student 1 if that pair is chosen; the second number tells the payoff to student 2. For example, if student 1 chooses

TABLE 22-4
Payoff matrix for class exercise

	STUDENT 2	
STUDENT 1	C	D
C	(2, 2)	(−5, 5)
D	(5, −5)	(−2, −2)

TABLE 22-5
Payoff tabulation for class exercise

DECISION NUMBER	STUDENT 1'S CHOICE (C OR D)	STUDENT 2'S CHOICE (C OR D)	YOUR PAYOFF (FROM TABLE 22-4)
1			
2			
3			
4			
5			
6			
7			
8			
9			
10			
11			
12			
13			
14			
15			
16			
17			
18			
19			
20			
		Your payoff total:	

D and student 2 chooses C, then together their choices select the pair (5, −5); as a result, student 1 gets a payoff of +5 and student 2 gets a payoff of −5. For both student 1 and student 2, the object of the choice is to obtain as high a payoff as possible; however, each choice of C or D is to be made without any discussion with the other student.

After each of you has independently (and secretly) made a choice, write it in the appropriate column of Table 22-5. Then show it to the other student. Record his or her choice also. Using the pair of choices, turn to Table 22-4 to find your payoff; record this also in Table 22-5. Repeat this procedure 20 times: secretly made a choice; simultaneously reveal your choices; tabulate payoffs.

Your assigned goal in this exercise is to gain for yourself as high a *total* payoff as possible. Do not discuss your choices with the other student, and do not concern yourself with the total payoff that he or she is accumulating. Your goal is not to get a higher total than your counterpart but to get as high a total as possible for yourself. After completing the sequence of 20 decisions, find your total payoff. Share it with your classmates and compare results.

Ask those class members who achieved the highest totals to describe their thinking as they made the sequence of decisions. Similarly, ask those with the lowest totals to describe their thinking as they made their decisions.

After thinking about your own experience and that of others, write a paragraph (Exercise 4, below) that summarizes your advice to yourself on how to obtain a higher total if you were to start over on the exercise, paired with a different student.

If the rules of the exercise were changed to permit discussion of the decision between you and the other student, what differences would this make?

PRISONER'S DILEMMA AS A MODEL FOR OTHER DECISION PROBLEMS

The prisoner's dilemma is not just an interesting hypothetical problem. It also is a model for a general type of decision situation. The characteristics of such a decision problem—which will henceforth be called a *PD problem*—are these:

PD 1 There are two decision makers, each with the same pair of choices designated, for simplicity, C ("cooperate") and D ("defect").

PD 2 Each decision maker, acting alone, gains more from choice D no matter what the other chooses.

PD 3 Each decision maker gains more if the other chooses C no matter what choice he himself or she herself makes.

PD 4 Both decision makers gain more if they both choose C than if they both choose D.

Examination of Table 22-1, now recast as Table 22-6, confirms that the prisoner's dilemma story has the four characteristics of a PD problem. The choice "keep quiet" corresponds to the choice C, "cooperate." (By making this choice one prisoner "cooperates" with the other for their mutual benefit.) The choice "rat" corresponds to choice D, "defect." (With this choice a prisoner "defects" and abandons loyalty to the other prisoner, instead pursuing his or her own interests.)

Likewise, Exercise 1, with payoffs given in Table 22-4, can now be recast as Table 22-7, and can be seen to have the four characteristics of a PD problem. (*Check* that this is so.)

TABLE 22-6
Payoff matrix for prisoner's dilemma

	PRISONER 2	
PRISONER 1	Keep quiet (C)	Rat (D)
Keep quiet (C)	(−3, −3)	(−10, 0)
Rat (D)	(0, −10)	(−7, −7)

TABLE 22-7
Payoff matrix for class exercise

	STUDENT 2	
STUDENT 1	C	D
C	(2, 2)	(−5, 5)
D	(5, −5)	(−2, −2)

When the consequences of a PD problem are expressed in terms of numerical payoffs and displayed in a matrix, then the following terminology is useful:

- r *reward* payoff to each decision maker when both make choice C, "cooperate."
- p *penalty* payoff to each decision maker when both make choice D, "defect."
- t payoff that creates a *temptation* for one decision maker to choose D while the other cooperates and chooses C.
- s payoff to the *sucker* who chooses C when the other defects and chooses D.

In Table 22-1, $r = -3, p = -7, t = 0, s = -10$. In Table 22-4, $r = 2, p = -2, t = 5, s = -5$. In these, as in all PD matrices, $t > r > p > s$.

A PD problem is perplexing because it seems to contain a trap. Each decision maker, despite logical reasoning, seems trapped into choosing an outcome that yields less than the best possible payoff. Rational decision makers, who decide and act so as to achieve greatest satisfaction, choose D—and by so doing deprive themselves of a greater gain.

The next section of the chapter will suggest a way out of this dilemma. But first you are invited to become acquainted, through some exercises, with a wide variety of decision situations that satisfy the characteristics of a PD problem.

Exercise 2

Verify that each of the tables on page 396 is a matrix for a PD problem by correctly labeling each decision maker's choices with C and D and by showing that characteristics PD 2, PD 3, and PD 4 are satisfied.

- a Table 22-8.
- b Table 22-9.
- c Table 22-10.
- d Table 22-11.
- e Table 22-12.

TABLE 22-8

	DECISION MAKER 2	
DECISION MAKER 1	?	?
?	(0, 0)	(−10, 5)
?	(5, −10)	(−5, −5)

TABLE 22-9

	DECISION MAKER 2	
DECISION MAKER 1	?	?
?	(−3, −3)	(8, −5)
?	(−5, 8)	(5, 5)

TABLE 22-10

	DECISION MAKER 2	
DECISION MAKER 1	?	?
?	(20, 20)	(0, 500)
?	(500, 0)	(100, 100)

TABLE 22-11

	DECISION MAKER 2	
DECISION MAKER 1	?	?
?	(−4, −4)	(−12, 0)
?	(0, −12)	(−8, −8)

TABLE 22-12

	DECISION MAKER 2	
DECISION MAKER 1	?	?
?	(0, 7)	(5, 5)
?	(2, 2)	(7, 0)

Exercise 3

For each of the PD matrices given in Exercise 2, identify the temptation (*t*), reward (*r*), penalty (*p*), and sucker (*s*) values.

Exercise 4

a Reflect on your experience in completing Exercise 1. Describe what you would try to do to obtain a higher total if you were to repeat the exercise, paired with a different student.

b If the rules of Exercise 1 were changed to permit discussion at each step between you and the other student, what difference would this make? Discuss.

Exercise 5

Tina and Chuck plan to get together to review for a coming mathematics test. Each has two choices: (1) to study beforehand and thus make their review more beneficial, and (2) not to prepare for the review. Each prefers that the other prepare but would prefer not to prepare himself or herself. The review session will be more beneficial to both if both prepare than if both do not. A matrix that gives the utility values of the various outcomes is shown in Table 22-13 (page 398).

a Verify that this is a PD problem by checking that characteristics PD 1, PD 2, PD 3, and PD 4 are satisfied.

b What are the temptation, reward, penalty, and sucker values (t, r, p, s) for this PD problem?

Exercise 6

Adams and Barnes both own stores that sell records and tapes in Collegetown. From time to time they engage in a price war, drastically reducing their prices to attract student customers. Experience has suggested to Adams and to Barnes that they should stop cutting each other's throat with competition and make an agreement to cooperate. Their profit amounts for different decision choices are given in Table 22-14 (page 398).

a Verify that this is a PD problem by showing that the four characteristics are satisfied.

b What are the temptation, reward, penalty, and sucker values (t, r, p, s) for this PD problem?

TABLE 22-13

	CHUCK	
TINA	Prepare	Don't prepare
Prepare	(3, 3)	(1, 5)
Don't prepare	(5, 1)	(2, 2)

TABLE 22-14

	BARNES	
ADAMS	Cooperate	Defect
Cooperate	($600, $600)	($100, $1000)
Defect	($1000, $100)	($300, $300)

Exercise 7

The students in an algebra course all find it difficult to ask questions in class because they do not want to risk embarrassment by asking stupid questions. Yet each class member finds it beneficial if others ask questions.

Consider a special case of this situation—a class with two members. Each must choose between two alternatives:

C: *Cooperate* with the other by asking some questions about confusing material.

D: *Defect* by keeping quiet, assuming no responsibility for asking questions.

a Discuss whether this situation satisfies the characteristics of a PD problem.

b Suppose we have decided that this is indeed a PD problem and have started to supply utility values for the different possible outcomes. Our matrix so far is as shown in Table 22-15. What are the ranges of possible values for x and y if this is to be a PD matrix?

TABLE 22-15

	STUDENT 2	
STUDENT 1	C	D
C	(5, 5)	(x, y)
D	(y, x)	$(-1, -1)$

Exercise 8

Two nations competing in an arms race may either persist or stop. If both nations stop, they can devote the resources they would otherwise spend on armaments to socially useful projects that would make their people better off; and the balance of power between the nations would still be preserved. If both persist, both will pay a high price for their socially useless weapons systems and will be comparatively no stronger militarily. If one nation persists and the other does not, the nation that persists will have the military superiority to defeat its adversary and thereby realize its best outcome; the nation that stops will lose control over its fate and thereby suffer its worst outcome.

a Discuss whether this situation satisfies the four characteristics of a PD problem.

b Using a utility scale with a range of -100 to 100, assign utility values to each possible outcome and construct a matrix that summarizes this decision situation.

Exercise 9

Suppose that two African tribes share a grazing land for their cattle. As the herds grow in size, the tribes begin to realize that their common grazing area will not support a continual increase: grass will not be replenished as fast as it is eaten.

Supply additional details that might fit this situation. Choose the details in such a way that the characteristics of a PD problem are satisfied.

Exercise 10

Consider the relationship of wages in a specific industry to the total national economy described in the matrix in Table 22-16 (page 400).

a Union A must decide whether to bargain for a wage increase or not. It is, of course, traditional for unions to bargain for increases for their members. But do such increases, in fact, make things worse by fueling an inflation spiral that actually will raise the cost of living and decrease their purchasing power? Discuss.

b Discuss whether this situation satisfies the four characteristics of a PD problem.

TABLE 22-16

	ALL OTHER EMPLOYEES IN THE NATION	
UNION A	No raise	Raise
No raise	Prices remain stable; present level of purchasing power is maintained.	Purchasing power for members of union A decreases by $5 a week.
Raise	Purchasing power for union A members increases by $5 a week.	Unstable economy; cost of living rises; Everyone's purchasing power decreases $1 a week.

Exercise 11: Invent

Consider the following situation: Two individuals who have purchased a building together are formulating plans for the maintenance of the building. Invent details of a scenario that describes this situation as a PD problem.

Exercise 12: Journal

Make a list of decision situations in which you are or have been involved, where two (or more) people have a choice of cooperating or not. For each situation decide whether it has the characteristics of a PD problem. Select one of the PD problems and analyze it using what you know about the prisoner's dilemma. Evaluate the results: in what ways was your analysis helpful?

Note The Appendix contains complete answers for Exercises 3*b* and 7*b*. It contains partial solutions or hints for Exercises 2*b*, 3*c*, 7*a*, and 11.

CIRCUMVENTING THE DILEMMA

The Prisoner's Dilemma seems unsolvable. Each person is tempted to defect in order to obtain a better outcome for himself or herself. But when both persons do this, they obtain an outcome that is worse instead of better.

As with many problem situations, one good solution for the PD problem is to prevent its occurrence. This solution can be illustrated by a pollution-control problem. The problem has been simplified to include only two decision makers; realistic pollution-control problems, of course, involve numerous decision makers, but the difficulties are like those of the simple example.

Example 1 ▶

Suppose that there are two manufacturers with large plants on Columbia Lake. Both have chosen this location because water is needed as a coolant in the production process and because the lake is convenient for waste disposal. Since the lake is large, the companies' heating of the water and discharge of soluble waste originally had negligible effects. But now that both manufacturers have grown and prospered, their effect on the lake is noticeable.

The two manufacturers have begun to talk about what to do. Shall they, at considerable expense to each, cooperate and end their damage to the lake? Or shall they continue to destroy the lake, each gaining what it can in the process?

It is reasonable to suppose that each decision maker, reflecting on the choices available, sees two possibilities:

C: Cooperate in cleaning up the manufacturing process.

D: Defect from any agreement to cooperate and continue to destroy the lake.

These choices probably satisfy the characteristics of a PD problem.

Now suppose that the manufacturers, facing this important decision, have made cost estimates for each alternative. These are shown in Table 22-17. The entries in Table 22-17 are multiples of $1 million. Negative numbers designate decreases in annual profits; positive numbers designate increases.

To face such a decision is unpleasant. The cost of cooperation is high. However, the cost of defection—the eventual destruction of the lake—is even higher. There is a temptation, after first agreeing to cooperate, to defect. (A sense of obligation to employees and stockholders may even make this choice seem to be a moral one.) How can this decision dilemma be resolved?

One way to alter the problem is this:

> Suppose that there is an organization, Columbia Lake Preservation, committed to maintenance of environmental quality in the lake region. Suppose further that this organization has the power to impose a reward-penalty system on the two manufacturers. Specifically, suppose that it can, through efforts of various sorts, impose punitive measures resulting in reduced profits of $2.5 million for a manufacturer which persists in pollution. And suppose that it can take measures to generate increased profits of the same amount, $2.5 million, for a manufacturer which ends pollution. In such a case, the values given in Table 22-17 would no longer describe the outcomes. Instead, the manufacturers face decisions described by the matrix in Table 22-18.

TABLE 22-17
Payoff matrix of first-year effects (in $ millions) of pollution-control choices

	MANUFACTURER 2	
MANUFACTURER 1	Cooperate	Defect
Cooperate	(−2, −2)	(−6, 1)
Defect	(1, −6)	(−4, −4)

TABLE 22-18
Payoff matrix for pollution-control choices, revised to include rewards and penalties of $2.5 million

	MANUFACTURER 2	
MANUFACTURER 1	Cooperate	Defect
Cooperate	(0.5, 0.5)	(−3.5, −1.5)
Defect	(−1.5, −3.5)	(−6.5, −6.5)

Careful examination of Table 22-18 shows that we no longer have a PD problem. With the payoffs changed by the reward-penalty amounts, the most profitable choice for both decision makers is C, "cooperate." Neither will be tempted to abandon that choice, because to do so would result in a lower payoff.

Thus the imposition of a reward-penalty system has changed the nature of the decision. The decision is transformed into one where cooperation is the dominant choice; there is no longer a PD problem. ◀

A summary of the steps in Example 1 can be useful in applying a similar technique to other situations.

1 A certain decision is found to be a PD problem.

2 People who believe that cooperation is the only defensible alternative alter the decision structure to allow their belief to be fulfilled. By introducing a reward-penalty system that increases the payoff for cooperation and decreases the payoff for defection, these people change the problem into one where cooperation is the dominant choice.

Use of rewards or penalties to induce cooperation is not a novel idea, as is illustrated by Examples 2, 3, and 4.

Example 2 ▶

There are many more people who do business in the downtown area of a community than there are parking spaces. It is beneficial to all when there is cooperation and parking spaces are shared—with each person making only brief use of them—instead of monopolized by a few. To enforce cooperation, meters and parking fines are used. ◀

Example 3 ▶

Americans agree that education of the young is a worthwhile national investment and have agreed to cooperate and pay taxes to support it. But some are tempted to defect and reap the benefits of a system supported by others without bearing a share of the cost. To enforce cooperation, penalties such as fines and imprisonment are imposed for tax evasion. ◀

PRISONER 1	PRISONER 2 Keep quiet (C)	Rat (D)
Keep quiet (C)	(0, 0)	(−7, −3)
Rat (D)	(−3, −7)	(−10, −10)

TABLE 22-19
Payoff matrix for prisoner's dilemma, incorporating reward and penalty values

Example 4 ▶

An organized crime syndicate recognizes that clever prosecutors may provide suspects with decision choices like the prisoner's dilemma. Syndicate leaders thus institute a reward-penalty system that changes the payoffs. For example, the syndicate may threaten a member who rats with harassment or even death. And it may offer esteem and perhaps money as a reward for keeping quiet. A prisoner who regards these rewards and penalties as equivalent in value to, say, 3 years' imprisonment, can no longer view the problem as originally presented by the prosecutor. The matrix in Table 22-19 gives the new view of the decision. The syndicate's penalties and rewards have removed the dilemma; now there is a dominant choice: keep quiet. ◀

It seems best to consider the prisoner's dilemma as a problem to *avoid* rather than solve. Whenever a group of people identify a goal that can be achieved through cooperation, they are in danger of creating a PD problem unless, as they decide to cooperate, they also consider how to enforce cooperation.

As Americans who prize individual freedom, we have a long tradition of reaction against limits to this freedom. We do not, for example, like to limit a manufacturer's enterprise by saying that it may not use the resources it finds or that it may not pollute. But we live in a different world from that of our ancestors. In the past, resources appeared unlimited and the greed of one did not seem to penalize others. Now, open space, clean air, water, and other natural resources all can be seen to be limited—and some are even scarce.

A long-dominant pattern of thought (first popularized by the economist Adam Smith in *The Wealth of Nations* in 1776) has been that decisions reached by individuals will result in what is best for an entire society. The prisoner's dilemma makes us realize that this need not be so. Because many actual decisions resemble the prisoner's dilemma, this realization is important.

Garrett Hardin, a biologist, in an often cited paper, "The Tragedy of the Commons," argued eloquently that individual decisions concerning population growth and conservation of natural resources do not promote the common good unless the situation is designed so that they will.* Hardin maintained that continued freedom to enjoy and share scarce resources requires the establishment of social arrangements of "mutual coercion, mutually agreed upon." The hypothetical re-

*Garrett Hardin, "The Tragedy of the Commons," *Science,* vol. 162, Dec. 13, 1968, pp. 1243–1248.

ward-penalty system imposed by Columbia Lake Preservation, discussed above, is an example of such social arrangements. Cooperation should not be voluntary, Hardin argued—for voluntary cooperation favors only those who lack conscience.

The possibility of solving decision dilemmas by avoiding them—by restructuring so that the choices which are best for the group are also most profitable for the individual—is appealingly simple. It was done on paper in Examples 1 and 4 simply by adding and subtracting numbers. But in practice the reward-penalty system is not so easy. Tough questions include these: Where does the reward come from? Who enforces the penalties? (Exercises 13 to 23 include questions to stimulate your critical thinking and evaluation of reward-penalty systems.)

SUGGESTIONS FOR FURTHER READING

The prisoner's dilemma has been studied extensively since the 1950s. Theoretical studies have been carried out as part of the branch of mathematics called *game theory,* which was born in 1944 with the publication of *Theory of Games and Economic Behavior* by John Von Neumann and Oskar Morgenstern (Princeton University Press). Experimental studies have been carried out in psychology. For example, see:

Anatol Rapoport and Albert M. Chammah, *Prisoner's Dilemma,* University of Michigan Press, Ann Arbor, 1965.

Much of this study has involved a form of prisoner's dilemma that differs in an important way from the dilemma presented here; instead of a one-time-only choice, the same decision is repeated again and again. In such a context, the desire for cooperation and fear of retaliation from the other player in the future provide an incentive for cooperation now.

An easy-to-read introduction to primarily nonquantitative aspects of game theory may be found in:

Henry Hamburger, *Games as Models of Social Phenomena,* Freeman, San Francisco, 1979.

An interesting discussion of game theory in the analysis of decisions described in the Bible is:

Steven J. Brams, *Biblical Games: A Strategic Analysis of Stories in the Old Testament,* MIT Press, Cambridge, Mass., 1980.

Robert Axelrod describes the results of his investigation of how cooperation can evolve in society despite individual self-interest in:

Robert Axelrod, *The Evolution of Cooperation,* Basic Books, New York, 1984.

One interesting aspect of his study is a computer tournament among programs written by experts in decision making. Each computer program was designed to carry out its author's strategies and play a repeated prisoner's dilemma game (similar to Exercise 1 in this chapter) against another computer program.

Exercise 13

Verify the claim that in Table 22-19 cooperation ("keep quiet") is the dominant choice for each prisoner.

Exercise 14

What is the meaning of the phrase (used in summarizing Hardin's argument) "voluntary cooperation favors only those who lack conscience"? Do you agree with it? Explain.

Exercise 15

a Verify that the situation in Example 1 (Table 22-17) is a PD problem.

b For each of Examples 2, 3, and 4, identify the choices C and D faced by the decision makers (before any rewards or penalties are imposed). Are the conditions for a PD problem satisfied?

c Select Example 2 or 3 and describe details of a reward-penalty system that provides an incentive for cooperation. How effective is it likely to be? Do you see drawbacks to such a system? Explain.

Exercise 16

In Example 1, the decision dilemma was removed by adding a reward of $2.5 million for each choice of C and subtracting a penalty of $2.5 million for each choice of D in the matrix of Table 22-17. The results were summarized in Table 22-18. In this exercise you are to consider the effects of different reward-penalty amounts.

a What if the reward and penalty amounts had been only $2 million? Determine the payoff matrix and analyze the decision.

b What if the reward and penalty amounts had been only $1.5 million? Determine the payoff matrix and analyze the decision.

c What if the reward and penalty amounts had been only $1 million? Determine the payoff matrix and analyze the decision.

d What is the minimum amount for rewards and penalties that would be effective in removing the dilemma and making C the only rational choice?

e What are some of the practical difficulties of implementing a reward-penalty system? Can they be overcome? Discuss.

Exercise 17

Example 4 assumed that criminals want the syndicate's approval enough to suffer 3 years imprisonment to get it and that they fear harassment enough to suffer 3 years imprisonment to avoid it. Experiment with Table 22-6 to find reward-penalty amounts other than 3 years imprisonment that also would enforce cooperation. What is the lowest such value?

Exercise 18

An abstract version of a PD matrix looks like Table 22-20, where t, r, s, and p stand for temptation, reward, sucker, and penalty payoff values and where $t > r > p > s$.

The goal of a reward-penalty system is to alter the payoffs so that choice C offers as much or more gain to each decision maker no matter what the other decides—that is, so that choice C dominates choice D.

In a PD matrix, $t > r$ and $p > s$; these relationships cause the decision makers to choose D. To remove the dilemma, t and p must be decreased, or r and s must be increased, or both.

The following steps describe how to calculate a minimum reward-penalty amount *RP:*

Step 1: Calculate $\frac{1}{2}(t - r)$; calculate $\frac{1}{2}(p - s)$.

Step 2: Choose the larger of the two values calculated in step 1; this is the minimum reward-penalty value RP. That is, RP = maximum of $\{\frac{1}{2}(t - r), \frac{1}{2}(p - s)\}$.

a Return to Exercise 16 and calculate RP from Table 22-17. Check that this is the same value you found in Exercise 16*d*.

b Return to Exercise 17 and calculate RP from Table 22-6. Check that this is the same value you obtained by experimentation in Exercise 17.

c Return to Exercise 6 and calculate RP from Table 22-14. How might Adams and Barnes make use of this information and establish a business arrangement that enforces cooperation?

d Return to Exercise 5 and calculate RP from Table 22-13. Is RP of use in this case? How might Tina and Chuck enforce cooperation?

TABLE 22-20

	DECISION MAKER 2	
DECISION MAKER 1	C	D
C	(r, r)	(s, t)
D	(t, s)	(p, p)

Exercise 19: Invent

One type of situation that may lead to a PD problem is scarcity of a resource that must be shared by the members of a group.

a Develop a scenario that describes in detail a situation of this type which has the characteristics of a PD problem.

b Describe a reward-penalty system to enforce cooperation in your situation. Discuss whether the reward-penalty system is a practical solution to the problem.

Exercise 20: Invent

In the discussion of rewards and penalties for the original prisoner's dilemma, esteem and harassment were mentioned as a reward and a penalty. More often than not, when we encounter a PD problem, we will find that payoffs are intangibles and cooperation (if it exists) is enforced in intangible ways. We may, for example, cooperate in sharing scarce resources because we desire friendship or admiration from our associates rather than because it is profitable.

a Develop a scenario that describes a PD problem where the payoffs are intangibles.

b Describe a reward-penalty system to enforce cooperation where the rewards and penalties are intangibles. Discuss the strengths and drawbacks of the system.

Exercise 21

a The method of solving a PD problem by instituting a reward-penalty system that removes the decision dilemma has been criticized as follows:

If cooperation is the *right* thing to do, then people should be expected to cooperate for that reason alone and not because they are rewarded for doing so or penalized for not doing so.

Discuss this criticism of a reward-penalty system; state whether you agree or disagree with it and why.

b Look back over the examples in the text and exercises of reward-penalty systems. Select one; list any drawbacks to the system that you think are significant. Decide whether the system is on balance a good idea or a poor one.

Exercise 22

In the prisoner's dilemma computer tournament described in Axelrod's book *The Evolution of Cooperation,* the strategy that worked best overall was called *tit-for-tat.* To use this strategy a player chooses C (that is, cooperates) at the first step and after that always mimics the decision made at the last step by the other player.

a If two players use the tit-for-tat strategy and do Exercise 1, what will be the outcome?

b Compare tit-for-tat with the strategy you described in Exercise 4*a.* Which is better? Why?

Exercise 23: Journal

Return to Exercise 12, for which you listed examples of prisoner's dilemma situations. For one (different from the one analysed earlier), describe carefully the choices involved. Was cooperation the eventual decision? How were rewards and penalties involved? How could the result have been improved?

Note The Appendix contains complete answers for Exercise 13 and 16*a* and *d.* It contains partial solutions or hints for Exercises 16*e* and 20*b.*

MULTIPERSON PRISONER'S DILEMMA SITUATIONS

Some of the examples of PD problems in this chapter would be more realistic if they were not limited to two decision makers. This section will consider briefly an extension of the prisoner's dilemma to include more than two people. The choice remains the same: to cooperate and, if others cooperate also, obtain a favorable outcome for all in the group; or to defect in favor of individual self-interest because there is no basis for trusting others to join in the cooperative decision.

A format similar to that for the PD problem can be used for the characteristics of a multiperson prisoner's dilemma decision (MPD) problem:*

MPD 1 There are n decision makers each with the same pair of choices designated, for simplicity, C ("cooperate") and D ("defect").

MPD 2 Each decision maker, acting alone, gains more from choice D no matter what the others choose.

*This formulation of the MPD problem is adapted from Thomas C. Schelling, *Micromotives and Macrobehavior,* Norton, New York, 1978.

MPD 3 Each decision maker gains more as the number of others that choose C increases, no matter what choice he himself or she herself makes.

MPD 4 There is some minimal coalition size, k ($1 < k \leq n$), such that if k or more people choose C and the rest choose D, those who choose C are better off choosing C than D.

These characteristics are illustrated by the dilemma faced by people in Oakwood, Michigan, during its annual fund-raising drive for United Charities.

- Each of the 1000 adult and employed residents of Oakwood faces the same choice: C, cooperate by making a contribution; D, defect by failing to contribute (MPD 1).
- Although United Charities is viewed as an honest and efficient organization, each Oakwood citizen can see more immediate benefit from his or her own use of the money than from the fund's use (MPD 2).
- Oakwood's United Charities provides benefits to the community in direct proportion to the amount of contributions: the more people contribute, the better (MPD 3).
- If a small number of people (5, 10, or 20) contribute, and if the amount is also small, then the contributors probably will feel disappointed with the outcome—for there will not be enough money to carry out any significant projects. On the other hand, it is not unreasonable to suppose that there is some minimum number—possibly 700—of people whose contributions will fund enough projects so that the contributors will be satisfied that they are better off by having given. Unfortunately, however, if only 700 contribute there also will be 300 freeloaders who enjoy the benefits of the contributions of others with no cost to themselves (MPD 4).

When there are a large number of decision makers in an MPD problem it is generally not necessary for all to cooperate for benefits to be realized. On the other hand, cooperation is least costly for each *one* when all cooperate; every group member is penalized for each person who defects. (In the paying of taxes, for example, a system of taxation can work even though there are a few who refuse to pay; however, each payer pays a bit extra for each freeloader.) When a group has goals that its members agree to, then the group needs to consider how to bring about cooperation to attain these goals.

This investigation of the MPD problem has been brief and will end after a few exercises; but you can be sure that you will meet the problem again, in your classes and in your reading. Talk about it with your friends, especially with those who are biologists, economists, philosophers, psychologists, or politicians. In addition to the consideration of possible reward-penalty systems, there are bigger questions to face. As we tackle problems like allocation of scarce resources, pollution control and disarmament, we must face questions such as:

What is economical?
What is possible?
What is right?

These questions can take us far from mathematics, but we can expect that our mathematical knowledge will help us in the same basic ways as elsewhere: in organizing information and evaluating alternatives as we seek the best solution.

Exercise 24

Ten people have formed an investment club. Each has agreed to contribute an amount—1 investment unit—per month. The gross return on each unit invested is 3 units. The members have agreed that the gross gain is to be shared equally by all club members.

At the outset, all members contribute the agreed-on unit per month and each realizes 2 units profit per month from that investment. Later, however, some members become unable (or unwilling) to make monthly contributions. Still the equal division of profits persists. Table 22-21 shows the various profit amounts, depending on the number of active investors.

a Each member of the investment club faces a choice; C, cooperate by making a monthly payment; D, defect by failing to contribute. Verify that the four characteristics of an MPD problem are satisfied by this situation. What is the minimal coalition size k of club members for whom choice C is more profitable than choice D?

b Suppose that you are a member of this investment club. What changes would you suggest? What different types of external situations might lead to different concepts of "fair" rules for investments and profit distributions?

TABLE 22-21

NUMBER OF MEMBERS WHO CONTRIBUTE	NUMBER OF FREELOADERS	GROSS GAIN	GROSS GAIN TO EACH MEMBER	PROFIT TO CONTRIBUTORS	PROFIT TO FREELOADERS
10	0	30	3	2	—
9	1	27	2.7	1.7	2.7
8	2	24	2.4	1.4	2.4
7	3	21	2.1	1.1	2.1
6	4	18	1.8	0.8	1.8
5	5	15	1.5	0.5	1.5
4	6	12	1.2	0.2	1.2
3	7	9	0.9	−0.1	0.9
2	8	6	0.6	−0.4	0.6
1	9	3	0.3	−0.7	0.3
0	10	0	0	0	0

Exercise 25

a Repeat the analysis of Exercise 24 for a situation in which the gross return on each investment unit is 2 units. (As an aid to your analysis you may want to develop a table like Table 22-21.)

b Repeat the analysis of Exercise 24 if the gross return on each investment unit is 10 units.

c What if the gross return on each investment unit is only 1 unit?

d How would external conditions affect your concept of "fair" investment club practices?

Exercise 26

a Some people give the following argument against a reward-penalty system to enforce cooperation:

If the group really wants a given cooperative outcome, its individual members will reflect that by cooperative behavior. If individuals do not choose to cooperate, then group regulations should not force them to do so.

Discuss this argument. What are situations to which it applies?

b In "The Tragedy of the Commons," Hardin argued that voluntary cooperation favors only those who lack conscience. In what situations is Hardin's view more appropriate than the argument above?

Exercise 27

Make a list of societal decision problems for which the MPD problem serves as a model. What solutions have been tried on these problems? Which ones have worked? Which problems can be circumvented by a reward-penalty system?

Exercise 28: Journal

Examine your own experience—both past and anticipated—to find decision dilemmas that resemble the MPD problem. Analyze one of these problems and make suggestions about future handling of such dilemmas.

Exercise 29: Test Yourself

Develop a list of new terms and key ideas presented in this chapter. Supply an example of each item on your list.

Note The Appendix contains partial solutions or hints for Exercises 24*a* and *b*, 25*a*, and 27.

AFTERWORD

I wrote this book to help you fill a "kitchen drawer."

When I was a child growing up, I lived in an old farmhouse. The kitchen had a tall, built-in dish cupboard with doors that swung wide to reveal large stacks of plates, bowls, and cups—the remnants of many different sets that had belonged to grandparents, great-grandparents, great-aunts, and great-uncles. The cupboard was grand, but the real treasure chest was the big wide drawer beneath it.

The drawer, which had two sturdy handles, was 4 feet wide and 1 foot deep and extended a full 2 feet back into the wall. Inside were bits of string (rolled neatly), discarded eyeglasses, an ancient pair of compasses, a fever thermometer, a bottle of crumbly aspirin, ragged newspaper clippings offering household hints, a deck of cards with aces missing, a choke collar, chair casters, a door stopper, pieces of stone-hard chewing gum, safety pins, Popsicle sticks, a pocket edition of the Psalms, a jar of assorted nails and tacks and nuts and screws, at least seven screwdrivers, sugar cubes that advertised the Sunrise Motel and Restaurant, Canadian pennies, a conch shell for producing ocean noises, a handwarmer—I have hardly begun to inventory its treasures.

This drawer was my family's problem solver. When something broke, when someone ached, when someone was bored, the solution always was, "Look in the kitchen drawer."

The drawer did not have all the answers, but its varied resources always were enough to get us started toward a solution. Sorting through its odds and ends stimulated a recollection of family history mixed with new ideas. A rainy summer afternoon was never an occasion with nothing to do; we could always browse through and reorganize the kitchen drawer.

Hindsight shows me that my education—and especially my study of mathematics—has been like filling a kitchen drawer. From here and there and elsewhere, I have collected odds and ends of ideas and insights that I have stuffed away in my memory to sort through and use when needed.

I have written this book to share with you the mathematical ideas that I find most useful—useful in personal matters (in organizing and estimating and deciding) as well as professional ones. I hope that our examination of these ideas has given you many bits and pieces that you will stash away in memory for later use in solving problems; that is, I hope you have begun to fill a mathematical "kitchen drawer."

REFERENCES

Everything has been thought of before but the problem is to think of it again.

—Johann W. von Goethe

	Relevant Chapters*
Andree, Josephine, and Richard V. Andree: *Cryptarithms,* Mu Alpha Theta Mathematics Club, Norman, Okla., 1978.	2
Averbach, Bonnie, and Orin Chien: *Mathematics: Problem Solving through Recreational Mathematics,* Freeman, San Francisco, 1980.	General
Axelrod, Robert: *The Evolution of Cooperation,* Basic Books, New York, 1984.	22
Bedian, Vahe: "Evaluation of Diagnostic Tests and Decision Analysis," *UMAP Journal,* vol. 1, no. 2, 1980, pp. 15–59.	4, 11, 16, 17
Billings, Karen, and Alice Kaseberg Schwandt: "New Techniques for Thinking Clearly: Brain Twisters Can Help Straighten Things Out," *MS,* vol. 8, no. 9, March, 1980, pp. 50, 54, 82.	1, 2, general
Brams, Steven J.: *Biblical Games: A Strategic Analysis of Stories in the Old Testament,* MIT Press, Cambridge, Mass., 1980.	22
———: *Game Theory and Politics,* Free Press, New York, 1975.	22
———: "The Network Television Game: There May Be No Best Schedule," *UMAP Journal,* vol. 1, no. 1, 1980, pp. 104–114.	17
Bronowski, Jacob: *The Ascent of Man,* Little, Brown, Boston, 1973.	General
Bursztajn, Harold, Richard I. Feinbloom, Robert M. Hamm, and Archie Brodsky: *Medical Choices, Medical Chances,* Dell, New York, 1981.	15, 17
Chartrand, Gary: *Graphs as Mathematical Models,* Prindle, Weber, and Schmidt, Boston, 1977.	13, 14

*References that are pertinent to the text in attitude or general content rather than (or in addition to) relating to a specific chapter are listed as *general.*

Relevant Chapters

Relevant Chapters	Reference
5, 6	Davis, Philip J.: "Mathematics by Fiat," *Two-Year College Mathematics Journal,* vol. 11, no. 4, September 1980, pp. 255–263.
General	Davis, Philip J., and Reuben Hersh: *The Mathematical Experience,* Birkhauser, Boston, 1981.
13, 15, 16	Eppen, Gary D., and F. J. Gould: *Quantitative Concepts for Management Decision Making without Algorithms,* Prentice-Hall, Englewood Cliffs, N.J., 1979.
1, 2, general	Ewing, David W.: "Discovering Your Problem-Solving Style," *Psychology Today,* vol. 11, no. 7, December 1977, pp. 69ff.
7, 12	Falk, Ruma, and Maya Bar-Hillel: "Magic Possibilities of the Weighted Average," *Mathematics Magazine,* vol. 53, no. 2, March 1980, pp. 106–107.
21	Farquharson, Robin: *Theory of Voting,* Yale University Press, New Haven, Conn., 1969.
1, 2, general	Frauenthal, James C., and Thomas L. Saaty: "Foresight—Insight—Hindsight," *Two-Year College Mathematics Journal,* vol. 10, no. 4, September 1979, pp. 245–254.
11, 12, 15, 16, 18	Freedman, David, Robert Pisani, and Roger Purves: *Statistics,* Norton, New York, 1978.
1, 2, general	Gagné, Robert M.: "Learnable Aspects of Problem Solving," *Educational Psychologist,* vol. 15, no. 2, 1980, p. 84–92.
20	Granovetter, Mark: "Threshold Models of Collective Behavior," *American Journal of Sociology,* vol. 83, no. 6, May 1978, pp. 1420–1443.
15	Gray, Mary: "Statistics and the Law," *Mathematics Magazine,* vol. 56, no. 2, March 1983, pp. 67–81.
6, general	Growney, JoAnne Simpson: "Contrariness: A First Step in Applying Mathematics," *UMAP Journal,* vol. 4, no. 4, 1983, pp. 381–88.
19	———: "Controlling the Effects of Interruptions," UMAP Instructional Module 657, *UMAP Journal,* vol. 5, no. 3, 1984, pp. 301–336.
20	———: "I Will If You Will . . . A Critical Mass Model," UMAP Instructional Module 539, *UMAP Journal,* vol. 11, no. 4, 1981, pp. 81–112.
5	———: "Measurement Scales," UMAP Instructional Module 546, Consortium for Mathematics and Its Applications, Lexington, Mass., 1981.
19	———: "Planning for Interruptions," *Mathematics Magazine,* vol. 55, no. 4, September 1982, pp. 213–219.
15	Hacking, Ian: *The Emergence of Probability,* Cambridge University Press, New York, 1975.
22	Hamburger, Henry: *Games as Models of Social Phenomena,* Freeman, San Francisco, 1979.
22	Hardin, Garrett: "The Tragedy of the Commons," *Science,* vol. 162, Dec. 13, 1968, pp. 1243–1248.
12	Hemenway, David: "Why Your Classes Are Larger Than 'Average,'" *Mathematics Magazine,* vol. 55, no. 3, May 1982, pp. 162–164.
7, 17, general	Hill, Percy H., et al.: *Making Decisions: A Multidisciplinary Introduction,* Addison Wesley, Reading, Mass., 1979.
4	Hoffman, Dale T.: "Monte Carlo: The Use of Random Digits to Stimulate Experiments," *UMAP Journal,* vol. 1, no. 2, 1980, pp. 61–85.
12, 13, 16, 17	Hogg, Robert V., Anthony J. Schaeffer, Ronald H. Randles, and James C. Hickman: *Finite Mathematics with Applications to Business and to the Social Sciences,* Cummings, Menlo Park, Calif., 1974.
9, 10, 13, 14, 15, 16, 21	Hunkins, Dalton R., and Thomas L. Pirnot: *Mathematics: Tools and Models,* Addison Wesley, Reading, Mass., 1977.

	Relevant Chapters
Kahneman, Daniel, and Amos Tversky: "The Psychology of Preferences," *Scientific American,* vol. 246, no. 1, January 1982, pp. 160–171.	15
Kogelman, Stanley, and Joseph Warren: *Mind over Math,* Dial, New York, 1978.	General
Krulik, Stephen (ed.): *Problem Solving in School Mathematics* (1980 Yearbook of the National Council of Teachers of Mathematics), NCTM, Reston, Va., 1980.	General
LaBrecque, Mort: "On Making Sounder Judgments: Strategies and Snares," *Psychology Today,* vol. 14, no. 1, June 1980, pp. 33–42.	15, 17
Lave, Charles A., and James G. March: *An Introduction to Models in the Social Sciences,* Harper and Row, New York, 1975.	3, 4, 8, 11, 13, 17
LeBoeuf, Michael: *Working Smart: How to Accomplish More in Half the Time,* Warner, New York, 1979.	2, 5, 19, general
Malkevitch, Joseph, and Walter Meyer: *Graphs, Models and Finite Mathematics,* Prentice-Hall, Englewood Cliffs, N.J., 1974.	7, 11, 12, 13, 14, 15, 16, 21, 22
Merrill, Samuel: "Approval Voting: A 'Best Buy' for Multicandidate Elections?" *Mathematics Magazine,* vol. 52, no. 2, March 1979, pp. 98–102.	21
Moore, David S.: *Statistics: Concepts and Controversies.* Freeman, San Francisco, 1979.	5, 12, 15, 16
Payne, Thomas A.: *Quantitative Techniques for Management: A Practical Approach,* Reston, Reston, Va., 1982.	13, 15, 16
Polya, George: *How to Solve It: A New Aspect of Mathematical Method,* Doubleday, Garden City, N.Y., 1957.	General
———: *Mathematical Discovery: On Understanding, Learning and Teaching Problem Solving,* Wiley, New York, 1981.	General
Pomeranz, Janet Bellcourt: "Confidence in Confidence Intervals," *Mathematics Magazine,* vol. 55, no. 1, January 1982, pp. 12–18.	18
Roberts, Fred: *Discrete Mathematical Models,* Prentice-Hall, Englewood Cliffs, N.J., 1976.	5
Rubinstein, Moshe F., and Kenneth Pfeiffer: *Concepts in Problem Solving,* Prentice-Hall, Englewood Cliffs, N.J., 1980.	5, 7, 9, 11, 15, 16, 17, 22, general
———: *Patterns of Problem Solving,* Prentice-Hall, Englewood Cliffs, N.J., 1975.	5, 7, 9, 11, 15, 16, 17, 22, general
Schelling, Thomas C.: *Micromotives and Macrobehavior,* Norton, New York, 1978.	3, 4, 20, 22
Shufelt, Gwen (ed.): *The Agenda in Action* (1983 Yearbook of the National Council of Teachers of Mathematics), NCTM, Reston, Va., 1983.	General
Schulte, Albert P. (ed.): *Teaching Statistics and Probability* (1981 Yearbook of the National Council of Teachers of Mathematics), NCTM, Reston, Va., 1981.	4, 15, 16
Simon, Herbert A.: *The Sciences of the Artificial,* 2d ed., MIT Press, Cambridge, Mass., 1981.	19, general
Singleton, Robert R., and William F. Tyndall: *Games and Programs: Mathematics for Modeling,* Freeman, San Francisco, 1974.	22
Slovic, Paul, Baruch Fischhoff, and Sarah Lichtenstein: "Risky Assumptions," *Psychology Today,* vol. 14, no. 1, June 1980, pp. 44–48.	15, 18
Stokey, Edith, and Richard Zeckhauser: *A Primer for Policy Analysis,* Norton, New York, 1978.	4, 5, 15, 16, 17, 21
Straffin, Philip D., Jr.: *Topics in the Theory of Voting,* Birkhauser, Boston, 1980.	21
Tobias, Shelia: *Overcoming Math Anxiety,* Norton, New York, 1978.	General
Trueman, Richard E.: *An Introduction to Quantitative Methods for Decision Making,* 2d ed., Holt, Rinehart and Winston, New York, 1977.	13, 15, 16, 17

APPENDIX

ANSWERS AND SUGGESTIONS FOR SELECTED EXERCISES

CHAPTER 1

1 (a) Play first and gain control of the game by removing 2 objects. Choose subsequent moves so that the sum of the numbers taken in your opponent's last move and your move is 3.

(b) Play first and remove 1 object; continue as in *a*.

(c) Have your opponent play first; choose subsequent moves as described in *a*.

2 (a) and (d) Let your opponent play first; follow each of his or her moves with a countermove which makes the total number of objects taken in his or her move and yours equal to 3.

3 Any number that is 1 more than a multiple of 3 is a losing number; the largest number in your list should be 49.

4 (a) Divide n by 3; if the remainder is 1, insist that your opponent play first.

(e) For (999; 1, 2) Nim, play first and take 2 objects (and leave your opponent with a losing number). Follow each of your opponent's moves with a countermove chosen to make the total for the pair of moves equal to 3.

6 Play first and remove 1 object. Each time your opponent takes *x* objects, you follow by taking *y* objects with $x + y$ equal to what fixed total?

7 (c) For (13; 1, 2, 3) Nim, have your opponent play first. Follow each of his or her moves *x* with a countermove *y* so that $x + y$ equals what fixed total?

8 (a) and (b) In a Nim game of the type (N; 1, 2, 3), to decide whether a number is losing, divide the number by 4; if the remainder is 1, it is a losing number.

9 (c) If dividing *n* by 4 leaves a remainder of 3, then *n* is 2 greater than a losing number; in such a case you should play first and take 2 objects.

10, 11 For each of these games identify the losing numbers and develop a strategy for play that will avoid losing numbers when it's your turn.

12 The game of 3, 4, 5 Nim is harder to analyze than the one-pile games. Working backward from the goal (leaving 1 object for your opponent) you can, however, identify certain pile combinations as losing ones. Four of the losing configurations are: 1, 1-1-1, 2-2, and 3-2-1. Check that these are losing configurations and find the others. Once you have identified all the losing pile arrangements, you can win at 3, 4, 5 Nim by choosing each move so that you leave your opponent with a losing pile combination.

13 A campus computer may be able to provide challenging opposition for your Nim endeavors. There are widely available computer programs that play endless variations of one-pile Nim and others that play extensions of 3, 4, 5 Nim which allow any number of objects in each of any number of piles.

14 Using your calculator, check the final entry in Table 1-7 by entering $4500 and multiplying by 1.09 ten times (each time taking the result of the last multiplication by 1.09). Why does this work? (There is an even quicker way to calculate that final entry if your calculator has a y^x key. If it does and if you don't see quickly how to use it, ask your instructor for help.)

15 The required investment is between $4200 and $4300.

16 Others, especially persons unknown to us, cannot select goals for us as well as we can ourselves.

17, 18 Be realistic here. Stick to goals that you really want and plan to achieve. For instance, you might have goals such as organizing a study schedule, completing an assignment, saving for a vacation weekend, and finding an apartment.

CHAPTER 2

1 (a) (b) (c) (d) For each of these problems, organization and persistence are important. Keep track of your efforts, so that when a trial proves to be incorrect, you don't waste time repeating it.

2 If you dislike mathematical problem solving, focus on ways to minimize your dislike and frustrations. Does it help if you think of practicing problem solving as a mental fitness program?

3 Think of your abilities as interrelated. Strategies that are successful in the development of one skill can be useful in the development of others as well, but the transfer is not automatic. Building on our strengths requires a conscious effort.

TABLE A-1

HOURS INVESTED (AND GRADES EARNED)			
PHYSICS	PSYCHOLOGY	FRENCH	AVERAGE GRADE
9 (A)	2 (C)	1 (C)	8/3 = 2.67
9 (A)	0 (D)	3 (A)	8/3 = 2.67
6 (B)	4 (B)	2 (B)	9/3 = 3.00

4 There are two important steps to follow in each part of this exercise: (1) Try several examples and guess what the pattern is. (2) Check your guess by predicting an answer for an example not yet tried. If your guess is incorrect, formulate another guess; if it is correct, try to develop an argument to show why the guessed pattern will always hold.

5 Make a table with five rows and three columns. List the explorers' names along the left; use "height," "fur" and "tail" as column headings. Fill in the table with the explorer's descriptions. Next, guess what Bigfoot looks like. Check each guess against the given information. Keep careful track of the possibilities that you eliminate, and soon you will discover the correct description of the monster.

6 Table A-1 lists several possibilities for Terry. It may be continued to find his best time allocation.

7 Sample problem Elaine is careful to buy only sweaters that look well with all her pairs of pants. If she has m sweaters and n pairs of pants, how many different outfits does she have to choose from? (*Answer:* $m \times n$)

10 (b) The largest area that can be enclosed by 100 feet of fence is 625 square feet, obtained by enclosing a region that is a 25 feet square.

(c) Finding this answer requires finding the square root of 1000.

11 (c) The four numbers are 3, 3, 2, 2. Their product is 36.

(d) The five numbers are 2, 2, 2, 2, 2. Their product is 32.

(f) The largest possible product is 1458.

(g) The largest possible product exceeds 1 quadrillion.

14 Errors are not low-cost mistakes in all courses. If you need to arrive at class with correct assignments, then you need something such as a study group with several other students in which experimentation can occur and mistakes can be made and discovered and eliminated. Mistakes can be costly, but fear of making mistakes can be even costlier. Making mistakes can be used to advantage if, after making them, we investigate why they occurred and learn ways to identify and correct them or to avoid them.

16 (a) It is not sensible to try to schedule the installation of ceiling and paneling at the same time and while carpeting or pool table assembly is going on. The workers will get in the way of each other and be unable to complete their work efficiently.

(b) If the requirement that the Winston-Salem representative "visit each other city once" is removed, lower-cost travel is possible.

19 Remembering "*P*lease *e*xcuse *m*e, *d*ear *A*unt *S*ally" reminds us that in evaluating complicated calculations, one does first what is in *p*arentheses; the order of operations is then to evaluate exponents, then perform *m*ultiplications, then *d*ivisions, then additions, and finally subtractions. Applying this rule gives 4 as the answer to the following calculation string: $14 \times (3 + 2)/(6^2 - 1) + 4 \times 3/6$.

20 *Vernon's decision method A:* Take the first job offer that comes along. (He might follow this rule if he needs money badly and is unsure of his chances of getting a job he likes.) *Vernon's decision method B:* Hunt for a job until getting an offer better than the one you just turned down. (He might follow this rule if he is worried about getting stuck with the worst possible job offer.) *Vernon's decision method C:* Keep hunting for a job until you get an offer that's better than all those that came before it. (If Vernon has plenty of time to wait for the right job, it may suit him to follow this rule.)

22 Be careful to round correctly. Entries for the last two rows of Table 2-5 are: 41.7 percent, 33.3 percent, 20.8 percent, 4.2 percent; then 45.8 percent, 29.2 percent, 16.7 percent, 8.3 percent.

24 The first column of entries for Table 2-7 is: 20.0 percent, 34.2 percent, 41.7 percent.

26 Surely a *real* Mr. Dogood would have some idea of the qualities that a good secretary possesses. (If not, he should seek advice from someone who does.) To make his selection, he should not only compare applicants with each other but also compare their qualifications with a list of desirable qualities.

CHAPTER 3

1 The *background constants* are as follows:
(a) Amount of money in Bill's pocket.
(b) Total population of Coaltown.
(c) Total number of calls.
(d) The aunt's investment.
(e) Amount of pizza for nine couples.

4 If the numbers of those traveling in each direction are the same, then one-way collection and two-way collection will yield the same revenue. Reduced cost of collection and fewer traffic tie-ups are advantages of one-way collection. Decreased revenue of an amount even greater than savings from eliminating collection is a possible disadvantage.

7 (a) Choppy is giving customers a better deal than was advertised.

8 20 percent.

9 (b) For (1), the religious group R_1; for (2), the whole population; for (3), the religious group R_2.

10 (c) The total number of males is not the same as the total number of females.

14 The portion of males Chris's father's age with positive test results who actually have tuberculosis is 16.1 percent.

17 (b) The final entry in Table 3-4 is $107.76.

18 (b) The final entry in Table 3-5 is $149.99.

CHAPTER 4

1 (a) The rule of 70 predicts a doubling time of 5 years. The population does not quite double in 5 years; however, the 5-year figure is closer to 10,000 than the 6-year population is.

2 (c) The bottom-line entry for Table 4-6 is $2073.60. The bottom-line entry for Table 4-7 is $2143.59. The bottom-line entry for Table 4-8 is $2182.86.

(d) The rule of 70 is a more accurate predictor of doubling time when there are more interest periods.

3 (b) See Table A-2.

5 Speedy Construction Company should first try to estimate d; if $d \geqslant 5$, the flat fee is better.

6 (b) Aunt Rachel needs more information about the degrees of risk. Once this information is known, the decision could be made to invest at the highest rate of interest available with an acceptable level of risk.

8 (a) A growth rate of 7 percent would lead to 3505 students in 5 years.

(b) A decline of 7 percent per year would leave 1739 students in 5 years.

9 (a) The first two columns of Table 4-10 show the pattern of decline indicated in Table A-3.

TABLE A-2

WITH THIS GROWTH RATE	THIS RULE IS A BETTER PREDICTOR OF THE DOUBLING TIME
20 percent	Rule of 72
14 percent	Rule of 72
8 percent	Rule of 72
5 percent	Rule of 70
2 percent	Rule of 70
1 percent	Rule of 70

TABLE A-3

	INITIAL POPULATION	
YEAR	LONDON	RURAL ENGLAND
1	200,000	1,000,000
2	199,000	995,000
3	198,005	990,025
4	197,015	985,075
5	196,029	980,150
6	195,049	975,249
7	194,073	970,373
8	193,103	965,521
9	192,138	960,693
10	191,176	955,890

TABLE A-4

ROW	BOTH HEADS	BOTH TAILS	ONE HEAD, ONE TAIL
161	4	5	16
162	7	7	11
163	8	6	11

10 (a) In a community of 100 adults with two of each age in the range 21–70, the community center started by the twenty 21–30-year-olds will undergo a gradual membership shift that will stabilize with a membership of the forty-two persons aged 50–70.

14 Your simulation will require use of rows 161–168 of Table 4-11. Partial results are shown in Table A-4.

16 Of the 300 simulation experiments, 54 result in 1; 49 result in 2; 49 result in 3; 47 result in 4; 50 result in 5; 51 result in 6.

17 The last column of Table 4-18 should contain these values: 49/100, 23/100, 11/100, 17/100.

22 Of the 100 simulation experiments, 50 choose an applicant of quality level 1; 26 choose an applicant of quality level 2; 20 choose an applicant of quality level 3; and 4 choose an applicant of quality level 4. The mean number of interviews needed for the 100 hiring simulation experiments is 2.44.

CHAPTER 5

1 Nominal scale: (b). Ordinal scales: (a), (f), (h), (i), (j), (k). Interval scale: possibly (d), units are points. Ratio scales: (c), units are milligrams per cup; (e), units are minutes; (g), units are miles; (h), units are games won. Think about *d*; explain how *d* might be any one of the four types of scales.

3 (b) There is no unit of measure for air quality.

5 When grades are used to measure knowledge in this way, the grade of 0 ordinarily does not correspond to a complete absence of knowledge; thus the scale is not a ratio scale. Unless each point corresponds to a fixed unit of knowledge, we do not even have an interval scale. Moreover, the simple ability to convert grades to points does not mean that knowledge is so easily measured. There are many difficulties. For example, does a grade of 3.2 describe a person with more knowledge than a grade of 3.1? Does each point measure the same amount of knowledge?

8 Other answers than the following are possible.
(a) College entrance examination scores.
(b) Number of mathematics courses completed.
(c) Number of times that I buy each in a year.
(d) This one is hard.

11 Don't let the lengthy story cause you to overlook the host's error. He is treating ordinal rankings as if there were known and uniform differences between rankings—that is, as if there were a unit of measure.

16 In this case we might expect the host to add his two guests' utility ratings and to serve coffee. But because the two guests probably do not attach the same meaning to *util,* the host still cannot be sure he has chosen the more preferred beverage. If it is very important to the host to know which beverage to serve his guests, he might devise some questions to try to measure the strengths of their preferences. For example, he might ask guest A, "Do you prefer tea to coffee and cookies, to coffee and cake, and so on?" Likewise, he might ask guest B, "Do you prefer coffee to tea and cookies, to tea and cake, etc.?"

17 None of these scales used 0 to designate the absence of all satisfaction.

Even if this decision is not one that particularly interests you, put yourself in the place of someone concerned about the decision and go through the effort of assigning utilities. If you practice by devising utility assignments in cases that aren't important to you, you will have more confidence in using them when an important decision comes along.

20 This may be a good time to reread the quotation from Lord Kelvin at the beginning of the chapter. His view—that when we cannot measure something, we have unsatisfactory knowledge of it—is by no means universal. Keep an open mind about the value of assigning utility values until you have experimented enough to feel comfortable with the process. At that point you will be able either to make use of utility scales in decision making or to forgo their use, whichever works better for you.

CHAPTER 6

1 (a) Sum (24 terms): \$0.35 + \$0.35 + · · · + \$0.35 = \$8.40. Is there a discount for bulk buying?

(b) Sum: ½ + ½ = 1. Does Rhonda have one *whole* tire?

(c) Sum: 100°C + 40°C = 140°C. What does your experience with combining water at different temperatures suggest as a reasonable temperature for the mixture?

(i) Sum: 10 utils + 15 utils = 25 utils. Might a candy bar satisfy Julie's desire for sweets and thus decrease the satisfaction she would get from a sundae?

(l) Sum: 80 points + 60 points + 85 points = 225 points (out of 300). Are the tests of equal importance?

(m) Are the criteria of equal importance? Does the same size util serve as the unit of measure for all three criteria?

(n) Can Steve do more than one activity at the same time?

(o) Does Dennis need any time between activities?

7 If you agree with the general idea behind statements (a) and (b) but not with the statements exactly as written, state what you disagree with in each and then rewrite them so that they correctly give your views.

8 (a) If cleaning the house and setting the table can be done while the other activities are going on, then the men may start as late as 3:15 P.M. Tuesday.

(b) The sum of 460 counts some homes more than once.

(c) We are not told that if Laura is not selected for one activity, she will be selected for the other. If indeed there is a chance that she won't be selected for either hockey or cheerleading, her chances of selection can't be 100 percent.

10 Here's something to think about as you assess the stated characteristics of a good problem solver: Does it make sense to approach mathematical problem solving as a game? Is learning to be a good problem solver a process that resembles learning to be a good tennis or basketball player, a good golfer, a good chess or bridge player? Are knowledge, daring, caution, and persistence qualities of someone skillful at a game? How can you approach problem solving as that same kind of skill? How can you develop the weaker aspects of your game?

CHAPTER 7

1 Matrix 1: Applicant C's ratings dominate the others; if the stated criteria are the only important ones, then choose applicant C.

Matrix 3: The ratings for applicant C dominate those for applicant A; eliminate A. To decide between B and C, consider the relative importance of the criteria "experience" and "pleasant disposition." Selecting the criteria and weighting their importance is probably a more important part of decision making than assigning ratings. Careful thought needs to be given to that aspect of the process.

2 (c) The same weighted average that Carl obtained will be obtained by using any list of weights obtained by multiplying Carl's list of weights by some fixed number. For example, if we multiply Carl's weights (0.3, 0.1, 0.1, 0.3, 0.1, 0.1) by 40, we obtain the list 12, 4, 4, 12, 4, 4, which leads to the same weighted averages as Carl found.

5 (b) When the weighted average is computed for a dominant row, each weight is multiplied by a number as least as large as its multiple in any other row. Thus these products, and their sum (the weighted average), will be at least as large as that for any other row. Assigning weights and computing weighted averages is not necessary when the choice matrix has a dominant row, for the highest weighted average always will belong to the dominant row.

6 (a) Using the given weights and utilities, the Renault is the choice with the highest weighted average (6.8).

(b) It would probably be wise for Betsy to consider future maintenance of the car as well. Do certain models have a likelihood of frequent repairs? Are parts and reliable service readily available? The emphasis in Chapter 7 has been on rankings and weights. However, in making any decision probably the most important aspect is the formulation of lists of alternatives and criteria. Only when these lists are carefully chosen and complete can a good decision be assured.

7 Raymond should talk with accountants and others—and perhaps do some reading—to learn as much as he can about the firms and cities he's considering. After gathering a lot of information, he might imitate Betsy's procedure in Exercise 6 to analyze his decision.

9 Examine the given matrix for dominance.

10 (b) There is no dominant choice in this case; thus the decision is not an easy one.

CHAPTER 8

1 (b) The purchase of 4 hamburgers and 4 books offers Patricia the highest utility (150 utils); thus she will choose that purchase.

2 (a) Probably not. Eating candy bars will probably reduce Sam's satisfaction from marshmallow sundaes, and vice versa.

3 (a) It is reasonable to add utilities in this case; doing so is based on the supposition that the satisfactions Jeff gains from studying and leisure do not detract from each other.

(b) After 6 hours of leisure per day, Jeff gains no additional satisfaction from more leisure time.

(c) 5 hours of study and 3 hours of leisure (73 utils).

5 (a) 4 hamburgers, 1 book; 95 utils.

(c) To purchase 10 books and 1 hamburger costs $42. A satisfaction of 150 utils could have been obtained at a cost of only $24 by purchasing 4 hamburgers and 4 books. (In fact, Patricia could have purchased 6 hamburgers and 3 books with $24 and gained 155 utils of satisfaction. How can we be sure that $24 is the best she can do?)

7 (c) From *a* and *b* we know that Terry gains 60 utils of satisfaction from a C in French and a B in physics, and 55 utils of satisfaction from an A in French and a C in physics. Thus Terry's utility values reveal the "surprising" result that he values a lower grade average more than a higher one. Could this have been predicted from examination of the first row and column of his utility matrix?

(d) One type of variation is this: change the hourly requirements for improving grades. For example, suppose that physics and French both require 2 hours of additional weekly study time to raise the grade one letter.

10 Donna's analysis can lead to efficient use of her time, but it ignores the value or importance of learning a particular subject. (For example, knowledge of statistics may be important to a person who will eventually be engaged in deciding whether certain drugs are safe enough to prescribe.)

12 (b) The persons who voted against funding for an annex to the county jail may think that current efforts to control crime are adequate, and that the marginal utility of additional money spent in this area is less than the marginal utility of expenditures elsewhere.

13 (b) Perhaps Bob works many hours because he needs the money. Or perhaps he doesn't need the money but is accumulating it toward something else that he really wants. Or perhaps he gains more satisfaction from his work than from any other activity on which he can spend his time.

17 In assessing the cost of thinking, here is an idea to examine: sometimes the time invested in thinking through one situation carefully results in the development of a thought pattern that can be used again and again. In such cases, thinking now saves time later. Can you identify situations where this idea applies?

CHAPTER 9

1 Since the total from the two separate lists is 40 names, and the number on the combined lists is only 25 names, 15 names must have been listed twice ($40 - 25 = 15$).

2 10, 22, 25, 3, 13, 7, 10, 10.

6 The mutually exclusive categories are: (1) graduates with salaries exceeding $30,000; (2) graduates with salaries exceeding $25,000 but not exceeding $30,000; (3) graduates with salaries exceeding $20,000 but not exceeding $25,000; (4) graduates with salaries exceeding $15,000 but not exceeding $20,000; (5) graduates with salaries of $15,000 or less.

7 The number of friends in the eight different categories are: 4, 2, 2, 2, 1, 1, 1, 0. List descriptions for the eight categories and pair the correct number with each.

10 Trial-and-error experimentation is needed to produce an example here. One way to start is to make a Venn diagram and to guess numbers for each region of the diagram. Then check to see whether your numbers are consistent with the information given in the exercise. If not, adjust your numbers until all the given conditions are satisfied. (There are many different correct answers possible.)

11 Examine the given information to determine how many people watch television for their news and to determine how many people don't watch television for their news. What's wrong with this pair of numbers?

CHAPTER 10

1 Yes.

2 (a) Here is a partial description of one design scheme. Adams: The letter A followed by 5 digits. Bedford: The letter B followed by 5 digits. Dover: The letter D followed by another letter and then four digits.

(b) Such a plan seems difficult to administer. For one thing, extra plates must be ordered for all designs. And what happens when a resident moves to another county? If having county identification is indeed desirable, wouldn't it be simpler to issue each vehicle owner a sticker for his or her plate that indicates the home county?

3 Yes. Eliminate either "oh" or "zero" and either "eye" or "one." The result is a set of 34 symbols, and 34^4 exceeds 1 million. (If this design scheme is used, what should be done about combinations of symbols that form offensive words?)

4 (b) Only design scheme 3 will work in this case.

6 $6 \times 4 \times 4$.

8 Describing a category of students requires specification of a sex (which can be done in 2 ways), a class (which can be done in 6 ways) and a major (which can be done in 67 ways).

10 (b) 5040.

(g) $\frac{12 \times 11 \times 10 \times 9 \times 8 \times \cdots \times 2 \times 1}{9 \times 8 \times \cdots \times 2 \times 1} = 12 \times 11 \times 10 = 1320$

11 (a) For a true-false test with 10 questions there are 1024 possible answer lists.

13 (a) Consider the tosses one at a time. Tom tosses a penny; there are two ways for it to turn up. Then Dick tosses; there are two ways for his penny to turn up. Finally Harry tosses; there are two ways for his penny to turn up.

15 (a) 8!

(b) $2 \times 4! \times 4!$

(c) The answer to *a* counts as different some arrangements that are nearly identical. For example, if the eight people are seated in one arrangement and get up and reverse their order in the seats, the new seating is counted as different even though each person is sitting beside exactly the same persons as before. Persons sitting together might not consider two arrangements different unless each of them was sitting beside a different person in the two arrangements. The total number of arrangements that are different in this latter way is much smaller than 8! and much more difficult to determine. (Can you find three such different arrangements? Can you find four? More than four?)

16 Is the answer 4×3 or $4 \times 4 \times 4$ or $3 \times 3 \times 3 \times 3$?

17 42 minutes.

22 Forty has binary representation 101000.

23 (e) 63 and 64. Compare *e* and *f*. What number corresponds to a binary representation consisting of *k* ones? What number corresponds to a binary representation consisting of a 1 followed by *k* zeros?

24 (d) 1100100.

25 Row 8 has entries 256 and 255. (255 has binary representation 11111111). The 256 numbers representable with eight digits range from 0 to 255.)

27 Experiment with products of more and more twos until you find a product that exceeds 1 million.

28 (b) 2^{22} bits.

29 (a) The decimal number 30 has quinary representation 110.

CHAPTER 11

2 There are 12 categories of students.

4 Start your tree with a pair of branches with labels such as "have knowledge" and "lack knowledge." Follow each of those branches with a pair of branches labeled "daring" and "not daring," and so on.

6 (b) Imitation of the process described by the given tree diagram will find you posing questions that divide the available numbers into two equal sets.

8 (b) Yes. The statement that each positive integer has a unique set of prime factors is such an important fact about numbers that it is called the *fundamental theorem* of *arithmetic*.

10 (b) How does one judge whether a decision was correct?

12 One-half of the families would be expected to have one child, one-fourth to have two children, one-fourth to have three.

13 (a) All together, 57,000 children will be born to the 27,000 families.
(b) Of the 27,000 families, 70 percent will achieve the goal of having a blond child.

14 The tree diagram should have 16 terminal vertices, each labeled with the fraction $1/16$.

15 (b) The diagram drawn in *a* should allow you to see that a customer will win $5 only $1/16$ of the time.

16 (b) The percent (10) of high school graduates who are female and graduate from college is higher than the percent (9) of high school graduates who are male and graduate from college.

18 The percent of male applicants admitted is 36.6 percent.

20 (b) The percentage of persons like Chris's father whose x rays indicate tuberculosis and who actually have tuberculosis is 16.1 percent.

21 Tree diagrams are useful in a greater variety of situations than Venn diagrams. The left-to-right ordering of branches corresponds nicely to the frequent need to organize information in a sequential manner. To represent the factoring of an integer (Exercise 8) or the steps of a decision (Exercise 10), the tree diagram seems natural. (Could a Venn diagram be used in these cases?) On the other hand, in logic, in the analysis of arguments, the Venn diagram is very useful. Its ovals represent categories of objects that are in some way related; their positions describe the relationship. For example, the relationship between men and mortals in "All men are mortal" can be described as shown in Figure A-1, and the relationship between women and tall people in "Some women are tall" can be described as shown in Figure A-2.

FIGURE A-1
"All men are mortal."

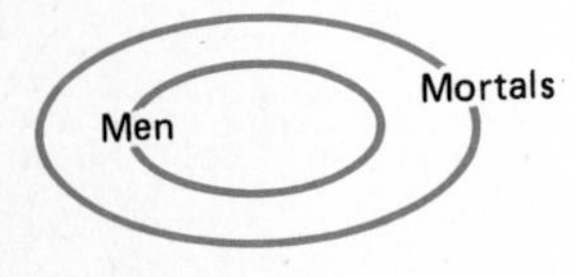

FIGURE A-2
"Some women are tall."

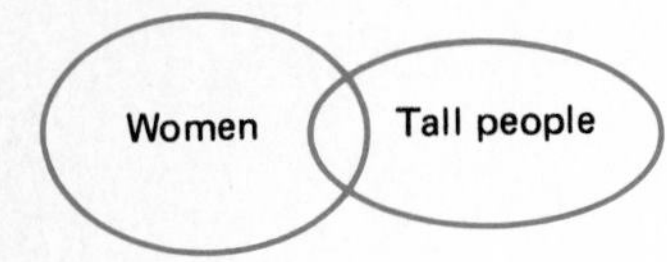

CHAPTER 12

2 (a) Each of the terms *white, married,* and *female* designates a mode.
(b) The average is likely to be the median (though is possibly the mean); the type of average should always be specified.
(c) Check this with an example. If the population and the mean are known, can the total income be calculated? What if the population and the median are known?

5 (a) "Mean word length" is the number obtained by adding the lengths of all words and dividing by the number of words. Although word size ranges from 1 to 13, probably there will be more short words than long ones, and the mean is likely to be less than 7.

(b) The modal word length is 3, and the median is 4. By inspection one can see that the mean will be greater than the median; why? The median is the best measure of the center. Why?

8 Nancy's average speed is less than 40 miles per hour because she spent a longer time driving at 30 than at 50 miles per hour. Her total distance is 20 miles; her total time is greater than ½ hour.

9 (a) Myron's mean weekly car sales is 3.37 cars.

10 The term *law of averages* refers to the concept that if the same event occurs over and over again, then the relative frequency of a given outcome will tend to stabilize at its true probability. (For example, if a die is perfectly balanced, so that each face is equally likely to turn up, then the law of averages predicts that in a large number of tosses the proportion of fives that turn up will be close to ⅙.)

20 The mean and the standard deviation will increase; the median and mode will be unchanged.

21 (a) Data set 1 has a standard deviation of 2.83. Data set 2 has a standard deviation of 7.07. Data set 3 has a standard deviation of 21.21.

22 For data set 2 the first quartile is 15, the median is 25, and the third quartile is 35.

25 (d) Add the same number to each data value.

27 Use Tchebyscheff's theorem.

CHAPTER 13

1 A digraph is more appropriate. Let vertices denote teams; draw an arrow from vertex T_i to vertex T_j if T_i beats T_j.

2 Let each member of Morton's family be represented by a vertex. Possible relationships to represent by edges are "is a parent of," "is a child of," "is a descendant of," and "is related to." All but the last will involve a digraph. Graphing the relationship "is related to" will result in a "complete" graph—that is, a graph in which each vertex is edge-joined to every other vertex.

5 Although road maps look like graphs (dots, like vertices, indicate cities and towns; and line segments, like edges, designate roadways), a typical road map conveys considerably more information than a typical graph. It has its dots and lines placed carefully so that locations and distances may be estimated. Vertex size indicates population, and color and thickness of lines indicate types of roads.

6 A solution may be obtained from Figure 13-10 by modifying the labeling so that each label is used at least three times.

10 (a) There are (at least) two choices for what the vertices should represent: either students or languages. To decide which, first see this problem as similar to the Grimm Zoo problem in Example 2. The goal of these was to assign animals to enclosures. The goal here is to assign courses to time periods. Identify elements of this problem similar to those of Example 2 and imitate the steps used in its solution.

14 Vertices labeled "inflation" have the greatest total number of outgoing edge arrows; from this we conclude that they rate inflation as most important.

18 (a) Add edges EF and FG to Figure 13-15.
(b) There are two possible schedules that satisfy the given condition.

19 (a) One convenient way to lay out the vertices has vertex L (Lockport) at the western end of the graph and vertex D (Dunkirk) at the eastern end.

20 (b) Delaying (or shortening) any activity will cause a delay (or shortening) of the entire project.

21 (a) This change in activity B will have no effect on the total length of the schedule, because B is not a critical activity. (However, B cannot be lengthened indefinitely without effect. How long can B take without delaying the total length of the schedule?)

22 (d) Work should begin first thing in the morning on August 13.

23 (c) All activities except C and E are critical.

25 (b) The minimum completion time is 20 weeks.

26 The given time requirements are only estimates. Which activities seem least likely to be finished on time? How would this uncertainty affect your scheduling?

27 Only six of the activities are critical. Among the other activities, some (like G, H, I, and J) are nearly critical and must be completed almost as soon as possible, and others (like F, M, N, and O) are far from critical and allow a lot of leeway in completion.

CHAPTER 14

1 (b) One possible Euler circuit for the star graph is A, D, E, H, J, G, I, F, B, C, D, H, G, F, C, A.

2 (c) You will get stuck at either vertex F or vertex H and be unable to go on without covering some edge a second time. Why does getting stuck occur at one of these vertices?

3 (a) Add edges that join the two pairs of odd vertices.

4 (b) Any route that covers the entire graph must go over a minimum of three street sections twice.

5 Think of rooms as being like vertices of a graph and doorways as edges.

10 (a) The sum of the degrees of the vertices is equal to twice the number of edges, and taking twice any positive integer gives an even number.
(b), (c), and (d) Whenever any collection of objects is counted twice, the result is an even number.

11 Put one lump in the first cup, one lump in the second cup, and twelve lumps in the third cup.

16 (a) The smallest number of edges is 5.
(c) Yes.

19 If you are not able to find a circuit that has all edges numbered between 1 and 30 outside the congested section, then relax that goal but come as close to it as you can. What is the best you can do?

20 (a) The minimum number of edges is 3.
(b) and (c) The minimum number of edges to be traveled twice in *c* is less than the minimum number in *b* even though *c* has more odd vertices.

21 (a) Yes, 11 is the minimum number.
(b) For the larger graph (5 blocks by 7 blocks) the number of blocks that must be traveled twice is smaller than that for the 5-block by 6-block graph, even though the larger graph has more odd vertices.

22 (b) This exercise requires some patience, erasing, and trying again. If the mail carrier crosses once to a diagonally opposite corner, count this as 2 crossings. It is possible to find a mail route with as few as 32 crossings. Is this the best possible route?

CHAPTER 15

1 Probability values that are reasonable matches for the given statements are:
(a) $p_E = 0$
(b) $p_E = .5$
(c) $p_E = .99$
(d) $p_E = .6$
(e) $p_E = .03$

2 Here is one list of interpretations of the traveler's statements:
(a) At 1 P.M. I seldom find stop-and-go traffic on the bridge.
(b) At 4 P.M. on about two days out of three I expect stop-and-go traffic on the bridge.
(c) At 6 P.M. the bridge is free of congestion about as often as not.
(d) Bridge traffic is almost never congested at 9 P.M.

10 (a) If only one can be promoted then the list "S_1: Simpson, Thompson, Wilson" will serve as a sample space.
(b) The list "S_2: MNO, MN, MO, NO, M, N, O, no Republican winner" includes all possible combinations of Republican winners and is one example of a sample space.

12 (a) One possible sample space is "S_3: MMM, MMF, MFM, MFF, FMM, FMF, FFM, FFF," which lists the sexes of each child, in order from oldest to youngest.

16 $p_0 = .25$

17 (a) The probability of at least 3 aircraft waiting is .3.

18 (a) The fact that the sum of Kerry's three values is greater than 1 is *not* a problem here, because the three events he uses are overlapping events.

FIGURE A-3

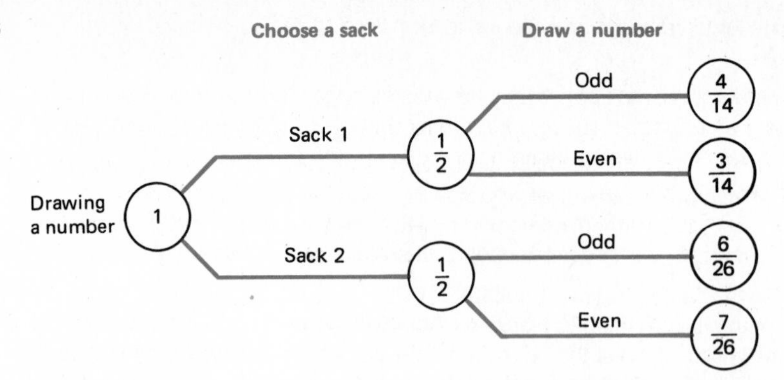

19 The best arrangement of odd numbers can be found by using a tree diagram (see Figure A-3). Suppose, for example, that sack 1 contains the numbers from 1 to 7 and sack 2 contains the numbers from 8 to 20. In this case the chance of drawing an odd number is $4/14 + 6/26 = 47/91 \approx .52$. To complete the exercise, experiment with different arrangements of the numbers in sacks 1 and 2. Most people guess that the solution is near to .5. Prove them wrong. (For extensive work with problems of this sort it is useful to have a working knowledge of *conditional probability,* a concept not included in this text.)

20 (a) and (b) The probabilities for Flogg and Mugg are identical.

(d) Flogg and Mugg might both go free; thus the events in *a* and *b* are not mutually exclusive.

23 Probability assignments 2 and 3 satisfy the necessary mathematical rules; that is, all values are nonnegative; and the sum, for all elementary events in the sample space, is 1.

24 (b) The outcomes 0 heads and 3 heads would be expected about the same number of times. (Why?) The outcomes 1 head and 2 heads would be expected about the same number of times. (Why?) Each of the latter two events would be expected to occur about three times as often as either of the former two. (Why?)

28 (a) Using the multiplication rule (Chapter 10), we calculate that the total number of New Sylvania license plates is $26 \times 10 \times 10 \times 10$. The number that begin with B is $1 \times 10 \times 10 \times 10$. An estimate for the probability that the license plate of a car chosen at random begins with B is

$$\frac{1 \times 10 \times 10 \times 10}{26 \times 10 \times 10 \times 10} = \frac{1}{26}$$

30 (a) Use a numbered tree diagram.

31 (a) The total number of ways that the slot machine can come to rest is $5 \times 5 \times 5 = 125$. Only one of these shows all bumblebees. If the wheel turns smoothly and all outcomes are equally likely, then a reasonable estimate of the probability of three bumblebees is $1/125$.

(d) $64/125$

34 The probability of failure for the system that includes duplicate backup control is .000001.

35 The probability of winning nothing is .997779.

37 The given information is inadequate for accurate predictions. In addition to historical data, information about recent flood-control measures and changes in runoff patterns need to be considered in estimating the likelihood of flooding. On the basis of the historical data alone, we could say that there is a low probability (near zero) that the Susquehanna River will flood next year and a high probability (near 1) that it will flood at least once during the next 50 years.

38 (c) Probability estimates from a tree diagram predict the patterns into which outcomes from a large number of experiments will fall if a perfectly fair coin is tossed in a manner that favors neither heads nor tails. Differences between the estimates and the outcomes of real tosses can occur because of an insufficient number of tosses, unbalanced coins, or a tossing method.

41 In 7 tosses of a coin it would not be unusual to have 2 or fewer heads or to have 5 or more heads, and thus to lose. (In fact, the actual probability of a loss in this case is about 45 percent; calculation of this probability requires knowledge beyond the scope of this text.) In 70 tosses it would be unusual to have 23 or fewer heads or to have 47 or more heads. (One of these outcomes will occur less than 10 percent of the time.)

48 (b) On what evidence are the prosecutor's probability estimates based?

CHAPTER 16

2 The expected value of the policy is −$30; this means that the insurance company takes in, on the average, $30 more per policyholder than it pays out. Martin should consider carefully whether it is worth $50 per year to ensure a payment of $10,000 in the unlikely event that he dies; he should shop around to learn rates from several competing companies; he should find out if there are important exclusions from the policy under consideration and from other policies with which he compares it.

3 The expected net gain to Middlecreek Construction Company is $70,000.

5 (a) The probabilities needed to calculate the expected value can be obtained from a numbered tree. The expected number of girls is two. (However, the probability of having two girls is only 3⁄8.)

(b) This question has the same structure as *a*.

7 (a) The expected value is the same as the weighted average calculated in Chapter 12: 3.37 cars.

(d) This problem is similar to *c*. However, it is possible to calculate an answer for *c*, but it is not possible to do so in this case. Why?

(e) If the ticket is a gift, so that you have no purchase cost, its expected value for you is 21 cents. If, however, you purchase the ticket for 50 cents, then its expected value is −29 cents.

8 (a) When the probability of rain is 50 percent, the expected value for Mark of carrying an umbrella is 1.5, and the expected value of not carrying an umbrella is 0.

(b) When the probability of rain is 70 percent, Marilyn should carry an umbrella.

14 To calculate the expected value of selling with a realtor, first deduct the 5 percent commission from the selling price.

15 The expected time for the shorter route is 17.5 minutes. Perhaps Paula prefers the predictability of the time needed for the longer route to the time savings of 7.5 minutes per day. It may be very important for her to avoid ever being late.

16 If you expect the caller to be in only 25 percent of the time (and if someone else will answer the phone in other cases, so that there will be a charge for the call), then it will cost less, on the average, to make operator-assisted calls.

18 Consider carefully the question, "Does Steve's reasoning make sense?"

20 $EV_{30,000}$ = \$3000; $EV_{35,000}$ = \$3375; $EV_{40,000}$ = \$3375; $EV_{45,000}$ = \$2875; $EV_{50,000}$ = \$2250. How should Roger decide between order sizes of 35,000 and 40,000?

CHAPTER 17

1 (a) The maximax rule selects Chapters 7 and 9.

(b) The maximin rule selects Chapters 8 and 9.

(c) The most-likely-value rule selects Chapters 7 and 8.

(d) The expected-value rule selects Chapters 7 and 8; the expected values are: $EV_{7,8}$ = 19.6; $EV_{7,9}$ = 18.2; $EV_{8,9}$ = 19.2.

(e) The minimax-regret rule selects Chapters 8 and 9. Roberta's regret matrix is shown in Table A-5.

(f) In this situation, with question B twice as likely as any other, the author favors the most-likely-value rule. The choice that it recommends is seconded by the expected-value decision, indicating that it not only is most likely but offers the best result on the average.

2 (f) The given information and the results of *a–e* select an order size of 40,000. Why? Before deciding, Roger could inquire whether he must order paper in lots that are multiples of 5000. If not, he may be able to find an order size between 35,000 and 40,000 that allows a greater expected profit than 40,000 copies allows.

TABLE A-5

	TOPICS			
CHAPTERS	A	B	C	D
7, 8	2	0	8	3
7, 9	6	7	0	0
8, 9	0	6	2	1

3 (a) If the decision is one of many similar decisions, then use of the expected value rule can be justified; it selects Groner Construction Company, the choice that is best on the average. However, for a one-time decision (especially, as in this case, when the funds of others are involved) the conservative maximin rule often is used; it selects Yingst. Trying all five rules reveals that the expected-value rule and the most-likely-value rule both select Groner; therefore Groner is a justifiable choice even for a one-time decision. Still another angle to consider is the pressure that will follow an incorrect decision. The author chooses the expected-value rule (and thus Groner), on the grounds that an administrator must make many decisions in which uncertainty plays a role; although they will not all turn out to be correct, it is defensible to take some risks, making choices leading toward a good average. The fact that the minimax-regret rule also selects Groner provides backup support here.

8 (a) Maximax decision: plant beans. Maximin decision: plant oats. Minimax-regret decision: since the maximum regrets for all choices are the same, this rule does not dictate a decision.

(b) If moderate rainfall is most likely, the most-likely-value rule selects beans.

(c) If probability of substantial rain = probability of moderate rain = probability of light rain = $1/3$, then the expected-value rule chooses beans.

(d) and (e) Use of the rainfall data leads to the following estimates: probability of substantial rain = $4/20$; probability of moderate rain = $11/20$; probability of light rain = $5/20$. When these probabilities are used, beans have the highest expected value.

(f) Unless there is some very important reason for having an income of at least $16,000 (instead of at least $15,000) the best choice is beans.

9 (f) To improve her position, Edwards could cut out the five patronage jobs promised to Fredericks if he supports Simmons. She also might reduce the number of patronage jobs for Fredericks if he remains neutral.

(g) Lurking in the background of this problem is the question whether political patronage is a desirable practice; should support of a political party ever be a basis for hiring? In a real situation, perhaps a problem like this should be eliminated rather than solved.

10 (c) Generally, jurors seem to use the minimax-regret rule to choose between "guilty" and "not guilty." If, over time, this rule seems to allow too many rascals to go free, regret over "condemning a possibly innocent person" decreases and regret over "allowing a rascal to go free" increases. When it begins to seem as if innocents are being wrongly incarcerated, the regrets move back.

13 Most of us try to follow Twain's advice most of the time. The trouble with a rule like "Always do right" is that we still face the question, "How do I know what is right?" To answer that question, and to deal with decisions that are morally neutral rather than right or wrong, we can benefit from systematic processes like those in the text.

CHAPTER 18

3 Estimation of quantities such as how much time a task will take and how much money is needed for weekly expenditures is an activity for which this exercise can be of practical assistance to you. A more frivolous activity—one that can, however, be used as a fund-raising game at a carnival or fair—is guessing ages, heights, or weights (or waist or cranial circumferences or whatever). With practice you can learn to guess so that a high percentage of your guesses fall within a small distance of the true value. (If you are actually using this as a carnival gimmick, you charge a fee for your guess and offer a prize whenever the true value does not lie within a certain promised distance—your confidence interval—of your guess.)

4 Interval estimates are useful because they indicate clearly our uncertainty of an exact value.

CHAPTER 19

2 (a) Accurate guesses require experience; however, your guesses should have the following relationship: Max's time > Marsha's time > Melanie's time.

(b) For Max, $t = 41.6$ minutes.

(e) Max is trying to complete a task that takes longer than the time that he expects between interruptions.

3 (a) $t = 15.4$ hours.

4 Some of the values of t in Table 19-4 shouldn't be taken seriously; in actual situations, human frustration will take over and prevent their occurrence.

(b) 10.6, 13.4, 18.7, 41.6, 67.0, 170, over 2000. (These are numbers of minutes.)

(e) No, no, yes, yes; no, no, yes.

(f) The observation is "roughly" true. More precisely: in *b*, when $U = i = 10$, $t = 1.87U$; in *c*, when $U = i = 20$, $t = 1.79U$. Intuitively, we might expect that if interruptions occur, on the average, every i minutes, then we would usually be able to accomplish tasks of length i. However, because the interruptions are random, they will not necessarily be spaced at intervals that will permit us to accomplish the whole task. In fact, sometimes we will have almost completed the task when an interruption occurs and requires us to start over. The random nature of the interruptions makes it possible that much time will be wasted through starting over.

(g) If $p = .25$, three 5-minute tasks will take 38.7 minutes; one 10-minute task will take 67 minutes.

6 (a) Might someone who believes Parkinson's law expect that organizing tasks to save time is futile, and that the job will take all the time that is available no matter what efforts one might make to change things?

(c) Henry Ford's statement ignores the difficulty of organizing a major project into small jobs.

7 (b) About 28 minutes.

8 (a) Forever. (Calculation using the formula for t gives a value in excess of 2 million time units.)

(b) Hora must assemble a total of 111 ten-part components; each takes an average of 10.57 time units. Thus assembly of an entire watch will take a total of 1173 time units, on the average.

9 (d) In this case, the time required for Jane to accomplish one half-hour segment of her task is $t = 53$ minutes. The total report (six half-hour segments) will thus take about 318 minutes, or 5 hours and 18 minutes. This calculated value is substantially less than the time of 9 hours calculated in *b* because of the different choice of time unit. In *b* the use of a 15-minute block as the time unit treated each interruption as if it wasted the entire 15-minute block in which it occurred. In *d*, however, each interruption is treated as if it wastes only the 1 minute in which it occurs. The accuracy of the predictions of formula 1 depends on the choice of time unit; sound results require using a time unit whose length agrees with the expected length of interruptions.

10 (a) 1: about 1 minute. 2: about 2 minutes.

(b) 1: about 9 minutes. 2: about 13 minutes.

(f) In *a* the suggested reorganization would require an average of 5.2 minutes and would not be worthwhile. In *b* the reorganization would require about 5.8 minutes; the savings of 3.2 minutes would be worthwhile, especially if it could be repeated by reorganizing many tasks into shorter subtasks. In *c*, *d*, and *e* the investment of organizational time is definitely a time saver; however, it is reasonable to ask whether even reorganization reduces the time to a tolerable length.

13 (a) Professors Kline and Lewis are responsible both for finding time to prepare effective classes and finding time to respond to the needs of students. Not surprisingly, they find it impossible to do both at once.

(b) Because many of an instructor's duties require uninterrupted intervals of time, and because the adequate completion of such duties (preparing lessons, grading papers, doing research) is as important to students as time spent on individual meetings with them, it seems reasonable for a professor to divide out-of-class time into two categories. Some time should be scheduled as *available for interruption* ("office hours"); at other times the instructor would be *unavailable for interruption*. When the instructor is available for interruption, the free time between actual interruptions can be used to accomplish brief tasks (such as reading mail and drafting memos) or to complete brief subtasks that result when lengthy "uninterruptible" tasks are reorganized.

15 In making your list, consider internally caused interruptions (such as fatigue, lack of interest, and short attention span) as well as externally caused ones (such as visitors and noise). After you have made your list, consider the following: Many people dislike the thought of investing extra time, before they begin a project, to get organized; is this true for you? Is this a costly attitude? Is it ever beneficial?

16 It may be hard to fit a problem of your own to the specific model discussed. You may, for example, have to guess at numerical values. Even if your values for U and *p* are simply guesses, working through the problem can be a useful experiment. At the end of your effort, one possible conclusion is that the analysis has little value or provides little insight in the situation you have described. The solution methods we learn do not always apply to the problems we encounter; a sensible procedure is to try a method if we think it might work, but to try it as an experiment whose results may need to be rejected.

CHAPTER 20

1 (a) One possible description: The person who favored protest even if no one else protested made his or her views public. One by one the others saw their thresholds met. Eventually, all thresholds were met and the group agreed to stage a protest.

(b) The one person who was willing to protest if someone else would found no one to join him or her, and so the possibility of protest died.

2 (a) When Mike announces his willingness to protest to the group, one other will join him. If the group is talkative, soon all the members will see that their thresholds are met, and a protest will emerge.

3 Other examples include cheering enthusiastically at a football game, attending a university reception for new students, and studying on Sunday afternoon or, oppositely, not participating in these activities.

4 (a) In both examples, $t > S_t$ for all positive data values; in both cases the predicted participation level is 0.

5 (c) The "Other Side of 30" group is likely to violate the restrictions. This follows from statement 2 in the text's amplification of the threshold criterion. When the group gets together to discuss the matter, the two members with threshold 1 will discover each other and will start a chain reaction that will eventually involve the entire group.

6 (a) The quantity S_t is a sum of nonnegative terms. As t increases, the number of terms added either stays the same or increases, and so S_t can never decrease.

(b) The graphical consequence of the fact that S_t is nondecreasing is that when we read across a graph of threshold data from left to right, each section is always either level or ascending.

8 (a) The final column of Table 20-6 should contain these entries: 0, 0, 0, 15, 25, 25. The final column of Table 20-7 should contain these entries: 0, 0, 0, 5, 15, 15.

(c) Group A is likely to maintain a full problem session with 25 students attending. If unforeseen circumstances change the number of attendees by a few, as long as 20 students attend, the problem session will continue. The combined attitudes of members of group B will result in a dying problem session. We might expect that 15 will show up at the first meeting; but, of these 15 students, only 5 have indicated satisfaction with an attendance level of 15. Only these 5 will show up at the second session. Since a group of 5 is insufficient to justify their attendance, they will not attend the third session. The group B problem session will thus be dead after the second meeting.

(d) The fact that group A can sustain a problem session while the whole class could not suggests a strategy to apply in a variety of situations. Even if a certain group cannot maintain a behavior in the group as a whole, perhaps a separated subgroup can. Sometimes religious and political subgroups adopt this strategy; they separate themselves from a larger societal group and within the separated group willingly adopt certain behaviors. (The Amish are an example of a separated subgroup; certain communes also are examples.)

11 (a) If the group members become well informed about each other's attitudes, then the 45 members who are satisfied with a threshold of 40 will discover each other and join together in behavior X.

(b) If the group members whose thresholds are described in Figure 20-5*b* believe that the number participating in (or willing to participate in) behavior X is 40 or less, then the number participating in behavior X will eventually decrease to an equilibrium with no members choosing the behavior. If the group members believe that the number of participants (or those willing to participate) is 60 or greater, then group behavior will eventually reach an equilibrium with all group members choosing behavior B. Since data values have been given in the graph only for thresholds that are multiples of 20, we may suppose that all thresholds have been rounded to these values and that group members will also round their estimates of the numbers of participants to multiples of 20.

(c) 1: ". . . the vertical coordinate of the nearest data point to the left of the intersection with the equilibrium line identifies a possible equilibrium situation for the group." 2: ". . . the intersection of the graph and the equilibrium line does not identify an equilibrium situation for the group."

12 (a) Data points (0, 0), (20, 20), and (70, 70) are equilibrium points; only 70 is a stable equilibrium.

13 (a) A graph of the threshold information is given in Figure A-4. Although both (0, 0) and (90, 90) are equilibrium points, only the latter is stable; as time passes and students get to know each other, we would expect that class behavior will tend toward a situation in which 90 of the 100 students cheat.

14 If guns are carried openly, behavior in Bowie would be expected to stabilize with 75 percent of the adult residents carrying guns.

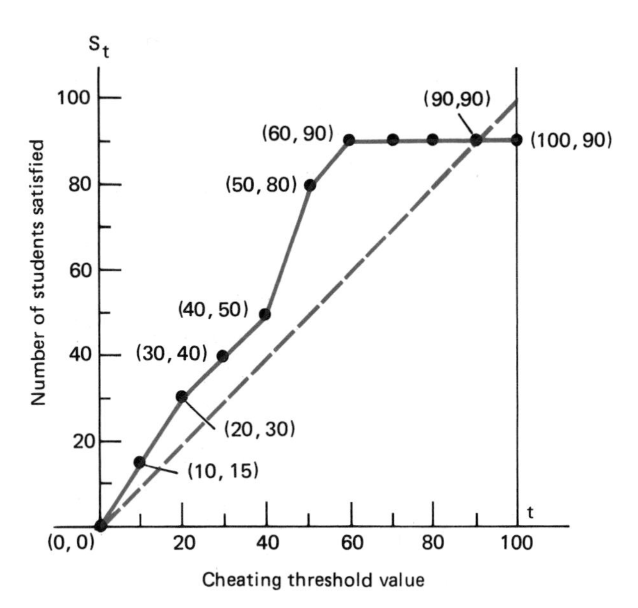

FIGURE A-4

TABLE A-6

NONCONSERVATION THRESHOLDS, PERCENTS	PERCENT OF COMMUNITY MEMBERS WITH GIVEN THRESHOLD
0	0
10	15
20	10
30	10
40	10
50	10
60	10
70	10
80	10
90	10
100	5

15 Possible samples of Disarm's reasoning are: No person can have a threshold of zero, since a person includes himself or herself in the threshold group. (Hence the first row of Table 20-16 must contain a pair of zeros.) The 10 percent of the people who would never carry a gun and the 15 percent who would carry a gun only if everyone was doing so would all be satisfied not to carry a gun if part of a group of 25 percent. (Thus in the second row of the table, corresponding to a threshold of 25 percent, we have 25 percent.) Consider the 40 percent of Bowie residents with gun-carrying thresholds of 50 percent. Members of this group would be unwilling to carry guns if only 25 percent carried them. Thus that same group of 40 percent surely would refuse to carry a gun if they would be part of a group of 75 percent. (Thus in the fourth row of the table, corresponding to a threshold of 75 percent, we have 40 percent.) If guns in Bowie are carried openly, behavior would be expected to stabilize with about 25 percent of the adult residents refusing to carry guns. Despite Disarm's effort, he has done little more than turn the graph for Exercise 14 upside down.

17 (a) One possible collection of threshold data for Watertown is shown in Table A-6. A graph for these data has two equilibrium points: (0, 0) and (100, 100). Only the latter is stable; thus we have a situation in which the likely outcome is 100 percent nonconservation. We may suppose that while 90 percent of the Watertown residents support conservation, many of them find it a sacrifice, and their willingness to conserve readily disappears when others fail to do so.

(b) One approach would be for the mayor of Watertown to urge the residents to raise their thresholds. A difficulty with this approach, however, is that it appeals only to those with a strong social conscience; others will still waste. Another approach would be for the mayor to urge the town to impose a fine for violators. The fine would serve as a way of enforcing the conservation that everyone wants.

CHAPTER 21

1 (a) Milhous will win with 28.3 percent of the vote.

2 (b) Ottinger would have won with the approval of 54 percent of the voters.

(c) Following are two contributions to a list of pros and cons are. *Pro:* Approval voting avoids the problem of splitting the votes of a majority between two (or more) popular candidates. *Con:* Even with approval voting, candidates might win without the approval of a majority of the voters.

3 (c) In the Alice-Ben-Carla-Donald example, the agenda order controlled the decision, since any one of the students could win if an appropriate agenda was used. On the other hand, the election below provides an example of an election in which the same candidates (P) will win no matter what agenda is used.

Voter 1	Voter 2	Voter 3
P	Q	S
Q	R	P
R	P	Q
S	S	R

4 In 18 of the 24 possible agenda involving Alice, Ben, Carla, and Donald, the last one on the agenda is selected. Thus the rule of thumb applies in this case. However, as noted for 3*c*, there are elections that defy the agenda effect and this rule of thumb.

5 (b) If "Atlanta" is replaced by "Alice," "Birmingham" by "Ben," "Chicago" by "Carla," and "Dallas" by "Donald," then the preferences in Raymond's problem turn in to the mathematics teachers' preferences.

9 Election 2

(a) In the author's view, S should win. Alternative S is preferred most by a greater number of voters and liked least by a smaller number of voters than either of the other choices.

(b) The plurality method selects V.

(c) The plurality-with-elimination method selects V.

(d) The Borda count method selects S.

(e) The Condorcet method does not select a winner.

Election 3

(a) Should the winner be A or B? Alternative A has a majority of the first-place votes. On the other hand, every voter puts B in either first or second position, and almost half of the voters put A last. The electorate needs to decide which it cares more about—choice by majority or overall preference—and then select an appropriate method before the election.

Election 7

(a) Should we refuse to answer this question? Except in the most obvious cases, how can we select a winner without having first chosen a voting method?

11 (c) The preferences expressed in election 9*a* leads to P as winner. However, two voters who changed their minds in favor of P caused P to lose in election 9*b*. Why did this occur?

13 The sequential pairwise voting method selected Donald, but every voter preferred Ben to Donald.

21 The group consensus—determined by the Borda method, since there is no Condorcet winner—is to fund the public library.

22 Would you recommend the Black method? Beside hearing of the good qualities of a new method, the legislative committee will need to hear about the flaws of other, more familiar or simpler methods. These flaws can perhaps be explained best by inventing examples of what can go wrong if a certain method is used.

23 Do you find that the groups to which you belong have a naive approach to voting? That is, do they believe that—if well-intentioned people think carefully about their choices and then vote—consensus will be properly measured, regardless of the voting method?

24 (a) Probably B and C are the least preferred choices of the prohibitionist and the alcoholic, respectively.

(c) If the alcoholic votes against the clubhouse—reflecting the second choice rather than the first—in the initial vote, then the outcome is better for him or her than it would have been had the alcoholic voted for the first preference.

CHAPTER 22

TABLE A-7

	D	C
D	(−3, −3)	(8, −5)
C	(−5, 8)	(5, 5)

TABLE A-8
Payoffs for decision maker 1

	D	C
D	−3	8
C	−5	5

2 (b) Labels C and D should be placed as in Table A-7. For decision maker 1, the decision payoffs are as shown in Table A-8. (The payoffs for decision maker 2 are identical for each combination of choices, but of course they occupy different matrix positions.) Since $-3 > -5$ and $8 > 5$, choice D is dominant and PD 2 is satisfied. Since $8 > -3$ and $5 > -5$, the decision maker prefers that the other choose C; thus PD 3 is satisfied. Since $5 > -3$, both decision makers are better off if both choose C than if both choose D; PD 4 is therefore satisfied.

3 (b) $t = 8, r = 5, p = -3, s = -5$

(c) Values for t, r, p, and s cannot be identified. Why?

7 (a) The exercise is stated in such a way that PD 1 is satisfied. For PD 2 to be satisfied, each student must gain more from keeping quiet, no matter what the other does. This would be true not for all students but only for those for whom shyness or fear of appearing foolish overrides the desire to gain information through asking questions. For PD 3 to be satisfied, each student must gain more from the class if the other asks questions, no matter what he or she does. This would not always be true but would probably be so if, for example, both students have similar levels of ability and interest. If understanding and learning are goals of both students and if the teacher is capable and responsive, then it is hard to imagine a situation when joint questions would not be more beneficial than joint silence; hence PD 4 would be satisfied.

(b) For the matrix to be a PD matrix it is necessary that $y > 5$ and $x < -1$.

11 One way to get started inventing a problem would be to identify among your friends or acquaintances those with whom you would consider (if you had the money and the inclination) entering into a joint purchase (perhaps for the purpose of earning income from rentals) and those with whom you wouldn't dream of doing

TABLE A-9

C	0	−7
D	−3	−10

TABLE A-10

	MANUFACTURER 2	
MANUFACTURER 1	Cooperate	Defect
Cooperate	(0, 0)	(−4, −1)
Defect	(−1, −4)	(−6, −6)

so. Out of your thoughts about the differences between situations in which cooperation seems likely and those in which it seems unlikely, develop a PD scenario.

13 The matrix of values for prisoner 1's decision is as shown in Table A-9, in which "keep quiet" (C) is the dominant choice. (Prisoner 2 faces a choice that involves the same payoffs.)

16 (a) If reward and penalty amounts of $2 million are added to and subtracted from the appropriate entries in Table 22-17, the matrix in Table A-10 results. In this case, "cooperate" is the dominant decision for both manufacturers.

(d) Why is $1.5 million the minimum? Examine the matrices obtained in *b* and *c* for evidence.

(e) Two difficulties are: Where will the reward money come from? How will the penalties be enforced? In a real environmental decision problem, some decision making body—possibly a private group like Columbia Lake Preservation, a group of county commissioners, or a state legislature—must decide at what cost the problem is worth rectifying. This figure must then be compared with the actual cost of doing so. Mathematical calculations can aid in determining the cost of a proposed solution; however, it then remains to decide whether the proposed solution is the best one—or at least good enough—and whether solving the problem is worth the cost.

20 (b) In the case of repeated prisoner's dilemma situations—treated, for example, by Axelrod in *The Evolution of Cooperation*—part of the reward or penalty for each decision is its effect on future decisions. Either cooperation or defection may generate more of the same from the other decision maker.

24 (a) As long as profits are divided equally, the noncontributors gain more than the contributors. This is a MPD problem in which choice D (not to contribute) is more profitable than choice C (contribute) for any fixed number of investors. Furthermore, each investor is better off the more others choose to contribute. The minimum size of a coalition that can benefit from choice C is 4. Their slight gain of 0.2 is better than nothing.

(b) Although this may not be as easy as it looks, it seems that a reasonable way to change the situation would be to deny a share of the gain to those who do not invest. A real situation that has characteristics similar to this problem is one where workers choose between joining and not joining a union. Benefits negotiated by the union may be extended to nonunion workers as well, and they gain without contributing dues or support to the union. While union members resent freeloaders, often they are powerless to exclude nonmembers from benefits. (Sometimes, however, the ill will expressed by fellow workers creates enough pressure to cause reluctant workers to join.)

25 (a) In this case the minimum size of a coalition that can benefit from choice C is 6.

27 Here are some situations to think about; how well is each modeled by the MPD problem? (1) Disease control: Residents of community must decide whether or not to be vaccinated. (2) Game hunting: Shall each hunter be allowed to decide whether to hunt deer or bear, or shall the amount of hunting be controlled? (3) Traffic congestion: Each driver in an area must decide whether to use a certain busy roadway at rush hour. (4) Community swimming pool: Each resident must decide whether or not to contribute. (5) Tuxedos: Each man attending a prom must decide whether or not to wear a dinner jacket.

INDEX

Acquaintanceship graph, conjecture, 210
Activity-analysis digraph, 217–226
  critical activity, 218
  critical path, 218
Addition, correct use of, 92–98, 135–140
Agenda for voting, 365–367
Agenda effect, 365–367
Air-quality scale, 80
Almanac questions, 309–312
Almanac survey, 309–312
Antinuclear protest, 338–339, 343
Approval voting decision method, 364
Archery-range warning, 15
Arrow, Kenneth J., results of, concerning voting methods, 387
Assigning weights to decision criteria, 102–114
Average, 183–190
  batting, 188
  law of averages, 188, 261, 270
  mean (arithmetic), 54, 78, 184, 193–194
  measure of "center," 183–184, 191
  median, 183, 191
  mode, 183
  weighted, 102–114, 188–189
    computing, 107
    reasonableness of, 104, 110
Average speed, 187, 189
Axelrod, Robert, 404, 408

Base 2, 157
Base 5, 160
Base 8, 161
Batting average, 188
Behavior threshold, 337–360
Belief in God, Pascal's decision about, 112–113
Berkeley, George, 313
Bernard, Claude, 40
Bernoulli, Jacques, 257
Bigfoot puzzle, 21
Bimodal data, 183
Binary number system, 157–160
Bit, 160
Black, Duncan, 380
Black's voting decision method, 380, 387
Blood-type example, 138–139
Borda, Jean Charles de, 372
Borda count voting decision method, 370, 372, 374, 376–388
Brams, Steven J., 404
Browning, Elizabeth Barrett, 146
Budget constraints, 36, 121

Calculation:
  of interest, 12–15, 37–38, 43–46, 50–52
  of numbers of possibilities, using multiplication rule, 146–161
  of population change, 48–50, 53–54
Calcutta population, 310–312
Carnival game, 179–180
Carroll, Lewis, 25
Cause-and-effect digraph, 210–211
"Center" of data, 183–190
Certainty, maximum, 252
Chain reaction, 337, 351
Chammah, Albert M., 404
Chang Tzu, 111
Chartrand, Gary, 26*n.*
Chinese proverb, 20
Choice matrix, 100–114, 295–309
"Christopher Robin," 329
Circuit in graph, 213, 228–242
Coffee-milk puzzle, 34, 36
Coins, tossing of:
  multiplication rule, 154
  probability assignments, 258–263, 269, 270
  simulation, 60–61, 65, 66
Communication flow, graph of, 204
Comparisons, meaningful, 82–83
Computer memory, 160
Condorcet, Marie Jean Antoine Nicholas de Caritat, marquis de, 372
Condorcet loser criterion, 379–381

Condorcet voting decision method, 372–373, 375–388
Condorcet winner criterion, 379–381
Confidence interval 275*n.*, 311–317
  experimentation with, 315
Connected graph, 228*n.*
Consensus, recognition of, 361–388
Consistency criterion, 379–381
Constants:
  background, 34–37
  percents, 37–46
Corneille, Pierre, 99
Criteria for judging voting decision methods, 379–381, 387–388
  Condorcet loser, 379–381
  Condorcet winner, 379–381
  consistency, 379–381
  majority, 379–381
  Pareto, 379–381
  independence-of-irrelevant-alternatives, 387
Critical activity, 218
Critical mass, 351
Critical path, 218
Cryptarithm, 17–18, 150–151

Darwin, Charles, 235, 245
Davis, Philip J., 92*n.*
Decision analysis, 170–171
Decision criteria, 99–114
Decision-making problems of individuals:
  candidates to support, 304–305
  car purchase, 110–111
  carrying an umbrella, 87–88, 282–283, 285
  college selection, 100
  crops to plant, 303–304
  God, belief in (Pascal), 112–113
  guilty or innocent, 305–306
  hiring a secretary (Dogood), 28–33, 70–71, 150
  job selection, 111, 167
  newspaper quantities to order, 290–300
  products to buy, 105–106, 116–127
    clothes selection, 105–106
    hamburgers or books, 116–127
  topics to study, 22, 52, 127–129, 294–299
Decision-making rules for individuals, 294–308
  expected-value, 297
  maximax, 296
  maximin, 296
  minimax-regret, 297–298
  most-likely-value, 296
Decision matrix, 103–114
Decision methods of Casper Dogood, 28–33, 70–71, 150
Decision tree, 170–171
Declining marginal utility, 118, 123–124, 129
Degree of a vertex, 234
Diagonal of a polygon, 20
Diagrams:
  digraph, 203–226
  graph, 201–226
  numbered tree, 171–182, 259
  tree, 162–182
Dice, tossing of:
  multiplication rule, 154
  probability assignments, 262–264
  simulation experiments, 62, 66–67
  tree diagram, 173, 180
Dictatorship voting decision method, 371, 372*n.*
Digraph, 203–226
  edge, 203
  vertex, 203
Digraph applications:
  cause-and-effect, 210–211
  communication flow, 204
  election results, 375–376
  family tree, 207
  project scheduling, 217–226
  tournament results, 206
Divide-and-conquer strategy, 346
Dodgson, Charles Lutwidge (Lewis Carroll), 25
Dogood, Casper, hiring decision problem of, 28–33, 70–71, 150
Dominance in a matrix, 100
Double-agent problem, 18
Drucker, Peter, 235

"Easy" problems, 141
Eccentric host, 83–84, 89
Einstein, Albert, 48, 51, 309
Electoral College voting decison method, 362, 363
Elementary events, 250, 251
Ellis, Havelock, 131
England, population of, 48–50, 53–54
Enmity, graph of, 202–206, 208
Environment, role of, in threshold analysis, 347–360
Equilibrium line, 349
Equilibrium point, 349
Erasmus, Desiderius, 115
Estimation, 309–312, 316, 317
  of probabilities, 246–249, 259–262
Euler, Leonhard, 228, 233
Euler, circuit, 228
Euler circuit rules, 229
Euler trail, 228
Even vertex, 228
Events:
  elementary, 250, 251
  independent, 273
  mutually exclusive, 249–251
Expected value, 275–293
  of carrying umbrella, 282–283, 285
  definition of, 277
  of gambling games, 278–279, 283
  of insurance purchase, 281, 285, 289

Expected value (*Cont.*):
of newspaper sales, 290
of operator-assisted calls, 288
of tosses of dice, 277, 285
of travel time, 283, 288
Expected-value decision rule, 297
Experimentation:
with confidence intervals, 315
to test speculations about random events, 48–55, 60–72

Factorial notation, 153
Factoring using tree diagram, 169
Fallacy:
gambler's, 269, 272
postmortem, 272
Fallon, Richard, 311*n.*
Family-planning rule, 62–64, 67–70, 175–176, 178–179, 264–265
Farquharson, Robin, 383*n.*
Feiffer, Jules, 201
Ford, Henry, 329
France, Anatole, 256
Freedman, David, 313*n.*
Friendship patterns, graph of, 201–203, 208, 209

Gambler's fallacy, 269, 272
Gambling games, 179–180, 275–279, 283
Game theory, 112, 404
Genetic code word, 155
Goal-directed problem solving, 3, 12–15
saving money, 12, 45–46, 52
Goethe, Johann W. von, 415
Grades, improvement of, 22, 52, 127–129
Granovetter, Mark, 350
Graph, 203
circuit, 213*n.*, 228–242
connected, 228*n.*
edge, 203
path, 213*n.*
of threshold data, 342–360
vertex, 203
degree of, 234
even, 228
odd, 228
(*See also* Digraph; Tree diagram)
Graph applications:
acquaintanceship patterns, 210
cause-and-effect, 210–211
communication flow, 204
friendship patterns, 201–203, 208, 209
highway pothole inspection, 233
ice cream route, 230–231
mail carrier, 240–241
milk route, 240
scheduling, 209, 211–226
arranging courses, 209

Graph applications, scheduling (*Cont.*):
ASM conversion, 212–215
project scheduling, 217–226
sales trip, 216–217
threshold analysis, 340–360
trash collection, 235–239
zoo design, 204–206
Gray, Mary, 272*n.*, 273*n.*
Group decision making:
behavior change at threshold, 337–360
prisoner's dilemma situations, 389–412
voting, 361–388
Growney, JoAnne S., 318*n.*
Guess a number, 168

Hamburger, Henry, 404
Hand, Learned, 389
Hardin, Garrett, 403, 405, 411
Hickman, James C., 222*n.*
Hiring decision, 28–33, 70–71, 150
Hogg, Robert V., 222*n.*
Holmes, Oliver Wendell, Jr., 92
House numbers (ordinal scale), 77

Imitate successes, 19, 98
Independence-of-irrelevant-alternatives criterion, 387
Independent events, 273
Independent interruptions, 322
Indifference, 123
Individual decision making (*see* Decision-making problems of individuals; Decision-making rules for individuals)
Insurance problems, 262, 268, 281, 285, 289
Interest earnings, 12–15, 37–38, 43–46, 50–52
Interruptions, 318–334
calculation of total time, 323–326
house of cards, 324
5-, 10-, and 20-minute tasks, 325–326
10-minute job, 323–324
desirability of, 332
devastating, 322
effects of (*see* Organization to reduce effects of interruptions)
independent, 322
probability of, 322
simulation of, 319–322
time for an interrupted task (formula 1), 323
Interval scale, 77
"Invent" exercises:
Chapter 1, 12; Chapter 2, 22, 27, 33; Chapter 3, 46; Chapter 4, 55, 72; Chapter 5, 89, 90; Chapter 6, 96; Chapter 7, 111; Chapter 8, 129; Chapter 9, 143, 145; Chapter 10, 156, 160–161; Chapter 11, 169, 170, 182; Chapter 12,

"Invent" exercises (*Cont.*):
189, 196; Chapter 13, 207, 210–211, 225; Chapter 14, 234, 241; Chapter 15, 248, 272; Chapter 16, 291; Chapter 17, 307; Chapter 19, 333, 334; Chapter 20, 358; Chapter 21, 378–379; Chapter 22, 400, 407

Jackson, Andrew, 361
"Journal" exercises:
Chapter 1, 15; Chapter 2, 18, 19, 22–23, 25, 27, 33; Chapter 3, 37; Chapter 4, 55, 72; Chapter 5, 84–85, 90; Chapter 6, 98; Chapter 7, 114; Chapter 8, 127–129, 131; Chapter 9, 145; Chapter 10, 156; Chapter 11, 169, 170, 182; Chapter 12, 190, 196; Chapter 13, 210–211, 215, 225, 226; Chapter 14, 235, 241; Chapter 15, 249, 271, 272; Chapter 16, 286, 291, 292; Chapter 17, 306–308; Chapter 18, 315, 316; Chapter 19, 334; Chapter 20, 347, 359–360; Chapter 21, 385; Chapter 22, 400, 408, 411

Keynes, John Maynard, 47
Kitchen drawer, 413
Koenigsberg bridge problem, 233

Laplace, Pierre-Simon de, 258, 272
Lave, Charles A., 40*n.*, 48*n.*, 116*n.*, 209*n.*, 304*n.*, 306*n.*
Law, Vernon, 23
Law of averages, 188, 261, 270
Law of large numbers, 261
Lessing, Gotthold Ephraim, 130, 227
Level of confidence, 312
License plates, counting of, 146–149, 151–152
Lincoln, Abraham, 337
Locating the "center" of data, 183–190
London, migration to, 48–50, 53–54
Lottery problems, 71, 265, 267–268, 280–281

Magic square problem, 17
Majority, lack of, 371*n.*
Majority criterion, 379–381
Majority rule voting decision method, 371
March, James G., 40*n.*, 48*n.*, 116*n.*, 209*n.*, 304*n.*, 306*n.*
Marginal utility, 118, 123–124, 129
Mathematics Magazine, 227*n.*, 318*n.*, 364*n.*
Mathematics quiz, probability of, 257
Matrix, 99–114, 116–131, 295–308, 391–408
- choice, 100–114, 295–309
- decision, 103–114

Matrix (*Cont.*):
- payoff, for prisoner's dilemma, 391–408
- of utility values, 110–114, 116–131

Maximax decision rule, 296
Maximin decision rule, 296
Mean (arithmetic), 54, 78, 183–190
Measurement of preferences, 86–91, 99–114, 115–131
Measurement scales, 75–91
- arithmetic of, 78
- and averages, 184, 189
- interval, 77
- nominal, 76
- operations on, 77–79
- ordinal, 77
- probability, 252–253
- ratio, 77
- temperature, 77, 79, 81
- utility (*see* Utility scale)
- versatility of, 78

Measures of "center," 183–190
- mean (arithmetic), 183
- median, 183
- mode, 183

Measures of variability, 193–197
- percentile, 191
- quartile, 191
- standard deviation, 192–193

Median, 183–190
Memory of computer, 160
Merrill, Samuel, 364*n.*
Migration to London, 48–50, 53–54
Milne, A.A., 329
Minimax-regret decision rule, 297–298
Mnemonic devices, 27
Mode, 183–190
Mohs' scale, 80
Morgenstern, Oskar, 404
Morse code, 154
Most-likely-value decision rule, 296
Multiperson prisoner's dilemma (MPD) problem, 408–412
- MPD characteristics, 408–409

Multiplication rule, 148
- applications of, 146–161

Mutually exclusive categories, 135–145, 163, 178, 249, 251
Mutually exclusive outcomes, 249, 251

NB-10, 313–315
Nim games, 3–12
- (12;1,2) Nim, 3–6
- (n;1,2) Nim, 7
- (n;1,2,3) Nim, 8–10
- losing number, 8, 10
- original version, 11
- winning number, 8, 10
- winning strategy, 6
- Wipe-Out, 11

Nominal scale, 76
Normal distribution of data, 193
Numbered tree diagram, 171–182, 259

Octal number system, 161
Odd numbers of lumps of sugar (riddle), 234
Odd vertex, 228
Ordinal scale, 77
Organization to reduce effects of interruptions, 326–334
  assistant vice president, 331–332
  dormitory counselor, 318, 329–330
  parable of the watchmakers, 330–331
  political science students, 326–328
  professors, 333
  20-minute task, 332
Organizing information:
  lists (Pareto's rule), 85
  matrices, 99–114, 116–131, 295–308, 391–408
  mutually exclusive categories, 135–145, 163, 178, 249, 251
  tree diagrams, 162–182
  Venn diagrams, 136–139, 182
Outcomes:
  of chance events, tree diagram for displaying, 173, 175–176, 178–181, 258–273
  mutually exclusive, 249, 251

Parable of the watchmakers, 330–331
Pareto, Vilfredo, 85, 379*n.*
Pareto criterion, 379–381
Pareto's 80-20 rule, 85
Parkinson, Northcote, 318, 329
Parkinson's law, 318, 329
Pascal, Blaise, decision of, about belief in God, 112–113
Path:
  in graph, 213*n.*
  in tree, 162
Payoff matrix for prisoner's dilemma, 391–408
Payoffs for prisoner's dilemma, 391, 395
PD (*see* Prisoner's dilemma)
People v. Collins, 272–273
People's Almanac, The, 81–82
Percent, 37–46
  base for, 38, 41
  dating and marriage, 40–41
  interest earnings, 12–15, 37–38, 43–46, 50–52
  safety statistics, 38, 41–42
  sale prices, 39–40
Percentile, 191
Persistence in problem solving, 97, 98, 167
PERT (program evaluation and review technique), 221

Pi, mnemonic device for, 27
Pisani, Robert, 313*n.*
Pizza problem, 35
Placid community center problem, 54
Plurality voting decision method, 369, 371–388
Plurality-with-elimination voting decision method, 370–388
Population change in England, 48–50, 53–54
Postmortem fallacy, 272
Prime factor, 169
Prisoner's dilemma (PD), 389–412
  circumvention of, 400–408
  MPD characteristics, 408–409
  multiperson problem (MPD), 408–412
  payoff matrix for, 391–408
  payoffs for, 391, 395
    penalty, 395
    reward, 395
    sucker, 395
    temptation, 395
  PD characteristics, 394–395
  PD problem examples: arms race, 399
    class participation, 398
    common grazing land, 399
    investment club, 410–411
    mathematics review, 397
    pollution control, 401–402, 405
    price war, 397
    shared maintenance, 400
    shared parking, 402
    shared taxes, 402
    support of charity, 409
    wage bargaining, 399
  repeated, 392–394
  reward-penalty system, 400–408
  two-person problem, 390–408
Probabilistic thinking, 245–249, 258, 292, 316
Probability, 245–274
  elementary event, 250, 251
  mathematical rules for, 250–251
  sample space, 249
  scale, 252–253
  two aspects of, 245–246
    relative frequency (interpretation 2), 246
    strength of belief (interpretation 1), 246
Probability assignments:
  bases of: degree of certainty, 246–249
    relative frequency, 260–262
    strength of belief, 246–249
    tree diagram, 259–262
  for outcomes of tossing coins, 258–263, 269
  for outcomes of tossing dice, 262–264
  for roulette wheel, 254
  for sex distribution of children, 264, 265
Problem-session attendance, 341–342, 346, 351–352
Problem solver, qualities of a good one, 97, 98, 167
Problem-solving strategy:
  focus on goal, 3, 12–15

Problem-solving strategy (*Cont.*):
  imitate successes, 19, 98
  kitchen drawer, 413
  learn from mistakes, 23–25
  observe others, 19–23, 98
  observe widely, 16–23, 47–48, 93, 98
  observe yourself, 16–19
  simulation, 55–72
  speculate freely, 47–48, 93
  take control, 25–27
  test carefully, 47–48, 93
  trial and error, 23–25
Program evaluation and review technique (PERT), 221
Project exercises for students:
  committee decision, 385–386
  threshold analysis, 359–360
Project scheduling, 217–226
Purves, Roger, 313*n.*

Qualities of a good problem solver, 97, 98, 167
Quartile, 191
Quinary number system, 160

Rabbit under the bush, 227–228
Randles, Ronald H., 222*n.*
Random-digit simulation, 55–72, 319–322
Random-digit table, 56–60, 320–321
  instructions for use of, 65
Range of data, 191
Range of uncertainty, 252, 309–317
Ranking items (ordinal scale), 77–85
Rapoport, Anatol, 404
Rating items, using utility scale, 86–91, 101–114, 116, 130, 398, 399
Ratio scale, 77
Rational decision maker, 390
Rational decision making, 390
Reading level, 196
Recognizing consensus, 361–388
Relative frequency of an event, 261
Rice, Peter, 385*n.*
Roethke, Theodore, 162
Rounding rule, 44*n.*
Rousseau, Jean-Jacques, 48–50, 53–54
Routes in graphs, 227–241
  applications of: ice cream vendor, 230–231
    mail carrier, 240–241
    milk route, 240
    pothole inspection, 233
    rabbit under the bush, 227–228
    trash collection, 235–239
    walking tour of Koenigsberg, 233
Rule of 70, 48, 50–51
Rule of 72, 51

Saint Ives riddle-rhyme, 12
Sample space, 249
Sartre, Jean Paul, 275
Satisfaction, maximization of, 115–131
Saturation, 117
Saving money, 12, 39–40, 45–46, 52
Saxe, John Godfrey, 16
Schaeffer, Anthony J., 222*n.*
Scheduling problems, 209, 211–226
  arranging courses, 144, 152, 209
  ASM conversion, 212–215
  project scheduling, 217–226
  sales trip, 216–217
  studying, 22, 52, 127–129, 302
Schelling, Thomas C., 350–351, 356*n.*, 408*n.*
School board budget decision, 368–370, 376
Scrabble, 155
Seating arrangements, 155, 208
Sequential pairwise voting, 365–367
Simon, Herbert A., 330*n.*
"Simple" problems, 141
Simulation, 55–72, 319–322
  of births, 62–64, 67–70
  of hiring a secretary, 70–71
  of interruptions, 319–322
  of tossing coins, 60–61, 65, 66
  of tossing dice, 62, 66–67
Slope of a line segment, 349*n.*
Smith, Adam, 403
Sophisticated voters, 383
Speculation about population patterns, 48–50, 53–54
Splicing to complete a circuit in a graph, 232
Standard deviation, 192–193
Straffin, Philip D., Jr., 370*n.*
Studying, 22, 52, 127–129, 302
Summarizing information:
  locating the "center," 183–190
  measuring variation from the "center," 191–197
Survey results used in problem solving, 140–141, 143–144, 181, 341, 356–360
Swift, Jonathan, 34

Table of random digits, 56–60, 320–321
  instructions for use of, 65
Tchebycheff's theorem, 192–193
Temperature scales, 77, 79, 81
Tennis-tournament problem, 19
Terminal vertex, 162
Testing probability values, 271
Testing speculations:
  by calculation, 48–55
  by simulation, 55–72
Thinking cost, 131
Thinking probabilistically, 245–249, 258, 292, 316
Thomson, William, Lord Kelvin, 75
Thoreau, Henry David, 3
Thought experiment, 271–272

Threshold analysis, 340–360
chain reaction, 337, 351
equilibrium line, 349
equilibrium point, 349
guidelines for application of, 349–350
role of environment in, 347–360
threshold criterion, 340
threshold value, 339–340
Threshold analysis applications:
antinuclear protest, 338–339, 343
cheating, 354–355
class participation, 345, 347–348
gun carrying, 356–358
head shaving, 354
problem-session attendance, 341–342, 346, 351–352
water conservation, 358
Threshold criterion, 340
Threshold value, 339–340
Tie in voting, 377, 380
Time required for interrupted task:
calculation of, 323–326
formula for, 323
simulation of, 319–322
Tit-for-tat, 408
Tossing coins:
multiplication rule, 154
probability assignments, 258–263, 269
simulation, 60–61, 65, 66
tree diagram, 259
Tossing dice:
multiplication rule, 154
probability assignments, 262–264
simulation experiments, 62, 66–67
tree diagram, 173, 180
"Tragedy of the Commons, The," 403, 411
Tree diagram, 162–182
decision, 170–171
numbered, 171–182, 259
rules for use of, 163
Tree diagram applications:
decision analysis, 170–171
displaying outcomes of chance events, 173, 175–176, 178–181, 258–273
guessing a number, 168
organizing information, 162–182
Truman, Harry S, 371
Tucker, A. W., 390*n.*
Turgenev, Ivan Sergeyevich, 135
Twain, Mark, 294, 307

UMAP Instructional Modules, 75*n.*, 94*n.*, 318*n.*, 337*n.*
Umbrella, utility of carrying, 87–88, 282–283, 285
Uncertainty, maximum, 252
Util, 86
Utility, marginal, 118, 123–124, 129
Utility scale, 86–91, 101–114, 116, 130, 398, 399
open-ended, 116
Utility theory, measuring strengths of preferences as, 86–91, 99–114, 115–131
Utility values, matrix of, 110–114, 116–131

Values, role of:
in decision making, 27–33, 105–106
in problem solving, 27–33
Vaughan, Bill, 96
Venn diagram, 136–139, 182
Vertex, 162, 203
degree of, 234
even, 228
in graph, 203
odd, 228
terminal, 162
in tree, 162
Von Neumann, John, 404
Voters' paradox, 384
Voting decision methods, 361–388
approval, 364
Black, 380, 387
Borda count, 370, 372, 374, 376–388
Condorcet, 372–373, 375–388
dictatorship, 371, 372*n.*
Electoral College, 362, 363
majority rule, 371
plurality, 369, 371–388
plurality-with-elimination, 370–388
(*See also* Criteria for judging voting decision methods)
Voting decision methods applications:
car purchase, 384
committee decision, 382, 385–386
job selection, 367–368
"mathematics student of the year," 362, 365–367
Mt. Olympus school board, 368–370

Watt, William W., 183
Wealth of Nations, The, 403
Weighted average, 102–114, 188–189 188–189
computing, 107
reasonableness of, 104, 110
Weights for decision criteria, 102–114
Whitehead, Alfred North, 55
Whitney, Joseph, 141
Wipe-Out (game), 11
Word:
in computer, 160
genetic code, 155

x-ray reliability, 42, 181